"This is one of the best one-volume works on the creation/evolution dialogue in print."

— **Kenneth Keathley,** Senior Professor of Theology, Southeastern Baptist Theological Seminary

"An excellent scientific treatment of the world of Genesis, using full-color illustrations and readily understandable explanations, that will be an essential textbook for any class on creation and an indispensable resource on the subject for lay readers."

— **Michael L. Peterson,** Professor of Philosophy, Asbury Theological Seminary

"This book does a much-needed service for all Christians troubled by the relation of their faith to science. While other writers have also taken the position that Genesis is to be understood from the standpoint of the worldview of its own time, Hill has spelled out just how to do that in nontechnical language and convincing detail."

— **Roy Clouser,** Professor Emeritus, The College of New Jersey

"Carol Hill's worldview approach may be a key way to unlock the meaning of an ancient witness like Genesis, and readers will benefit from this unique perspective in the interpretation of Scripture."

— **Scott Hoezee,** Director of The Center For Excellence in Preaching, Calvin Theological Seminary

"Carol Hill takes both Scripture and science seriously, affirming the inspiration of the Bible and the evidence for biological evolution."

— **Deborah B. Haarsma,** Astrophysicist and President of BioLogos

"Carol Hill's worldview approach brings the reader face to face with archeological, biblical, and scientific data that enable one to gain a new appreciation for what the Bible is trying to teach. This approach is a very helpful tool!"

— **James K. Hoffmeier,** Emeritus Professor of Near Eastern Archaeology & Old Testament, Trinity Evangelical Divinity School

"Carol Hill's book **A Worldview Approach to Science and Scripture** *is terrific! It provides an excellent integration of God's revelation in his written word to his revelation in his created world as seen through the eyes of science."*

— **Walter Bradley,** Emeritus Professor of Engineering and Material Science, Texas A&M University and Baylor University

"Building on her expertise in geology, Carol Hill argues attractively for a 'worldview' approach to the book of Genesis. Her common-sense approach shows the errors of many common views, setting out the context for the biblical records."

— **Alan Millard,** Emeritus Rankin Professor of Hebrew and Ancient Semitic Languages, University of Liverpool

(Complete endorsements and reviewer biographies are on pages 226-227.)

THE GOAL OF THIS BOOK IS TO PRESENT SOLID SCIENCE TO THE CHURCH – NOT TO DESTROY FAITH, BUT TO RECONCILE SCIENCE WITH FAITH AS A WAY TO TRUTH.

KREGEL PUBLICATIONS
GRAND RAPIDS, MICHIGAN

A Worldview Approach to Science and Scripture

by CAROL HILL

Published by Kregel Academic, an imprint of Kregel Publications,
2450 Oak Industrial Dr. NE, Grand Rapids, MI 49505-6020.

Cover photographs: (background dinosaur image): Image of *Archaeopteryx* fossil from a quarry in Germany. Photo: Bob Ainsworth; *(background starry images within the title circles):* Image of the Milky Way Galaxy, photographed from Earth. Photo: Zora Zhuang; *(upper left): Adam and Eve*, Peter Wenzel (1745-1829), © Roman Romanadze/Dreamstime.com; *(upper right): Sunrise over Earth*, © Pavel Chagochkin/Dreamstime.com; *(left middle):* The troglobite *Oliarus polyphemus* planthopper in Kazumura lava tube, Big Island of Hawaii. Photo: Peter and Ann Bosted; *(lower left): Noah's ark riding on a swell after the Great Flood*, 3D computer illustration, © James Steidl/Dreamstime.com; *(lower right):* Cuneiform tablet of the Sumerian King List, Babylonia, 1813-1812 B.C. The Schøyen Collection, MS 2855, Oslo and London.

Back cover photographs: (upper left): Tyrannosaurus rex (T. Rex) dinosaur of Late Cretaceous age. 3D illustration, © Orlando Florin Rosu/Dreamstime.com; *(upper right):* A completely dark-adapted, transparent, deep-sea turtle from the Kermadec Trench of New Zealand. Sea Creatures image. Public domain.

Title page image (reverse leaf): Joshua Commanding the Sun to Stand Still upon Gibeon by John Martin, 1816. National Gallery of Art, Washington, D.C. Open access.

Publisher for Kregel Publications: Catherine DeVries

Kregel Academic & Ministry Director: Laura Bartlett

Managing Editor, Kregel Academic & Ministry: Shawn VanderLugt

Copy Editor: Deb Helmers

Proofreader: Rachel Bono

Layout & editing: Susan Coman

ISBN 978-0-8254-4614-6

Printed in the United States of America

19 20 21 22 23 / 5 4 3 2 1

I dedicate this book to my husband,

Alan, for his love, patience,

and encouragement

during the twenty years

since I began this project.

CONTENTS

Foreword RECONCILING SCIENCE AND SCRIPTURE ---------- VIII

Chapter 1 A WORLDVIEW APPROACH ---------- 2

A Worldview Approach: What It Is Not ---------- 3
Three Main "Creationist" Positions ---------- 4
A Worldview Approach: What It Is ---------- 7
Cosmology of the Ancient Near East ---------- 9
Basic Premise of a Worldview Approach ---------- 12
Separating Worldview from Revelation ---------- 14
Timeline of Human History ---------- 15
Preview of What Is to Come ---------- 15

Chapter 2 THE SIX DAYS OF CREATION ---------- 16

Four Traditional Views ---------- 17
Worldview Approach to Genesis 1 ---------- 24
Objections to a Literary View ---------- 26

Chapter 3 THE GARDEN OF EDEN ---------- 28

Location of Garden of Eden ---------- 29
A River Rises *in* Eden ---------- 36
A Worldview Approach to a Garden in Eden ---------- 37
Garden of Eden on a Modern Landscape ---------- 38

Chapter 4 THE NUMBERS AND CHRONOLOGIES OF GENESIS ---------- 40

Two Christian Views on the Age of Earth ---------- 42
The Mesopotamians' Worldview of Numbers ---------- 42
Long Ages of the Patriarchs ---------- 47
Scriptural Problems with Biblical Genealogies and Chronologies ---------- 51
A Worldview Approach to Biblical Chronologies and Age of Planet Earth ---------- 53

Chapter 5 NOAH'S FLOOD: HISTORICAL OR MYTHOLOGICAL? ---------- 54

A Worldview Approach to Noah's Flood ---------- 55
A Time and Place for Noah ---------- 58
Ancient Mesopotamia ---------- 61
Noah's Worldview ---------- 64
Meteorology and Hydrology of the Flood ---------- 65

Chapter 6 NOAH'S FLOOD: GLOBAL OR LOCAL? ---------- 70

"Universal" Language of Genesis 6-8 ---------- 71
Noah's Ark ---------- 74
Landing Place of the Ark ---------- 78
Feasibility of a Local Flood ---------- 82
Local Flood Model and Route of Noah's Ark ---------- 86
The Nature of "Nature Miracles" ---------- 86

Chapter 7 FLOOD GEOLOGY ---- 88
Flood Geology Theology ---- 90
Worldview of the Biblical Author(s) ---- 91
Worldviews in Conflict ---- 92
Case Study #1: Mount Ararat ---- 98
Case Study #2: The Grand Canyon ---- 99
Worldview and the Young-Earth/Old-Earth Debate ---- 107

Chapter 8 EVOLUTION AND THE "NEW" GENETICS ---- 108
Evolution: What Does This Word Mean? ---- 109
After Its Kind: What Does This Phrase Mean? ---- 111
Developments in Evolutionary Biology ---- 114
Supposed Problems for Evolution in the Fossil Record ---- 117
Contemplating the Evidence: My Story ---- 120
Christian Views on Evolution ---- 124
A Personal Perspective ---- 129

Chapter 9 ADAM AND EVE AND ORIGINS ---- 130
What Is the Scientific Evidence? ---- 132
Migration of Humans around the World ---- 140
Four Christian Views ---- 147
A Worldview Approach to Origins ---- 151

Chapter 10 PUTTING IT ALL TOGETHER ---- 158
The Timeline of Biblical History ---- 159
A Worldview Approach to Creation ---- 161
A Worldview Approach to Evolution and Pre-Adamites ---- 161
A Worldview Approach to Adam and Eve ---- 164
A Worldview Approach to Noah's Flood ---- 167
A Worldview Approach to the Patriarchs ---- 172
A Worldview Approach to Moses and the Monarchy/Exile ---- 174
A Worldview Approach to Christ, the Church Age, and Present ---- 177
Final Thoughts ---- 179

References ---- 180

Index ---- 216

Full Endorsements ---- 226

About the Author ---- 228

FOREWORD

RECONCILING SCIENCE AND SCRIPTURE

This book is not a theology book, nor is meant to compete with theology books. It is an apologetics book modeled after Bernard Ramm's 1954 classic work *The Christian View of Science and Scripture* – only this new attempt of reconciling science with Scripture incorporates within it the explosion of scientific knowledge and advancement in biblical scholarship over the last 60 plus years. The purpose of Ramm's book was to promote harmony between Christianity and science, and the purpose of this book is the same. But to achieve harmony I found it necessary to take what I call a "Worldview Approach" to science and Scripture in order to reconcile and harmonize these two seemingly opposing subjects.

I am a scientist – specifically a geologist – but I am also well informed in other fields of science and in ancient biblical history. My background includes a number of published articles in *Perspectives on Science and Christian Faith,* the apologetics journal of the American Scientific Affiliation, of which I am a fellow. I am also the senior editor and one of eleven authors of *Grand Canyon: Monument to an Ancient Earth; Can Noah's Flood Explain the Grand Canyon?* also published by Kregel, a Christian publishing house. These published works form the foundation of this book; yet, my goal for the book surpasses the subject matter covered in these earlier publications.

My overall intent for this book is to make the world of Genesis *real* to people. How can the text of Genesis be understood without considering the worldview, culture, and times of the people who wrote it? How can it be visualized without seeing maps, tables, figures, and artwork representing the world in which these ancient people lived? Without this factual and illustrative information, the biblical world of Genesis remains foreign and "lost" to modern generations, making it difficult for us today to fathom the *worldview* (mindset, or different way of looking at life) of God's chosen people.

Before reading this book, two minor points of clarification need to be made: one in regard to the use of the terms "author(s)" and "scribes," the other in regard to why both the King James Version (KJV) and New International Version (NIV) of the Bible are quoted. Throughout this book I talk about the "author(s)" of Genesis, or sometimes about the "scribes" who wrote the stories handed down to them via oral communication. Authorship of Genesis is a major source of misunderstanding among Christians, as clarified by John Walton and Brent Sandy in their book *The Lost World of Scripture.* The "authors" of the Genesis stories are lost in antiquity, these ancient stories having been passed down for millennia to later generations – first orally, and then in written form. Therefore, when using the term *author(s)* with respect to Genesis, it is meant the *unknown* authors of Genesis. To maintain that Genesis was written word for word by Moses is incorrect: Moses was not the original "author" of Genesis, as these stories were composed thousands of years before Moses lived. However, Moses could have been the historian author who compiled these ancient stories into the book of Genesis. Also, "scribes" were not the "author(s)" of Genesis; they merely copied these stories from oral or written accounts, usually priding themselves on a faithful reproduction of the text. Sometimes I use the term "authors/scribes" to denote the process by which these stories were written: first they were oral stories, then written down and copied by scribes – sometimes over the course of hundreds of years – and then finally canonized into the book of Genesis that we now have in our Bible.

Today it is not popular to quote the King James Version of the Bible. That is because it is hard for us in the modern world to understand the old English in that version. However, since I often reference or expound on the ideas of theologians, historians, or biblical scholars and archeologists who quote the King James Version, I must also quote the KJV so there will not be a disconnect in the use and interpretation of specific words. For example, at the beginning of Chapter 1, I quote Genesis 1:6-7 (KJV) in order to later discuss the ancient meaning of the word *firmament*. The New International Version translates this word as "expanse," which does not convey the exact meaning of firmament, so I do not reference the NIV in my discussion of Figure 1.1.

ACKNOWLEDGMENTS

I want to thank Bob and Sharon Collins, Francis Collins, Dick Fischer, Gary Havens, Roy Hill, Randy Isaac, Peter Rust, David Snoke, and Davis Young for reviewing early renditions and sections of this book. After a lapse of five years while working on *Grand Canyon: Monument to an Ancient Earth,* the manuscript became significantly updated, and this involved a new group of reviewers: Larry Collins, Gregg Davidson, Dick Fischer, Earl Godwin, Alan Millard, Ken Touryan, and Ken and Helen Wolgemuth; theologians Roy Clouser and George Murphy who reviewed the theology of the entire book; biologists/geneticists Sy Garte and Joel Duff who reviewed Chapter 8 on evolution. A special thanks is owed to an anonymous donor who paid for most of the artwork in the book, and to Susan Coman who did an excellent job on layout and in identifying artwork from Dreamstime. Finally, I want to thank Kregel for its amazing support in publishing my book.

— *Carol Hill*

The King James Version of the Bible. The fingers are pointing to Genesis 1:20. *Photograph by Suzanne Tucker. Used with permission, Dreamstime.com.*

Genesis 1:20. And God said, Let the waters bring forth abundantly the moving creature that hath life, and fowl that may fly above the earth in the open firmament of heaven. (KJV)
Painting by Tintoretto, Wikimedia Commons.

A Worldview Approach to Science and Scripture

Sunrise over Earth.
3D Rendering © Pavel Chagochkin | Dreamstime.com.

CHAPTER 1

A WORLDVIEW APPROACH

Genesis 1:6-7. And God said, Let there be a firmament in the midst of the waters, and let it divide the waters from the waters. And God made the firmament, and divided the waters which were under the firmament from the waters which were above the firmament: and it was so. (KJV)

What picture comes to mind when you read these two verses of Genesis? Water vapor in the Earth's atmosphere being the waters above the firmament and liquid water in the oceans being the waters below the firmament? That is the picture that comes to my mind, and after over fifty years of reading the Bible it is hard for me to imagine that the biblical author(s) may have had another concept in mind. The book of Genesis stems from the Mesopotamian culture of about 5000-1500 B.C., and in the cosmological worldview of that time and place, the concepts of "firmament," "waters," and "earth" had different meanings than they do today. I will attempt to explain this ancient cosmology later in this chapter, but first I want to pose the question: What is a Worldview Approach to the science-Scripture debate? Before I say what it is, I will say what it is not.

A Worldview Approach: What It Is Not

More Than Hermeneutics

Worldview is *not* simply a hermeneutics approach, although the concept of worldview encompasses many aspects of hermeneutics. Hermeneutics is the branch of theology that deals with interpretation, or exegesis, of Scripture. It takes into account the ancient language of the biblical author; the historical, physical, political, and cultural setting of the biblical account; the context of the passage with respect to other passages before it, after it, or related to it; and whether the language is figurative, poetic, prosaic, prophetic, narrative, or spoken as parables, metaphors, or hyperboles. Theologians who study hermeneutics understand that the Bible should be interpreted from the *culture* of that time, but few extend that concept to the primitive *science* of that time, even though the understanding of nature (or lack of it) is part of culture.

Hermeneutics is an important tool in interpreting Scripture in that it tries to understand the Bible from all of these various viewpoints, but it cannot begin to solve all of the problems placed on the biblical text by modern-day science. A few years ago I was at a creation conference and one of the keynote speakers – based only on his hermeneutical studies – concluded that the biblical flood *had* to be worldwide, without any mention of the impossibilities of this conclusion based on the geologic and other scientific evidence. There needs to be a better way of reconciling science-Scripture issues in a manner consistent with good scientific, linguistic, and biblical scholarship, and this is the goal of this book.

More Than Culture

Culture is probably one of the most important factors to be considered when interpreting the Bible. In order to understand a passage in its entirety, it is sometimes essential to know the cultural tradition behind it, otherwise its real meaning can be diminished or misinterpreted. An example of such a passage where the meaning is diminished by not knowing the cultural background of the times is Matthew 5:41: "If anyone forces you to go one mile, go with them two miles" (NIV). The historical and cultural context behind this verse is that a Roman, by law, could compel a non-Roman to carry his pack for one mile (1,000 steps), but no more. We call this the "go the extra mile" principle, one of many taught by Jesus in the Sermon on the Mount. The meaning of the passage is that it is not the strict obedience of the law (going exactly one mile and no more) that matters to God, but the attitude of one's heart towards others. Knowing the cultural background of this passage adds to its meaning.

An example of a passage where the meaning can be misinterpreted without knowing its cultural connotation is found in Matthew 5:13: "You are the salt of the earth. But if the salt loses its saltiness, how can it be made salty again? It is no longer good for anything, except to be thrown out and trampled by men" (NIV). This passage has perplexed me for years. What does it mean by "if the salt loses its saltiness?" As a geologist and mineralogist, I know that halite (common table salt) cannot lose its salty flavor. Also, it is not clear what "trodden under foot of men" (KJV) has to do with salt. What I didn't understand about this passage is that in Jesus's time and locality, salt was mined mostly from the nearby Dead Sea, and even today this salt is not always pure halite but is naturally intermixed with gypsum (calcium sulfate) and other minerals. If the halite (salty) fraction is removed from this mixture, then what is left is tasteless gypsum, which has no value in seasoning food or in preserving meat. Therefore, in Jesus's time, the gypsum fraction would be thrown out and used in making footpaths, which would be "trodden under foot of men." The real meaning of this passage, then, from the cultural perspective of that time and place, is that we should remain pure and not be contaminated, lest we be trodden under by the outside world.

While culture can be important in interpreting the biblical text, as shown in the above two examples, culture is *not* the same as worldview, even though the distinction between the two can be subtle. Cultural aspects of a society can be seen or discerned; worldview is not readily seen or discerned, especially by the people who are molded by it, because it is basically an unconscious response to culture. Worldview is all aspects of a culture bound up into a different way of thinking about the world. It is a *mindset* that *stems from* a culture, but it is not the culture itself, and that mindset can differ significantly between subgroups within a culture.

Three Main "Creationist" Positions

Although a number of apologetic positions have been applied to the science-Scripture controversy over the years, these have generally fallen into three main categories: the Young-Earth Creationist position, the Progressive Creationist ("concordist") position, and the Evolutionary Creationist ("accommodationist") position. The Young-Earth and Progressive Creationist positions are held fairly consistently by their followers, but within the Evolutionary Creationist position, a number of people often hold intermediary sub-positions on various issues.

Young-Earth Creationist Position

The Young-Earth Creationist position is that the Bible is the inerrant Word of God, correct in every detail, including any science that it might include. The days of Genesis 1 are literal days, the patriarchal ages are literal ages, and based on these literal time periods, the Earth and universe are around 6,000 years old (10,000 years max). All science is interpreted within this literal framework, especially the time-related sciences of astronomy, geology, biology (evolution), anthropology, and archeology – all must fit within this 6,000-year time period and if they don't, Young-Earth Creationists accept Genesis and reject science. This reinterpretation of science is called "creation science" and is presented by its Young-Earth Creationist advocates as being an equally viable alternative to modern or mainstream science. Creation science is very popular today among many pastors and lay Christians, but it is rarely accepted by people trained in science.

The logic behind the Young-eEarth position with regards to science goes something like this: God knows science better than anyone because he is the creator of the world; we humans in science only describe what he has already created; therefore, his Word should be considered as scientifically accurate as the non-scientific, theological parts of the Bible. Wouldn't God have revealed modern science to the ancient authors of the Bible or, at the very least, wouldn't he have made sure that at least a simplified, but still correct, scientific version be written into the Bible? This is the theological position that God's revelation in Scripture was given in terms of his omniscient knowledge of history and science, and that the pre-scientific understanding of the biblical author(s) is *not* a prime factor in judging science-Scripture issues. In other words, God's knowledge of science was transmitted through the biblical author(s) to the biblical text.

Progressive Creationist Position

This is essentially the "Day-Age" or "concordist" position where long periods of time are acknowledged, but where the Bible is taken literally in most cases. This view considers Genesis 1 to be a remarkable record of actual creation events in that it supposedly "concords" with astronomy and geology *if* the days are interpreted figuratively for long periods of time rather than taken literally as twenty-four-hour periods. The Progressive Creationist position (also called the "scientific concordist approach") seeks to give modern scientific explanations for the biblical text and essentially tries to harmonize scientific data with some of the most important doctrines of the Christian faith. The main method for this harmonizing is through hermeneutics. For example, a hermeneutics (word study) approach is taken with respect to the word *day* in the Genesis 1 text in order to biblically justify that a day can equal a long-age.

The logic behind Progressive Creationism seems to be that the Bible is scientifically accurate and was given in terms of God's omniscient knowledge of

LITERAL OR FIGURATIVE?

The logic behind Progressive Creationism seems to be that the Bible is scientifically accurate and was given in terms of God's omniscient knowledge of history and science; however, some passages must be interpreted figuratively rather than literally to make them fit with science. Like Young-Earth Creationism, Progressive Creationism considers that the pre-scientific understanding of the biblical author(s) is not a prime factor in interpreting the Bible; also, God's knowledge of modern science was transmitted through the biblical author(s) to the biblical text and, therefore, today's modern scientific theories can be imposed on the ancient text.

history and science; however, some passages must be interpreted figuratively rather than literally to make them fit with science. Like Young-Earth Creationism, Progressive Creationism considers that the pre-scientific understanding of the biblical author(s) is *not* a prime factor in interpreting the Bible; also, God's knowledge of modern science was transmitted through the biblical author(s) to the biblical text and, therefore, today's modern scientific theories can be *imposed* on the ancient text. A typical method of imposition used by Progressive Creationists is: invoke the logic that since God's dual revelation involves both science and Scripture, they must fit together; choose a biblical passage that seems to support a modern scientific theory; and then fit the two together without necessarily regarding the biblical context. An example of this approach to interpretation is Job 9:8, which says that God alone spread out the heavens. This passage has been interpreted by some Progressive Creationists as describing the expanding universe of the Big Bang Theory, even though the poetic intent of this verse is evident from its context: "He alone stretches out the heavens and treads on the waves of the sea" (NIV). In other words, Progressive Creationists attempt to read science in between lines of Scripture, attributing to the text a meaning that the ancient human author(s) could never have intended.

Evolutionary Creationist Position

The Evolutionary Creationist position stems from the concept of "accommodation," which is the basic theological principle that underlies all of God's revelation to humankind. God condescends to interact with humans on any level – from becoming the incarnate Christ; to directly interacting with people in the Bible; to indirectly interacting with them through visions, dreams, angelic beings, or the Holy Spirit. Very few theologians would quibble with this general sense of accommodation, and in fact this concept has been applied to a number of theological topics for hundreds of years.

However, relatively recently – within the last few decades – there has been a movement to extend the concept of accommodation to science-Scripture apologetics and to use the term in the specific sense of God accommodating his revelation to the pre-ingrained cultural and scientific ideas of the people with whom he is interacting. This position – that God accommodates his message to fit a culture's worldview – is sometimes referred to as "divine accommodation." This position states that the Holy Spirit descended to the level of the ancients while accommodating their more-primitive views of geology, astronomy, biology, and other sciences. This meaning of *accommodation* is also the interpretation of a Worldview Approach, *except* if it means that God accommodates untruth in his Bible, such as: Adam and Eve were not real persons; the garden of Eden was not a real place; Noah was not a real person and the flood was not a real event; the patriarchs were not real people but were invented at the time of the Monarchy; and so on.

The Evolutionary Creationist position was defined by Denis Lamoureux in his book *Evolutionary Creation.* While the main theme of his book is evolution, Lamoureux also applies his views to other theological topics such as the non-historicity of Adam and Eve, the garden of Eden, and Noah and the flood. However, since Lamoureux wrote his book in 2008, the term *Evolutionary Creationist* has expanded to include views other than those explicitly expressed by him, and for many Christians it can now have a much broader meaning or it can mean different things to different people. For example, most members of *BioLogos* – an online Christian community – consider themselves to be Evolutionary Creationists who subscribe to the unifying theme of God having used evolution and common descent as his tool to carry out the creation of all life. But not everyone in this group believes that the early chapters of Genesis are unhistorical. Lamoureux's position on what Evolutionary Creationists should believe seems set, since he comments in his 2016 book *Evolution: Scripture and Nature say YES!:* "Evolutionary creationists… firmly believe that real history in Scripture begins roughly around Genesis 12 with Abraham." Since the term *Evolutionary Creationist* was originally applied to the ideas of Lamoureux, and since no other term has emerged to replace it, in this book

I will use the modifier *(strict) Evolutionary Creationist* to identify persons who strictly follow Lamoureux's theological position. Then in Chapter 9, when I talk about the different views on Adam and Eve, I will include the subcategory "Historical Adam" to identify more conservative Evolutionary Creationists who acknowledge the historicity of Genesis.

A Worldview Approach: Not a Position

A Worldview Approach favors some aspects of all three of these positions and it disagrees with other aspects. It sees no reason why it *has* to align itself with a specific position and assume a positional label because its theology spans across the divisions that separate the three positions. Instead, it tries to interpret Scripture with respect to *both* the scientific and biblical evidence, while keeping in mind the worldview of the ancient authors – at least as much as can be discerned from Scripture, biblical archeology, and ancient documents related to biblical history. Or, in other words, in this book the concept of "worldview" is considered an *approach to* topics in the science-Scripture debate – it is *not a position on* the science-Scripture debate, although it can influence certain interpretations of Scripture. Since my interpretation of Scripture as presented in this book is based on this approach, I come to certain conclusions about people and events mentioned in Genesis based on what I consider to be the scientific and/or biblical evidence. But there is no reason why, when using this approach, others cannot come to interpretations different from mine.

A Worldview Approach: What It Is

Worldview, or the different ways people groups around the world look at life, is a general term that can be applied to either biblical or non-biblical subjects, and for the history, philosophy, and theology of worldview, the reader is referred to the book *Naming the Elephant: Worldview as a Concept* by James Sire and the references therein. In this book, however, a narrower perspective of worldview is offered. That is, worldview will be discussed specifically as it relates to the Bible and to ancient ideas of how the world was made and functioned. While the cultural aspects of biblical interpretation have been investigated for centuries in the form of hermeneutics, the interpretation of scientific concepts from a worldview perspective has primarily emerged in the last few decades as a number of theologians, apologists, and biblical archeologists have used the concept of worldview to disavow ironclad literalism in favor of understanding the ancient pre-scientific context and literary forms found in the Bible. While many authors have intertwined the concept of worldview into their writings, none to my knowledge have defined, identified, and used a Worldview Approach as their foundation for interpreting the major controversies within the science-Scripture debate as is being done in this book. For example, *In the Beginning...We Misunderstood: Interpreting Genesis 1 in Its Original Context,* co-authors Miller and Soden commit to interpreting Scripture from the different mindset of the ancients; however, their book covers only the Genesis 1 creation controversy.

WORLDVIEW IS SUBJECTIVE

Most people have no idea of what worldview is and how it can affect our thoughts and actions. This is because we see life and the world through this lens without realizing we are doing so. Our way of looking at things seems right and natural to us because that is the way we, in our particular culture, were brought up. And because it seems right and is the foundation of our reality, we come to believe that our worldview is innate and superior to all other worldviews. This ethnocentric bias characterizes all societies, not just ours.

You Don't See the World As It Is, You See It As You Are (Talmud)

Most people have no idea of what worldview is and how it can affect our thoughts and actions. This is because we see life and the world through this lens without realizing we are doing so. Our way of looking at things seems right and natural to us because that is the way we, in our particular culture, were brought up. And because it seems right and is the foundation of our reality, we come to believe that our worldview is innate and superior to all other worldviews. This ethnocentric bias characterizes all societies, not just ours. It is why for hundreds of years the West has puzzled over the "inscrutable" mind of the Asian – and why the Asian has puzzled over the Western mind. Or why in today's world we in the West cannot understand the Arab mind. Robertson McQuilkin, in his book *Understanding and Applying the Bible,* quotes T. E. Lawrence (Lawrence of Arabia), who lived among the Arabs during World War I as saying:

> Semites [Arabs] had no half-tones in their register of vision. They were a people of primary colours, or rather of black and white, who saw the world always in contour. They were a dogmatic people, despising doubt, our modern crown of thorns. They did not understand our metaphysical difficulties, our introspective questionings. They knew only truth and untruth, belief and unbelief, without our hesitating retinue of finer shades.

To the Western mind, Arabs were (are) seen as dogmatic because they weren't (aren't) grounded in the same worldview system. Or, as McQuilkin continues: "It [the differences between them and us] seems almost to come from a basic way of thinking." They *do* come from a different basic way of thinking! Worldview, then, is all aspects of culture bound up into a different way of thinking about the world and about relationships between different members of a society.

In Western culture, individuals are often considered above groups, so what makes the individual happy and self-fulfilled is what is important. Western culture looks at things in a linear, cause-and-effect manner where actions have consequences. In many Eastern (Asian) cultures, the situation can be quite different. Individuals are not that important – it is group relationships that are paramount. The group defines the reality for the individual and, therefore, the individual does what the group expects.

In animistic societies, the spirit world controls one's reality and destiny. Thus the spirit world must be manipulated or appeased to avoid evil, nurture good, and gain power. For example, in some animistic cultures, behavior such as eye contact must be avoided (in case someone tries to give you the evil eye), speech must remain soft (so as not to offend the spirits), and dreams and visions are interpreted as meaningful messages from the spirit world. An animistic worldview is still prevalent in Africa, even among Christians. It is one of the reasons why it has been difficult to control AIDS on that continent. From a spirit-controlled perspective, it is not having unprotected sex that causes AIDS – it is bad spirits.

As will be discussed in Chapter 9, animism is the oldest form of religion, thought to go back tens of thousands of years in Africa and Europe, as portrayed by the shamanistic cave art of these ancient peoples. Remnants of this earliest form of religion are still subliminally layered within all of the major religions of the world today. Animistic layers are present in Buddhism as the notions of good luck and good fortune, as bad numbers and good numbers, and as the concept of "yin" and "yang" whose harmonious (or unharmonious) interactions influence the destiny of people. They are present in Islam as *beraka,* where an impersonal spiritual force is concentrated in objects or persons. They are also present in the Bible, such as in 1 Samuel 28, where Saul consults the witch of Endor, a forbidden medium into the spiritual world. Such "layered" vestiges of root religions such as animism or polytheism affect a culture's worldview, but that culture rarely realizes it is being subtly influenced by such factors.

Worldview of the Genesis Author(s)

Consider all of the differences in worldview that are present in the world today, and then factor in 4,000 years of time difference between our culture and that of the biblical author(s)/scribes of Genesis. We have a Western mindset; they had a Near Eastern mindset. We have a highly scientific mindset, where science defines reality; they had a pre-scientific mindset. They had a written language in the early stages of forming, one that had a "proper" literary convention within their culture. We have a highly evolved written language where nuances of meaning can differentiate between shades of thought. Our modern concept of history is that it be based on "just the facts." But that wasn't the worldview of the ancients. They interpreted history from their different cultures and religious ideas.

This fundamental difference in worldview between us and the ancients also applies to the following argument we so often hear today from Christians, that "there are two revelations: *special revelation* which comes to us in Scripture, and *general revelation* which comes to us in nature, and these two revelations *cannot possibly be in conflict with each other.*" A Worldview Approach considers this view of revelation to be a *false interpretation* of Scripture because nature (as understood by modern science) was *not* revealed to the ancient, pre-scientific world and, therefore, we should *expect* that this ancient-world's recording of Scripture should conflict with modern science.

So how do these differences in worldview affect our interpretation of Genesis? It means that the worldview of the peoples of the ancient Near East *must* (as much as possible) be factored into the reading of Genesis if one is to resolve the conflict of Scripture with modern science.

Cosmology of the Ancient Near East

We now return to the beginning of this chapter, to the cosmology of the ancient Near East. Regarding this cosmology, I will give two examples of how the worldview of the Genesis author(s) might contradict a traditional interpretation of the biblical text.

"Firmament," "Earth," "Fountains of the Deep"

The Mesopotamians' concept of the world was that there existed a solid dome of heaven above the earth called the *firmament* (Fig 1-1). The Hebrew word for firmament is *râqîya',* which comes from the root *rq'* and which suggests a thin sheet of beaten metal (Job 37:18) or an extended surface expanse. The firmament was thought to stretch to the circle of the horizon, which was the boundary of light and darkness (Job 26:10). Above the firmament were the "waters" where the storehouses of snow and hail, and the chambers of wind, were kept; the highest heaven was conceived of as an immense ocean that supplied the rains. The primary function of the firmament was to provide a firm barrier that prevented the waters above the firmament from crashing down upon the earth below and flooding the world. "Sluice gates" or "windows" in the firmament, when they were opened, allowed the rain and snow and hail and wind to descend onto the earth.

The Sun, Moon, and stars moved across, or were fixed in, the firmament. The Sun moved from horizon to horizon each day, entering and exiting in the morning and evening through gates at the edges of the firmament. Clouds existed below the firmament and rose up in order to be filled with water from above the firmament. In Mesopotamian cosmology, the "earth" was considered to be the dry land, a flat disc upon which people walked and worked (the ancients had no concept of Earth as a planet), and their known world was made up of a single continent fringed by mountains (e.g., the Zagros Mountains to the east and the mountains of Ararat in the north) and ringed by a cosmic sea. These mountains ("pillars of the sky;" Job 26:11, Fig 1-1) were believed to hold up the bowl-shaped firmament and to have roots in the netherworld below the earth ("pillars of the earth"). The "fountains of the deep" existed below the flat earth. These fountains tapped into the rivers of the primordial deep or netherworld. Within the earth lay Sheol, the realm of the dead (Num. 16:30-33).

Such a world as just described seems incredibly naïve to the modern mind. But stop and think –

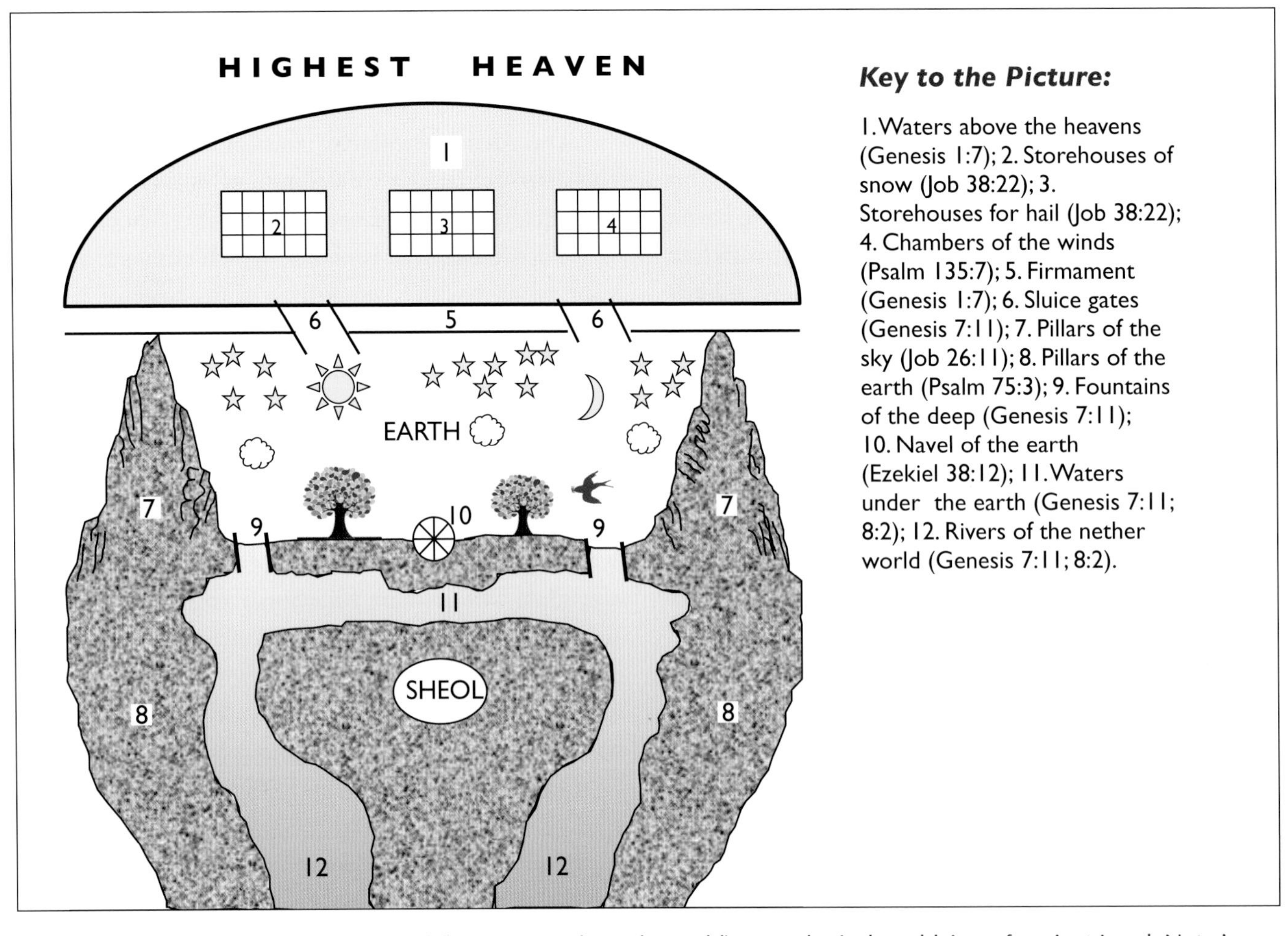

Figure 1-1. A drawing of the three-tiered (heavens, earth, underworld) cosmological worldview of ancient Israel. Note in the key that this cosmological worldview occurs in a number of Old Testament (KJV) books besides Genesis. This ancient view of the cosmos does not show any significant change or development throughout the Old Testament period. Modified from N. M. Sarna, *Understanding Genesis* (1966), The Jewish Theological Seminary of America, New York; then published by Schocken Books as *Understanding Genesis: The Heritage of Ancient Israel.* Permission granted by The Jewish Theological Seminary of America, copyright holder of the Sarna book.

what if you had lived before modern science influenced your concept of the world?

Imagine that you are living in southern Mesopotamia (what is now Iraq) more than 4,000 years ago, standing on a flat alluvial plain surrounded by mountains. You possess no scientific knowledge and observation of the world directly around you is the key to your interpretation of it. From your viewpoint, it looks like the blue sky stretches from horizon to horizon like an upside-down bowl (like the clay bowl you eat your food in), and touches the earth at the circular horizon. It also looks to you as though the surrounding mountains are holding up the sky at the edges of the horizon. During the day you see the Sun moving across the horizon, birds flying, and clouds floating below the surface of the solid, domed, sky blue bowl, and during the night you see the Moon and stars residing in what looks like the same region.

Most of the time, it is very hot and dry where you live, but sometimes a wind brings in clouds, causing it to rain or hail. Where do these waters mysteriously come from? There must be a storehouse of waters somewhere above the solid, bowl shaped, blue sky. And sometimes, this storehouse of waters is opened and then shut, like the sluice gates on the canals surrounding your city. Below the ground is another world, from which issues forth rivers and springs. You have heard that large rivers exit caves in the mountains surrounding your flat plain. You have also heard of thick, black water bubbling up from the ground in places like Hit, where pitch is collected in baskets and shipped down the Euphrates River to your city. There must be a vast reservoir of water somewhere below the earth (land) from whence these springs originate.

This three-tiered (heavens, earth, underworld and/or sea) cosmology was part of the Mesopotamians' worldview and reality because it could be *observed* from their ordinary, everyday experience. In fact, the Mesopotamians were not alone in having this cosmic worldview. Many primitive societies have conceived of the earth as a flat surface, the heavens as a dome, and the underground as the realm of the dead. Such a cosmological worldview (or varieties thereof) is still widespread in the tribal societies of Africa, Australia, North America, South America, and Borneo. This worldview was also prevalent in Egypt and the Near East in ancient biblical times, and it wasn't until the fourth century B.C. that the concept of the Earth as a planetary globe began to emerge in the Near East.

Joshua's Long Day

Another example of how the cosmological worldview of the ancients may have influenced the writing of the biblical text is Joshua's long day. Joshua 10:12-13 is a poem, as Bible translators have recognized since the Revised Version of 1885, so this poetic literary genre should be the reader's first clue that these passages should probably not be taken literally. John Walton has made the case for the interpretation of these passages from the viewpoint of the ancients' use of language describing celestial omens:

> Here I propose that Joshua 10 operates in the world of omens, not physics. Rather than ask what it would mean *to us* for the sun and moon to stop, we must ask what it would mean in the ancient context of celestial omens. It was considered a good omen if the full moon came on the fourteenth day of the month because then the month would be the "right" length and all would be in harmony. If in addition, the sun and full moon appeared simultaneously on opposite horizons on the 14th day of the month (a situation called "opposition"), this would be an especially propitious omen.

The Mesopotamian language, when it came to celestial omens, used active verbs like "wait," "stand," and "stop" to record the relative movement and position of celestial bodies. When the Moon or Sun did not "wait" or "stop," then the Moon would sink over the horizon before the Sun rose, and this would be a bad omen; i.e., one's enemies would prevail in battle. Thus, the Joshua 10:12 passage "Sun, stand still over Gibeon; and you, moon, over the Valley of Aijalon" (NIV) may describe the ancients' desire for the Sun and Moon to be in opposition, or the perfect time to defeat one's enemies (Fig 1-2). Using figurative language, Walton translates Joshua 10:12-15 as:

> O sun, wait over Gibeon and moon over the valley of Aijalon.' So the sun waited and the moon stood before the nation took vengeance on its enemies.

A different interpretation of Joshua's long day is that of Colin Humphreys and Graeme Waddington based on their astronomical knowledge and also on the Hebrew expertise of Semitic Languages Professor Alan Millard. These authors propose that a solar eclipse occurred on October 30, 1207 B.C. (the only known eclipse to pass over the land of Canaan and be visible from Gibeon when Joshua was in Canaan), and that the 1611 King James translators of the Hebrew Bible incorrectly concluded that the Sun and Moon stopped *moving,* while a plausible alternative interpretation of Joshua 10:12-15 is that the Sun and Moon stopped *shining* during a solar eclipse. Since a solar eclipse occurs only when the Moon is directly between the Earth and Sun, it blocks the Sun's light and also the Moon itself is not visible and so it is not reflecting sunlight to the Earth. Again, the idea is that if Joshua actually did encounter this known solar

Figure 1-2. *Joshua Commanding the Sun to Stand Still upon Gibeon* by John Martin, 1816. The Moon (left) and Sun (right) are rising and setting at the same time, a situation called opposition – considered an auspicious time to defeat one's enemies. *National Gallery of Art, Washington, D.C. Open access.*

eclipse as he was about to wage war on the Amorites, then that encounter would have been recorded from the pre-scientific and naïve cosmological worldview of people living then; i.e., it might have appeared like it had stopped moving to those viewing it. Also, what the Israelites would have witnessed during this solar eclipse was a "double dusk" effect, which may have been perceived as a long day. The appearance of the annular (partial) eclipse that these authors propose for 1207 B.C. is shown in Figure 1-3.

These two Joshua 10 interpretations, while different from each other, are basic to a Worldview Approach in that they show how biblical exegesis is not a static process but should be subject to an increased knowledge of science, history and linguistics that needs to be reconciled with Scripture. In the case of Joshua's long day, this solar-eclipse reconciliation does not negate the biblical premise that God performs miracles – only that the text does not insist on a physical miracle of the Sun and Moon stopping in their orbital movement. Nor does it negate the direct involvement of God with this battle or with any of the other battles describe in Joshua because the whole book of Joshua records the personal involvement of the Lord in the conquest of Canaan (e.g., *The Lord said to Joshua...*; Jos. 8:1). Rather, it shows that taking a Worldview Approach to Scripture can help with the interpretation of a difficult biblical passage like Joshua's long day, which for centuries has acted as an "incredibility" block to faith for millions of people.

Basic Premise of a Worldview Approach

We now have enough information in order to try defining a Worldview Approach. The basic premise of a Worldview Approach is that the Bible

Figure 1-3. The annular solar eclipse that passed directly over the land of Canaan in the afternoon of October 30, 1207 B.C. The light from the Sun started decreasing from its normal level at about 3:30 pm until 4:50 pm when it was approximately ten times less intense than normal and dusk set in. By about 5:10 pm, the level of illumination would have been somewhat restored before dusk fell again (a "double dusk"), and then the Sun finally set at about 5:38 pm. This event is not speculation but is known with great accuracy because astronomers can simulate long-ago eclipses using astronomical codes. Whether Joshua witnessed this eclipse is unknown, but possible, considering the correspondence of this 1207 B.C. solar eclipse with known biblical and Egyptian-pharaoh dates. Graeme Waddington and Colin Humphreys, with help from John Jarman, produced this simulation based on Stephenson et al. and on IAU codes and precepts.

in its *original context* records *historical* events *if* considered from the *worldview* of the biblical authors who wrote it. By *original context,* I do not mean the King James Version of the Bible or even the Hebrew Masoretic text, which is but a later translation of more ancient texts based originally on oral stories. I mean that the archeological and literary evidence relevant to the cultural context of that day should be considered along with what has come down to us as the written text. By *historical,* I mean not only history and prehistory in a traditional sense, but also the historical sciences such as anthropology, geology, and archeology when they relate to biblical issues. By *worldview,* I mean the basic way of interpreting things and events that pervades a culture so thoroughly that it becomes a culture's concept of reality – what is good, what is important, what is sacred, what is real. It extends to perceptions of time and space, of happiness and well-being – or to quote James Sires, it is "a fundamental orientation of the heart." The beliefs, values, and behaviors of a culture stem directly from its worldview.

Let us now apply this basic premise to two important theological topics – those of historicity and inerrancy – in order to compare a Worldview Approach to the three main Creationist positions described earlier.

Historicity

With respect to the subject of historicity, or the historical authenticity of the Bible, we will use Noah's flood as an example. We could take a Young-Earth

Creationist, global-flood stance based on the view that "all" and "earth" in Genesis 6-8 refer to the whole globe because God knew "earth" was a planet, and so in his revelation to the biblical author(s), he transmitted this intended meaning. Or we could take a local-flood stance that "all" and "earth" should be viewed from the cosmology of the Mesopotamians; that is, their whole "world" was perceived to be the Mesopotamian alluvial plain and surrounding area. Both of these views consider Genesis 6-8 to be the record of a historical flood and Noah to have been a historical person. Or we could take a third, Evolutionary Creationist stance (in the "strict" sense of Lamoureux): the flood account is an inspired version of an ancient tradition or myth that God accommodated in his Bible because of the theological importance of the Noah story. A Worldview Approach concurs with the second stance: it considers Genesis to be a historical rather than a non-historical account, but one based on the pre-scientific worldview of this ancient people group.

Inerrancy

Young-Earth Creationists would say that the Bible is the inerrant Word of God – correct in every detail – and that all science must be adjusted to fit with this interpretation. Progressive Creationists would say that the Bible is inerrant but that certain passages must be figuratively reinterpreted to fit with science. Some Evolutionary Creationists (again, in the "strict" sense of this position) would say that the Bible is the inspired Word of God, but since the early chapters of Genesis contain myth-like stories, these people, places, and events cannot be considered to be historical.

The issue of inerrancy raises a number of important theological questions. How does God interact with humankind and how has he revealed himself through his Holy Spirit over the millennia? Did God dictate his Word directly into the minds of the biblical authors/scribes, so that these authors wrote down his revelation verbatim? Or, did he meet them "where they were" and allow them to write down his revelation according to their own literary style and from their own cultural and worldview perspective? A Worldview Approach favors this second alternative, as does the *NIV Archeological Study Bible:*

> The Bible is rooted in history – it is not theology divorced from human events and cultures. It is also not science divorced from human understanding within a culture or surrounding cultures.

A Worldview Approach interprets the Bible as the *inspired* revelation of God, where the biblical authors were allowed to express their interaction with God from their own literary and pre-scientific knowledge base. Does this make the Bible untrustworthy, in that it has incorporated the naïve views of the ancients? No, because these pre-scientific views are part of *real history*. If the Bible does not conform to real history, it is not a historical document, and it is then that the stories of Genesis become mythological and thus untrustworthy. Fitting with real history makes the Bible reliable and gives substance and credibility to the basic doctrines of the Christian faith. Does this mean that the Bible is not addressed to all humans in all cultures at all times? No, it means that God's *revelation* is addressed to all people in all cultures at all times, but that the science is not. His timeless revelation will be the same for people 400 years hence, when scientific knowledge will certainly surpass that of today. God guided the biblical authors' thinking and writing according to his will ("All Scripture is God-breathed;" 2 Tim. 3:16, NIV), but he did not teach them modern science.

Separating Worldview from Revelation

The basic problem with a Worldview Approach to Scripture is, of course: How does one separate the "cultural" or worldview aspects from the "divine"? How can the culturally conditioned (scientific and otherwise) passages be distinguished from theologically important and revealed truths? The bottom line is that it is not easy to disentangle the essential theological material from its worldview influence, and setting out general principles for making the necessary distinctions is difficult, if not impossible.

Theologian George Murphy tried to explain this entanglement of worldview with the divine by comparing it to an equally difficult question: How could Jesus have been human and God at the same time? The Bible is the result of both human writing and the Word of God, just as Jesus is both fully human and fully divine. Another example of entanglement is the Evolutionary Creationist position that Jesus accommodated ancient science in order to reveal inerrant spiritual truths, implying that Jesus knew modern science, but pretended not to, and accommodated the untruths of ancient science anyway. A Worldview Approach would interpret this dichotomy in a different light and consider that Jesus, as a human who entered history at a specific time and place, took on and lived within the pre-scientific ideas of that time – he did not accommodate them from his God-like character.

Timeline of Human History

The intent of this book is to take a Worldview Approach to controversial biblical topics in the time frame of God's universe. In order to make this attempt easier for the reader to follow, I have constructed a timeline – starting from the beginning of the universe and going to the present – showing my conception of how biblical history fits within God's universal time frame (Fig 1-4). This timeline will be discussed in Chapter 10 after the relevant information of the next eight chapters is covered.

Preview of What Is to Come

This book is divided into ten chapters. There is a beginning chapter on what worldview is (this chapter), and a final summary chapter (Ch. 10) where we will again encounter the timeline diagram of Figure 1-4. The eight chapters between Chapters 1 and 10 will cover how a Worldview Approach impacts the most controversial and highly debated science-Scripture topics faced by the church today: the Six Days of Creation (Ch. 2), the Garden of Eden (Ch. 3), the Numbers and Chronologies of Genesis (Ch. 4), Noah's Flood: Historical or Mythological? (Ch. 5), Noah's Flood: Global or Local? (Ch. 6), Flood Geology (Ch. 7), Evolution and the New Genetics (Ch. 8), and Adam and Eve and Origins (Ch. 9). We will start biblical history with Chapter 2 where Figure 1-4 starts: with Genesis 1 and the six days of creation.

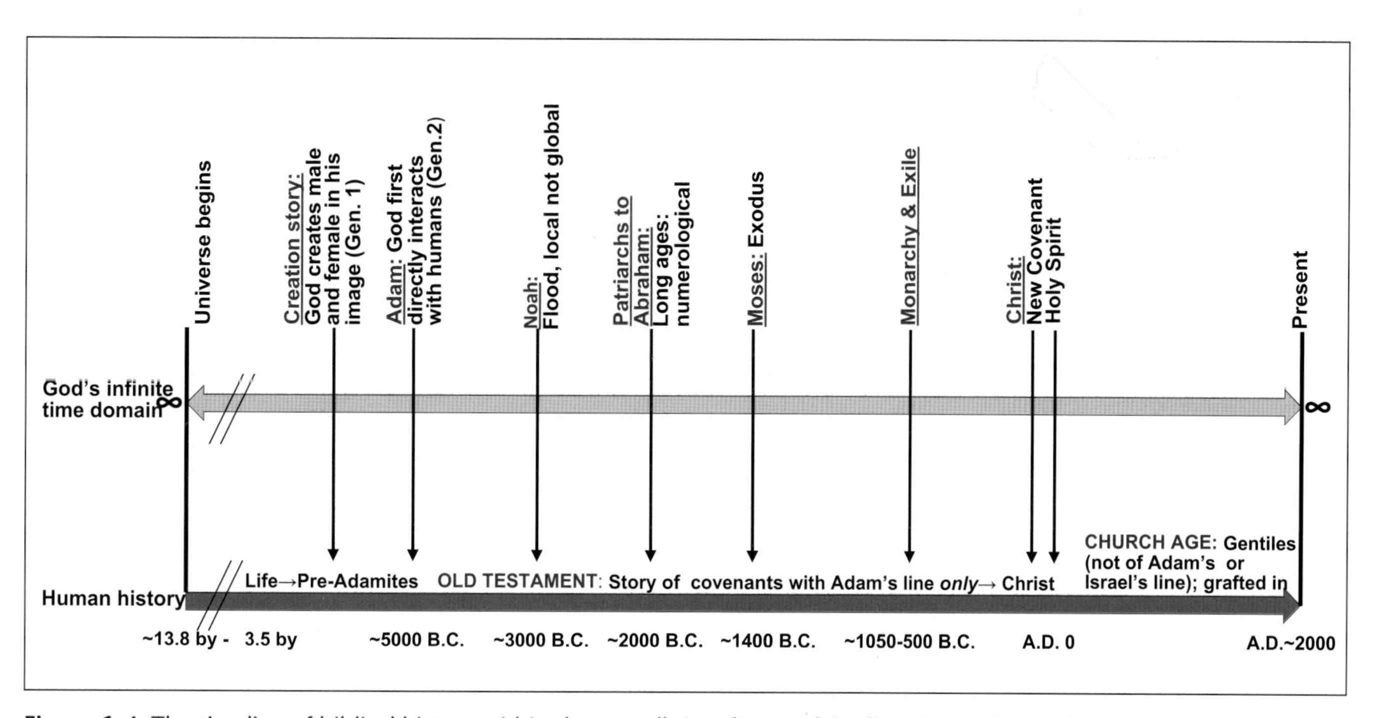

Figure 1-4. The timeline of biblical history within the overall time frame of God's universe (by = billion years).

GENESIS

Incipit liber bresith quē nos genesim
In principio creauit deus celū dicim⁹
et terram. Terra autem erat inanis et
vacua: et tenebre erāt sup faciē abissi:
et sp̄s dn̄i ferebat̄ sup aquas. Dixitq;
deus. Fiat lux. Et facta ē lux. Et vidit
deus lucem q̄ esset bona: ⁊ diuisit lucē
a tenebris: appellauitq; lucem diem ⁊
tenebras noctem. Factūq; est vespe et
mane dies unus. Dixit q̄z deus. Fiat
firmamentū in medio aquaȝ: ⁊ diui-
dat aquas ab aquis. Et fecit deus firmamentū: diuisitq; aquas que erāt
sub firmamento ab hijs q̄ erant sup
firmamentū: et factū ē ita. Vocauitq;
deus firmamentū celū: ⁊ factū ē vespe
et mane dies secūd⁹. Dixit vero deus.
Congregent̄ aque que sub celo sūt in
locū unū ⁊ appareat arida. Et factū ē
ita. Et vocauit deus aridam terram:
congregacionesq; aquaȝ appellauit
maria. Et vidit deus q̄ esset bonū: et
ait. Germinet terra herbā virentem et
facientē semen: ⁊ lignū pomiferū faciēs
fructū iuxta genus suū: cui⁹ semen in
semetipo sit sup terrā. Et factū ē ita. Et
protulit terra herbā virentē ⁊ facientē
semē iuxta genus suū: lignūq; faciēs
fructū ⁊ habēs unūqdq; sementē scdm
speciē suā. Et vidit deus q̄ esset bonū:
et factū est vespe et mane dies tercius.
Dixitq; autē deus. Fiant luminaria
in firmamēto celi: ⁊ diuidāt diem ac
noctem: ⁊ sint in signa ⁊ tpa et dies ⁊
annos: ut luceāt in firmamēto celi et
illuminēt terrā. Et factū ē ita. Fecitq;
deus duo luminaria magna: luminare
maius ut pesset diei et luminare min⁹
ut pesset nocti ⁊ stellas: ⁊ posuit eas in
firmamēto celi ut lucerent sup terrā: et
pessent diei ac nocti: ⁊ diuiderent lucē
ac tenebras. Et vidit de⁹ q̄ esset bonū:
et factū ē vespe ⁊ mane dies quartus.
Dixit eciā de⁹. Producāt aque reptile
anime viuentis ⁊ volatile super terrā:
sub firmamēto celi. Creauitq; deus cete
grandia: et omnē aīam viuentē atq;
motabilē quā pduxerāt aque ī species
suas: ⁊ omne volatile scdm gen⁹ suū.
Et vidit deus q̄ esset bonū: benedixitq;
eis dicens. Crescite ⁊ mltiplicamini: ⁊
replete aquas maris: auesq; mltipli-
cent̄ sup terrā. Et factū ē vespe ⁊ mane
dies quītus. Dixit quoq; deus. Pro-
ducat terra aīam viuentē in genē suo:
iumenta ⁊ reptilia: ⁊ bestias terre scdm
species suas. Factūq; ē ita. Et fecit de⁹
bestias terre iuxta species suas: iumenta ⁊ omne reptile terre ī genere suo. Et
vidit deus q̄ esset bonū: et ait. Facia-
mus hoīem ad ymaginē ⁊ silitudinē
nostrā: ⁊ psit piscib3 maris: et vola-
tilib3 celi ⁊ bestijs uniūseq; terre: omiq;
reptili qd mouetur ī terra. Et creauit
deus hoīem ad ymaginē ⁊ silitudinē
suā: ad ymaginē dei creauit illū: ma-
sculū ⁊ feminā creauit eos. Benedixit-
q; illis deus: ⁊ ait. Crescite ⁊ mltiplicamini ⁊ replete terrā: et subicite eā: et dn̄a-
mini piscib3 maris: et volatilib3 celi:
et uniūsis animātib3 que mouent̄
sup terrā. Dixitq; de⁹. Ecce dedi vobis
omnē herbā afferentē semen sup terrā:
et uniūsa ligna que hñt in semetipis
sementē generis sui: ut sint vobis ī escā
⁊ cunctis aīantib3 terre: omīq; volucri
celi ⁊ uniūsis q̄ mouentur in terra: ⁊ ī
quib3 est anima viuēs: ut habeāt ad
vescendū. Et factū est ita. Viditq; deus
cuncta que fecerat: ⁊ erāt valde bona.

Original by Johannes Gutenberg (printer), ca 1456. Scan by Jossi (public domain).

CHAPTER 2

THE SIX DAYS OF CREATION

Genesis 1:31. And there was evening, and there was morning – the sixth day. (N"

There has been more debate over the first chapter of Genesis than any other chapter in the Bible. How could the Sun, which was supposedly created on the fourth day, have come after plants on the third day when everyone knows that plants need the Sun to live? And, if there was no Sun until the fourth day, how could there have been an "evening and morning" in the first three days? Non-Christians rightly ask: If Genesis 1 is so scientifically inaccurate in these matters, shouldn't it be viewed as an ancient myth rather than the revealed Word of God?

Four Traditional Views

Because Genesis 1 is one of the major stumbling blocks to belief for non-Christians, it has been among the foremost apologetic issues of the Christian church for centuries. The main issue concerns the "days" of Genesis 1. Are these "days" to be taken as literal 24-hour days; are they to be taken for longer periods of time [as in 2 Pet. 3-8 after Ps. 90:4, "a day is like a thousand years" (NIV)]; or are they to be taken metaphorically [as in Joel 2:1, "the day of the Lord" (NIV)]? Unlike English, Hebrew has no word other than *yôm* ("day") to denote a span of time. Therefore, *yôm* can be used literally to mean a 24-hour day (as considered by the Hebrews, from sunset of one day to sunset of the next day), or it can be used figuratively, as when God told Adam that he will die on "the day" he eats the forbidden fruit (Gen. 2:17).

Although the Christian views of Genesis 1 have been many and varied over the centuries, they have usually fallen into four main categories:

Box 2-1. Four Christian Views of Genesis 1			
24-Hour Day	Gap Theory	Day-Age	Literary

Twenty-Four-Hour Day View

This view states that creation was instantaneous by fiat (Fig 2-1), and that the days of Genesis are 24-hour days, with each day marked by an evening and morning. The 24-Hour Day position is held by most Young-Earth Creationists, and some churches or Christian institutions even have this Young-Earth Creationist position incorporated into their constitutions and bylaws, or make it a faith requirement for membership. However, most of the major branches of the Christian church have deliberately avoided doing this.

Figure 2-1. God creating the Sun and Moon by fiat on the fourth day. Michelangelo, Sistine Chapel ceiling (1508-1512). *Wikimedia Commons.*

A 24-Hour-Day interpretation of Genesis 1 is based on a so-called "literal" reading of Genesis. This view, with a few exceptions, has been the traditional position of the English-speaking Protestant church, especially after the King James Version of the Bible was published in 1611, and it wasn't until the eighteenth century that this interpretation was challenged by the findings of science. While traditional interpretations should not be discarded without ample justification, tradition often proves wrong.

Perhaps the strongest Young-Earth Creationist argument in favor of a 24-Hour-Day interpretation is that it is considered to be the most straightforward reading of the biblical text. This is one of the most important principles of hermeneutics: don't alter the plain sense of the text. According to Young-Earth Creationists, if our present-day understanding of science or ancient culture is allowed to alter the plain sense of the text, there is a danger of starting down the "slippery slope" of naturalism, where people interpret the Bible in any way they please. Another important hermeneutical principle is that Scripture must be interpreted within the context of other passages. In the case of Genesis 1, other biblical passages such as Exodus 20:11 seem to support a 24-Hour-Day interpretation of the word *day*: "For in six days the Lord made the heavens and the earth, the sea, and all that is in them, but he rested on the seventh day. Therefore the Lord blessed the Sabbath day, and made it holy" (NIV). However, there are serious objections to the Young-Earth Creationist view – both scientific and theological.

The main difficulty with the 24-Hour-Day view is that not only does it imply that creation took six literal days, but, combined with the chronologies of Genesis (refer to Chapter 4), it implies that the universe and planet Earth are only about 6,000 years old. And this interpretation is in direct conflict with *all* of modern science! According to the latest findings of astronomy, the universe is 13.82 billion years old, not 6,000 years old. This age comes from recent cosmic-background microwave radiation maps (Fig 2-2), but there are many other independent lines of astronomical evidence that support a very old universe in the neighborhood of 10 to 20 billion years. A discussion of this evidence is beyond the scope and intent of this book: the books of astronomer Hugh Ross, *The Fingerprint of God, The Creator and the Cosmos,* and *Creation and Time,* are recommended to anyone interested in a Christian apologetics view of this topic.

Corresponding with a 13.82-billion-year age for the universe is a 4.56-billion-year age for our solar

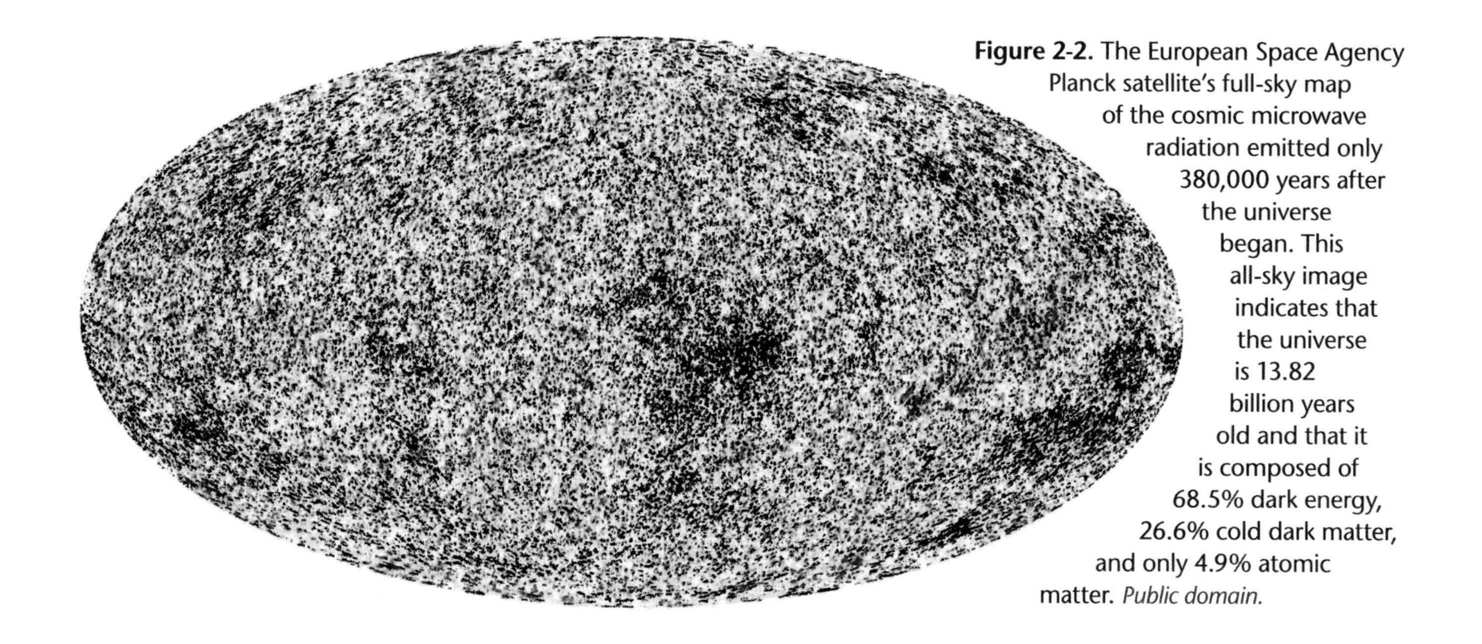

Figure 2-2. The European Space Agency Planck satellite's full-sky map of the cosmic microwave radiation emitted only 380,000 years after the universe began. This all-sky image indicates that the universe is 13.82 billion years old and that it is composed of 68.5% dark energy, 26.6% cold dark matter, and only 4.9% atomic matter. *Public domain.*

system and planet Earth. This age is also based on many independent lines of evidence, some of which are discussed in Chapter 7. From anthropological evidence, the date for early hominids (human-like) species goes back 4 to 6 million years, and the appearance of modern *Homo sapiens* goes back to about 200,000 years. This last topic will be covered in Chapter 9.

The reaction of 24-Hour-Day advocates to the scientific evidence for an old universe and Earth is to either deny the evidence or reinterpret it in light of their 6,000-year theology. This is no small matter because in addition to the denial of astronomy, geology, anthropology, and archeology, certain fundamental principles of physics, chemistry, and genetics also have to be discarded. Can all of science be wrong? Can the basic mechanisms of scientific discovery, on which all of today's technology is based, be so readily dismissed? And how can a 24-Hour-Day position be justified when it is in conflict with such an *overwhelming* amount of scientific data?

Besides the scientific evidence against the 24-Hour-Day view, there are also a number of exegetical problems with this interpretation. For one thing, the events of Genesis 1 seem to be internally conflicting. If the Sun was not made until the fourth day, how could there have been a "morning" and an "evening" on the first three days? Do the terms *morning* and *evening* refer to something other than a daily cycle – should they perhaps be interpreted as representing the beginnings and endings of longer periods of time? There is also the problem of the seventh day, where there is no "morning and evening" statement, and which is open ended and still ongoing. On the seventh day, God rested from his creation work and he is still resting (Heb. 4:1-5). Furthermore, in Genesis 2:3 the term *day* refers to the entire period of God's resting from creating the universe. This "day" began after he completed the creative acts and extends at least to the return of Christ. Therefore, if the seventh "day" is long, why can't the other six days also be longer than a 24-hour period?

Another clue that the "days" of Genesis 1 might not refer to literal 24-hour days is Genesis 2:4: "These are the generations of the heavens and of the earth when they were created, in the day that the Lord God made the earth and the heavens" (KJV). This verse points back to the preceding text of Genesis 1:1 through Genesis 2:3, and it tells how creation ("the heavens and the earth") emerged from what preceded Genesis 2:4. The word *generations* comes from the Hebrew word *tôlĕdôt*, which pertains to descent over a long time period (i.e., narratives, records, family histories). Thus, if "in the day" is taken metaphorically rather than literally, and if "generations" is taken as an extended time period, six literal days for Genesis 1 is a questionable interpretation. The subject of *tôlĕdôt* will be discussed further in Chapter 9 with respect to the correlation of "male and female" in Genesis 1 with Adam and Eve in Genesis 2.

Gap Theory View

The Gap Theory view (also called "The Pre-Adamic Ruin and Restoration Theory") was a nineteenth- to early twentieth-century attempt to reconcile the six days of Genesis 1 with an accumulating amount of scientific evidence for an old Earth and universe. This view envisions an indeterminate "gap" in time between Genesis 1:1 and 1:2. Genesis 1:1 describes God's original perfect universe, which could have been created billions of years ago; Genesis 1:2 describes some undefined catastrophic happening (such as Satan's fall from heaven); and then Genesis 1:3 on describes a re-creation in six literal days about 6,000 years ago.

The main advantage to the Gap Theory view is that it allows for long astronomical and geological ages without sacrificing a literal six-day interpretation of Genesis 1. This theory was immensely popular among Christians in the first part of the twentieth century – so popular that it was favored by the prominent theologian J. Vernon McGee and was included in the Scofield Reference Bible between 1909 and 1945. The Gap Theory view is essentially a compromise between the 24-Hour-Day and Day-Age views.

However, this theory has a number of major difficulties: (1) the Bible verses used to defend it are taken out of context or syntax, (2) the view is so theologically nebulous that a number of variations unsupported by Scripture have sprung out of it, (3) it is not substantiated by either the biblical or scientific evidence, and (4) the theory is not exegetically sound – it simply reads too much into the biblical text. For example, the syntax of Genesis 1:1-3 does not allow a gap: in Hebrew "and the earth was formless and void" is a circumstantial clause which cannot be separated from the preceding clause. Also, it takes a very imaginative interpretation of Genesis 1 to put Satan's fall between Genesis 1:1 and 1:3. In reality, Scripture does not give the slightest hint as to the time and place of Satan's fall, except that Satan was a fallen creature when humankind was created. The other main verses used to defend the Gap Theory view (mainly Jer. 4:23-26; Isa. 24:1, 19) do not refer to creation at all, but to future prophecy.

None of the biblical references that supposedly support the Gap Theory view even so much as alludes to science, yet great leeway is exercised with fitting scientific theories of fossils and fossil humans – who supposedly belonged to pre-Adamic life – into this view. One variation of the theory states that pre-Adamites were wiped out before the onset of the events recorded in Genesis 1:2; in other variations, two different Adams are supposed. There are still a number of variations on the Gap Theory floating around today. For example, one current variation of it goes something like this: the time prior to Adam was the "first creation," from Adam onward was the "second creation," and the time after Christ returns will be the "third creation." Operating from a Gap Theory point of view, this variation interprets the word *window* in Isaiah 24:18 (KJV) to mean "port hole" because it allows movement from one time frame to another. Such fanciful interpretations illustrate the main "con" to the Gap Theory view: theological persuasions of any kind can be incorporated into it. Because of these theological difficulties, and because of the rise in popularity of Young-Earth Creationism, this theory today has become almost extinct.

Day-Age View

In this view, the "days" of Genesis are not literal 24-hour days but six sequential ages of unspecified, although finite, duration. It is alternatively called the *concordist* (Progressive Creationist) view of Genesis 1 because the Genesis days are supposedly in agreement (concord) with geologic periods of time. The Day-Age view is probably second in overall popularity to the 24-Hour-Day view, being the position favored by a number of scientists who are Christians.

As with the Progressive Creationist view, the main attraction of the Day-Age view is that it is in agreement with an ancient universe and Earth as determined by astronomy and geology. In defense of the "days" of Genesis being long periods of time, the Day-Age view uses the argument that the Hebrew word *yôm* can have more than one meaning and can be used for a long period of time as well as for a 24-hour day. In addition, it attempts to fit these days with geologic "ages" or periods as defined by geology. For example, it notes a progressive creative sequence in Genesis 1: the creation of the universe (heavens) and Earth (Gen. 1:1-2); the creation of light (the Sun?) on Day 1 (Gen. 1:4-5, page 2); the formation of water vapor in the atmosphere and liquid water in the seas and on land (Day 2, Gen. 1:6-8); the creation of dry land in one place, and the creation of plants (Day 3, Gen. 1:9-13); the creation of birds and sea creatures (Day 5, Gen. 1:20-23); and the creation of land mammals and humans (Day 6, Gen. 1:24-31). On Day 4 the progressive sequence seems to be interrupted by the creation (or re-creation?) of the Sun and Moon and stars.

Like the 24-Hour-Day and Gap Theory views, the Day-Age view has many weaknesses. It could be argued, as with the Gap Theory, that this view was devised to accommodate modern science in its ideas of an old Earth/universe. It could also be argued that it is supported by verses taken out of context. For example, let's take the phrase "a day is like a thousand years" in context and fit it with the passages directly preceding it:

> Lord, you have been our dwelling place throughout all generations. Before the mountains were born or you brought forth the whole world, from everlasting to everlasting you are God. You turn people back to dust, saying, "Return to dust, you mortals." A thousand years in your sight are like a day that has just gone by, or like a watch in the night. (Psalm 90:1-4, NIV)

These passages, taken together in context, put a different slant on things. The whole idea of Psalm 90 is that God is eternal and man is mortal. Time to God is meaningless – a thousand years are but as yesterday or a watch in the night because he is from everlasting to everlasting. Seen in this contextual light, Psalm 90:4 does not seem to support the days of Genesis 1 as being long time periods. Rather, a "thousand years" is used as a hyperbole emphasizing how much more infinite God is compared to finite man.

This contextual objection is not nearly as damaging to the Day-Age view as the fact that, when inspected closely, this view is *not* in good concordance with either astronomy or geology. The Sun is presumably created on Day 1 since that is when light is created and there is a morning and evening; yet, on Day 4 the creation of the Sun is seemingly repeated. Some interpreters credit this "double creation" to the Sun and Moon and stars becoming visible on the fourth day; i.e., there was some kind of atmospheric cloud cover that obscured these heavenly bodies, so that from the point of view of the observer on the surface of the Earth, they did not appear visible until Day 4. But what observer? Humans were not created until the sixth day. Also, Genesis 1:14 says (KJV): "Let there be [exist] lights in the firmament of the heaven" – a creation command – not "let lights *appear* in the heavens," as if the Sun and Moon and stars were already there but then later somehow became visible.

With regard to geology, it is illuminating to compare the Genesis 1 sequence of events with the actual geologic and paleontological record to see how concordant it really is. In Genesis 1:11-12 (Day 3), the text indicates that plants were formed (before the Sun was visible?), and one might assume that these must have been very early forms of plant life such as algae. But instead, Genesis 1:11 continues: "... and the fruit tree yielding fruit after its kind, whose seed is in itself..." (KJV). But according to the geologic record, seed-bearing fruit trees (angiosperms) do not appear until the Late Cretaceous Period (~80 million years ago) – much, much later than primitive plants and also long after fish appeared in the Devonian Period (~400 million years ago). Yet, the Genesis text has fish appearing in Day 5 along with whales (Gen. 1:21, KJV). In addition, there is a reversed sequence in the appearance of birds and reptiles. Primitive reptiles first appeared in the Mississippian (~340 million years ago), whereas birds didn't appear until about the middle Jurassic (~155 million years ago). Furthermore, whales are mammals that didn't appear until much later in the Cenozoic Era (~50 million years ago). When the Genesis 1 "days" are carefully scrutinized with respect to the fossil record, the concordance is superficial at best.

From a Worldview Approach, the Day-Age view is also not in concordance with the Mesopotamians' cosmology when it comes to Day 2. The Day-Age view interprets the "waters which were above the firmament" in Genesis 1:7 to mean the atmosphere, and the "waters below which were under the firmament" to mean the seas and bodies of water on land. However, as discussed in Chapter 1, in the cosmological worldview of the Mesopotamians the "firmament" was the solid dome of the sky, the "waters which were above the firmament" existed in an immense reservoir above this solid dome, and the "waters which were below the firmament" were the underground waters of the netherworld (Fig 1-1). These are two completely different interpretations of the text – the Day-Age view being dictated by our knowledge of modern science, and the other by the pre-scientific views of the ancients. Or, as succinctly stated by theologian John Walton:

> If we try to turn it [ancient cosmology] into modern cosmology, we are making the text say something that it never said.... Since we view the text as authoritative it is a dangerous thing to change the meaning of the text into something it never intended to say.

Thus, the Progressive Creationist position appears to also be inadequate when considered from Scripture, from the correlation of science with Scripture, and from the worldview of the biblical authors/scribes who wrote the Genesis text.

THE LITERARY VIEW

In the Literary view, the "days" of Genesis 1 are figurative 24-hour days, where the divine works of creation are narrated in a topical order rather than in a strict sequential order. The narrative involves temporality (i.e., it starts "in the beginning" and works towards the creation of humans), but the narrative style is not constrained by the sequence of events. Importantly, the Literary view maintains that Genesis 1 was written following the convention of literary works prevalent in the ancient Near East about 4,000 years ago. A Worldview Approach favors the Literary view over the other three views because it considers the literary conventions of the ancient biblical authors/scribes to be essential to correctly interpreting Genesis 1.

Literary (Framework) View

So far we have examined three views on the six days of Genesis: the 24-Hour-Day view, which takes a strict "literal" interpretation of Genesis, but which denies or reinterprets science in concordance with its preconceptions of Scripture; the Gap Theory view, which compromises both science and Scripture; and the Day-Age view, which tries to force-fit the findings of modern science onto the ancient Genesis text. Most people today have not heard of a fourth option, although it is growing in popularity: the Literary view (also sometimes referred to as the "framework" view because the text is distributed over six days, with a seventh day of rest). Although a non-literal approach to the word *day* in Genesis 1 has been advocated by a number of scholars since the first century A.D., this particular view of a parallel construction for the days of Genesis was introduced to English-speaking, non-Jewish readers by N. H. Ridderbos in his book *Is There a Conflict Between Genesis 1 and Science?*

Now let's see what the Literary view is and how a Worldview Approach might help reconcile the differences between Genesis 1 and modern science. In the Literary view, the "days" of Genesis 1 are figurative 24-hour days, where the divine works of creation are narrated in a *topical* order rather than in a strict *sequential* order. The narrative involves temporality (i.e., it starts "in the beginning" and works towards the creation of humans), but the narrative style is not constrained by the sequence of events. Importantly, the Literary view maintains that Genesis 1 was written following the convention of literary works prevalent in the ancient Near East about 4,000 years ago – which convention not only included literary style, but the Mesopotamians' dual concept of numbers and their importance to the symmetry and harmony of a sacred text. (We will cover this dual concept of numerological and numerical numbers in Chapter 4; here it is discussed only in relationship to Genesis 1.)

As explained by Hebrew scholar Umberto Cassuto in his *A Commentary on the Book of Genesis,* the whole chapter of Genesis 1 is based on a system of numerical harmony. Not only is the number *seven* fundamental to its main theme (God created the world in six days and rested on the seventh), but it also serves to determine many of its details. It was considered a perfect period (unit of time) in which to develop an important work, the action lasting six days and reaching its conclusion and outcome on the seventh day. It was also customary to divide the six days of work into three pairs; alternatively viewed, as two parallel triads of days. So a completely harmonious account of creation, in accord with other ancient examples of similar schemes in literature of that time, and using the rules of style in ancient epic poetry and prose narrative of the ancient Near East, would be the parallel form of symmetry found in Genesis 1. In Genesis 1, the first set of three days represents a general account of creation, while the second triad is a more specific account of the first three days (Table 2-1).

Much debate has revolved around the Genesis 1 topics: (1) Are the days of Genesis long epochs of time or 24-hour periods? (2) How could the Sun have been created on the fourth day after plants? (3) Is modern science in concordance or discordance with the "days" of Genesis 1? But, if taken in the proper and intended context of literature written in the ancient Near East of around 2500-2000 B.C., there is no conflict with any of these topics. The Genesis author was simply writing in the customary prose-narrative style of that day. Thus, the Genesis 1 text was not meant to represent a sequential order of creation nor one that needs to fit with modern science. It was simply the way that writers of that day wrote narrative texts.

The fact that the Literary view is topical is important when it comes to the supposed conflict between Genesis 1 and modern science. The problem of the fourth day disappears because the structure of the narrative is parallel, not sequential. Note this parallelism in Table 2-1. The luminaries (Sun, Moon, stars) in Day 4 are specific to the general topic of light in Day 1. Therefore, the Genesis author was not trying to imply that the Sun was formed after plants – *we* just interpret the text this way from our Western, linear, cause-and-effect worldview. But this was not the biblical authors' worldview. *Their* mindset was to impose a sacred symmetry on the text.

Day 1. Light	**Day 2.** "Waters"; sea and heaven	**Day 3.** Earth or land; vegetation
Day 4. Light from luminaries (Sun, Moon, stars)	**Day 5.** Fish (whales) in sea and fowl in heavens	**Day 6.** Land creatures, land vegetation, and humans
Day 7. Rest		

Table 2-1. A Literary interpretation of the days of Genesis 1.

Look at the next two columns in Table 2-1. Here, all of the supposed geologic objections disappear. Why are "fish" and "whales" grouped together, when fish geologically appeared in the Devonian and whales in the Tertiary? Because both fish and whales (Day 5) are specific to the "waters" in Day 2. No time sequence is implied by these verses. The same applies to "fowl" and "heaven." Birds fly in the heavens and therefore they belong in Day 5 which is specific and parallel to Day 2. The parallelism of Days 3 and 6 is the same. Land creatures (Day 6) are specific to earth (*eretz,* or dry land), and these are land mammals (e.g., cattle; Gen. 1:24) that eat vegetation (Day 3). Such parallelism was the way the minds of ancient authors/scribes – from their literary and worldview perspective – compartmentalized and organized material to be written into a narrative text based on seven as a sacred number. Man is also created on the sixth day in Genesis 1:26, along with the other land creatures. But man was created to be the pinnacle of God's creation, ruling over the animals and eating the plants intended for food:

> So God created mankind in his own image;
> in the image of God he created them; male
> and female he created them. (NIV)

Worldview Approach to Genesis 1

A Worldview Approach favors the Literary (framework) view over the other three views because it considers the literary conventions of the ancient biblical authors/scribes to be essential to correctly interpreting Genesis 1. Let us now examine how ancient narrative texts were written in early biblical times in order to understand these literary conventions.

Figurative Language

A Worldview Approach to Genesis 1 incorporates the ancients use of figurative language, literary conventions, and the symmetry and harmony of numbers into the narrative text. For example, the words *day, morning,* and *evening,* taken figuratively, could apply either to a 24-hour period or a longer period of time. Take your pick, because it didn't matter to the biblical author(s) of Genesis 1. The perfect period of time in which to develop a work was considered to be seven days – whether literal days or figurative days. This does not mean that the biblical author(s) made up the story of creation, only that the divinely revealed story was literally incorporated into the framework of conventional epic prose prevalent at that time. And, since the language is figurative, then passages like Exodus 20 can be theologically reconciled with Genesis 1. Metaphorically, we are to *model* our human work week after the figurative work week of Genesis 1, working six days and resting on the seventh.

Literary Conventions

Literary conventions of the ancient Hebrews included analogy, carefully woven into language, and the use of repetition – this repetition included not only words, but also numbers, phrases, and structural elements such as parallelism. Prime examples of repetition in words, phrases, and structure are seen in Genesis 1:11-12; 1:27; and 2:1. Furthermore, in the worldview of the Mesopotamians, language not only stated facts, it could establish them. Thus in Genesis 1, when God said "Let there be light," by this statement (in the minds of the ancients) light was created. The ancient Hebrews also believed in the identity and essence between a name and what it meant and that a being or thing only fully came into existence once it was given a name; e.g., as in Genesis 2:19 where Adam names all of the animals. They also loved a play on words: for example, *adam* (generic humans) in Genesis 1 and *Adam* (a specific human) in Genesis 2. None of this play on words was gratuitous; it was the very basis of intellectual thought. And, while this type of thought, or worldview, is quite foreign to our way of thinking, it still needs to be considered because where the biblical author(s) were "coming from" is essential to the correct interpretation of Genesis.

Literary Harmony in Genesis 1 Text

An important part of the worldview of the ancient Mesopotamians was that harmony and balance be maintained in their everyday lives (kind of like "yin" and "yang" to the Chinese). We will discuss this topic in more detail in Chapter 4 when we talk about the numbers of Genesis, but it is especially appropriate here with regard to the construction of the Genesis 1 text because it shows the mindset of the ancient author(s) who wrote the text. One way harmony was achieved was by using symmetry and parallelism in prosaic writings and by using "good" numbers rather than "bad" numbers. A prime example of literary and numerical symmetry in an ancient Near Eastern text is the first chapter of Genesis (the Hebrew is read from right to left, and is in stanzas, not verses).

A close look at the Genesis 1 text reveals the carefully constructed and intricate harmony of the *Masoretic* Hebrew text in terms of the sacred numbers three and especially seven. The first stanza of Genesis has seven (7 x 1) Hebrew words in it, and the second stanza has fourteen (7 x 2) words (Box 2-2). After the introductory stanza, the section is divided into *seven* paragraphs, each of which pertains to one of the *seven* days. Each of the *three* nouns that occur in the first stanza ("God," "heavens," and "earth") is repeated throughout the chapter a multiple of *seven* times: "God" occurs thirty-five (7 x 5) times, "earth" is found twenty-one (7 x 3) times, and "heavens" appears twenty-one (7 x 3) times. Each stanza after the first contains *three* pronouncements that emphasize God's concern for man's welfare (*three* being the number of emphasis), namely the phrases "Let us make man," "Be fruitful and multiply," and "Behold I have given you every plant yielding seed." Thus, there is a series of *seven* corresponding dicta of triads (*threes*). The terms "light" and "day" are found *seven* times in the first stanza, and there are *seven* references to "light" in the fourth (parallel) stanza. "Water" is mentioned *seven* times in stanzas two and three; "beasts" *seven* times in parallel stanzas five and six; and the expression "it was good" appears six times – the *seventh* time "very good" for emphasis. To suppose that all of this is a mere coincidence is not possible. Unquestionably, the repetitions were introduced for the sake of parallelism in accordance with the customary stylistic literary conventions of that day. In other words, the text was *purposely* constructed this way by the biblical author(s) in

Box 2-2. A Hebrew translation of Genesis 1:1-2 illustrating its literary and numerical symmetry.

Genesis 1:1:

הָאָרֶץ׃	וְאֵת	הַשָּׁמַיִם	אֵת	אֱלֹהִים	בָּרָא	בְּרֵאשִׁית
ha'aretz	ve'et	hashamayim	'et	'Elohim	bara'	b^{e}reshit
the earth	and +	the heavens	(dir obj)	God	he created	In the beginning

Genesis 1:2:

תְהוֹם	פְּנֵי	עַל-	וְחֹשֶׁךְ	וָבֹהוּ	תֹהוּ	הָיְתָה	וְהָאָרֶץ
tohu	p^{e}ney	'al	v^{e}chosheq	v^{e}bohu	tohu	hay^{e}tah	v^{e}ha'aretz
deep	face	upon	and darkness	and void	formless	it was	and the earth

הַמָּיִם׃	פְּנֵי	תֹהוּ	מְרַחֶפֶת	אֱלֹהִים	וְרוּחַ
hamayim	p^{e}ney	'al	m^{e}rachefet	'Elohim	v^{e}ruach
the waters	face	upon	hovered	God	and Spirit (of)

order to attain a sacred symmetry and harmony commensurate with their ancient worldview of literary correctness.

Chapter 1 of Genesis is not the only section of Genesis to display numerical repetition and symmetry based on the number seven. It also applies to the texts of the garden of Eden, Cain and Abel, and Noah and the flood. This consistency gives a remarkable unity to the Genesis stories, or, as stated by Hebrew scholar Cassuto: "This numerical symmetry is, as it were, the golden thread that binds together all the parts of the section and serves as a convincing proof of its unity." Not only that, but the type of figurative and numerical narrative writing displayed in Genesis 1-11 is also evidence of its antiquity. Or, as put by archeologist Kenneth Kitchen: "Genesis 1-11 shared the view of the ancients, being ancient itself." Such antiquity (~2000-1600 B.C.) has been challenged by the so-called "critical scholars," who have proposed that many of the books of the Old Testament were written during the time of the Monarchy or Exile (~800-400 B.C.). This topic of how and when Genesis was written will be discussed in Chapter 10.

Objections to a Literary View

Let's now list some of the main objections that Young Earth Creationists and Progressive Creationists have against the Literary view of Genesis 1, and for each of these objections I will argue to the contrary from a Worldview Approach.

Not the traditional rendering of the church. I would argue: Genesis 1 was not written by or for the church. It was written by and for people of 4,000 years ago with a pre-scientific worldview before the Christian church existed.

Recent interpretation of the text. I would argue: While a Literary view interpretation of parallel structure is relatively recent, having been around only within the last one hundred years or so to English readers, and while this interpretation is partly based on the discovery of cuneiform texts that help reveal the literary style of the ancient Mesopotamians, the original intent of the text has not changed, since it was *never* intended to teach science but to teach God as creator.

Not the most straightforward rendering of the text. I would argue: This may not seem the most straightforward to us, but perhaps it was the most acceptable rendering of the text to an ancient Near Eastern audience. In their worldview, the way Genesis 1 was written down was the most honorable and straightforward way to write a sacred text. The parallelism, the figurative terms, and sacred numbers, as presented in this chapter, all show a "straightforward" text written by authors *not* familiar with modern science or in possession of a Western European literary worldview.

Non-chronological order to the text. I would argue: The text was not meant to be a sequential account of creation. It was written in the repetitive, parallel-construction, literary format of that day, following the then-rules of narrative style. The "days" of Genesis represent episodes of divine creativity stated in a literary framework that provides a teleological order rather than a chronological or causal order.

Non-literal interpretation of the text. I would argue: It is ironic that a figurative interpretation of Genesis 1 may be more "literal" than the 24-Hour-Day view because that is how the original author(s) *intended it to be.* Or, as Conrad Hyers has put it: "Unwittingly, 'literal' or 'concordist' views are secular rather than sacred interpretations of the text. To faithfully interpret Genesis is to be faithful to what it really means as it was originally written, not to what people living in a later time, or coming from a different worldview, desire it to be."

IMPORTANT LESSONS TO BE LEARNED FROM GENESIS 1

What are the most important lessons to be learned from Genesis 1?

(1) Genesis 1 was not written to teach astronomy, geology, or biology. It was written to show the relationship of a specific Near Eastern people group to a creator God and their unique status in God's divine covenantal program with them. Too much emphasis on either a scientific or "literal" reading of Genesis 1 takes away from this all-important truth.

(2) God revealed his creation story to a people group surrounded by nations practicing animism and polytheism – religions based on mythology. The whole chapter of Genesis 1 has a strong thrust against such mythology; for example, God created the Sun and Moon – they are not to be worshipped, God is. Also, humans are made in the image of God (Gen. 1:26) and are, therefore, not to worship images of wood and stone but the living God, who is creator of heaven and earth.

(3) Genesis needs to be interpreted from a Near Eastern, 2500-1500 B.C. worldview – not from a seventeenth century A.D., European "King James" worldview or a twenty-first century A.D., scientific worldview.

Part of great canvas, *Adam and Eve in the Garden of Eden* painted by Peter Wenzel in (1745-1829).
Photo taken in 2017 by Roman Romanadze in Vatican Musem's Pinacoteca, Rome.

CHAPTER 3

THE GARDEN OF EDEN

Genesis 2:8. And the Lord God planted a garden eastward in Eden; and there he put the man whom he had formed. (KJV)

The garden of Eden is another controversial topic. Was it a historical place or a mythological place? A Worldview Approach considers it to have been a real place, contrary to the (strict) Evolutionary Creationist position, which considers it to be non-historical. If it was a historical place, then where was it located?

Location of the Garden of Eden

The exact location of Eden was identified by the biblical author(s) in Genesis 2:10-14:

> [10] And a river went out of Eden to water the garden; and from thence it was parted, and became into four heads.
> [11] The name of the first is Pison [Pishon, NIV]; that is it which compasseth the whole land of Havilah, where there is gold;
> [12] And the gold of that land is good: there is bdellium and the onyx stone.
> [13] And the name of the second river is Gihon: the same is it that compasseth the whole land of Ethiopia [Cush, NIV].
> [14] And the name of the third river is Hiddekel: that is it which goeth toward the east of Assyria. And the fourth river is Euphrates. (KJV)

The Name of the First Is Pishon

The land of Havilah. The Bible mentions two Havilahs in the Table of Nations: Havilah the son of Cush (Gen. 10:7) and Havilah the son of Joktan (Gen. 10:29), grandson of Noah through the line of Shem. The "land of Havilah" has been interpreted by many biblical scholars to be Arabia, and Joktan to be the head of the tribes of Arabia, as most of his sons can be traced to places and districts within what is now Saudi Arabia and Yemen.

But within the land of Havilah where is the Pishon River? There is no river that flows from the western mountains of Saudi Arabia down to the head of the Persian Gulf. Although no river flows across Saudi Arabia today, evidence exists that such a river did flow there in the past. An average of four inches of rain now falls in Saudi Arabia each year, but the climate was much wetter than it is today during the periods from about 30,000 to 20,000 YBP (years before present) and from about 11,000 to 6,000 YBP, which period geologists refer to as "the Green Sahara." Even as late as 3500 B.C., ancient lakes are known to have existed in the "Empty Quarter" of Saudi Arabia, which is today the largest sand desert in the world (Fig 3-1). A drier, but somewhat still moist phase, existed from about 4000 to 2350 B.C., followed by a more arid phase from about 2350 B.C. to the present. The first great civilizations in Africa, Egypt, and Mesopotamia began during this approximate 6000-4000YBP time frame.

Has the Pishon River been found? Starting from the Persian Gulf at Umm Qasr, the now-dry Wadi al Batin (*wadi* means "dry river") can be followed upstream to the southwest past the borders of Kuwait and into Saudi Arabia, where it is incised into a Tertiary limestone-sandstone sedimentary rock terrain. Then, just past Al Hatifah, the dry riverbed is engulfed by immense sand dunes and disappears (Fig 3-1).

Although obscured by sand dunes, satellite photos can still detect the Wadi al Batin continuing to the southwest beneath the sand and emerging as the Wadi Rimah (that is, both wadis were part of the same river system in the past). It is this old river system – now an *underground* river bed detectable by satellite – that is the Pishon River of Genesis 2:11. This river flowed at a time when the climate was much wetter than it is today, which implies an extraordinary memory on the part of the biblical author(s), since the river dried up sometime between about 3500 and 2000 B.C. This is further evidence for

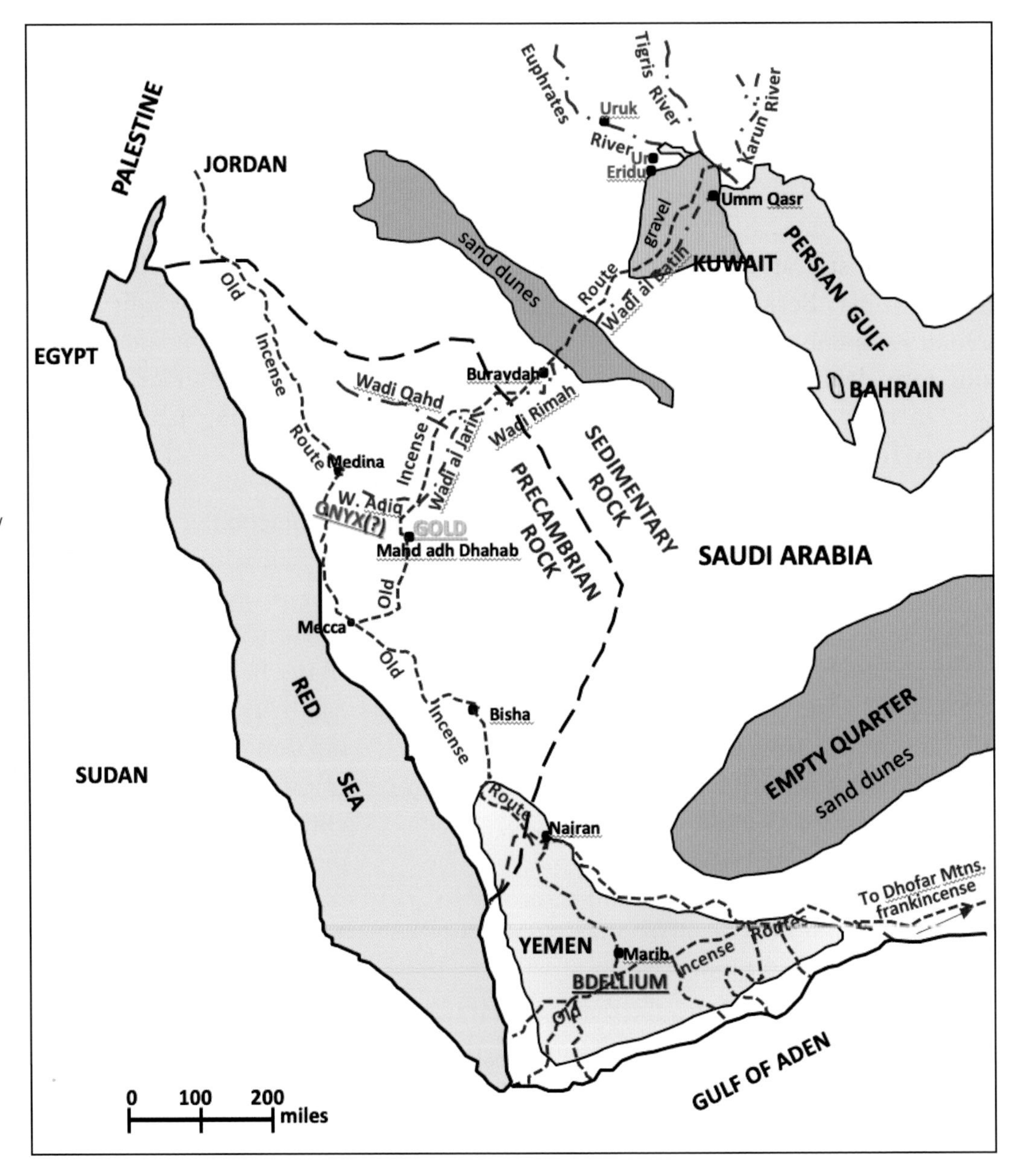

Figure 3-1. Map of the "land of Havilah" (Saudi Arabia and Yemen) showing the contact between Precambrian rock and sedimentary rock (long black dashes between the two areas), location of the old incense routes (short purple dashes) on the western side of the Arabian Peninsula, the area where bdellium was/is grown (green area), the gravel fan deposited by the once-flowing Pishon River (gray gravel area in Kuwait), where gold and onyx was mined in antiquity, and places mentioned in the text. The southeastern side of the Arabian Peninsula, in the Dhofar region (east of this map; purple arrow), is the best location for where frankincense was (and still is) grown.

the antiquity of the biblical text that we talked about in Chapter 2 and, in the case of the Pishon River, can be thought of as a "historical memory" of past events (and very rarely, of recent events).

Mahd adh Dhahab: cradle of gold. About 80 miles farther in the upstream direction, the Wadi Rimah forks into the Wadi Qahd on the northwest, and the Wadi al Jarir on the southwest (Fig 3-1). The Wadi al Jarir continues up-gradient to the area of the Mahd adh Dhahab gold mine exactly as Genesis 2:11-12 records that the River Pishon encompasses the whole land of Havilah, "where there is gold.... And the gold of that land is good" (KJV). The gold of that land is indeed good! Mahd adh Dhahab (literally meaning "cradle of gold") was one the largest and richest gold mines of the ancient world. It is the gold mine believed to be the fabled "Ophir" mine mentioned in the Bible (Ophir was another of Joktan's sons) and the source of King Solomon's gold (1 Kings 9:28; 10:11). Gold occurs at Mahd adh Dhahab mostly as electrum (gold-silver) within quartz veins. Besides gold, the mines have also produced a substantial amount of silver, copper, zinc, and lead. The quartz veins containing the gold intrude (cross-cut, or are younger than) the Mahd adh Dhahab Series of Precambrian volcanic and sedimentary rocks.

The onyx stone. Onyx, sardonyx, agate, and carnelian are all varieties of chalcedony, a very fine crystalline quartz. In antiquity, all of these (and other) varieties of fine-grained quartz were sometimes referred to by the general label "onyx." Chalcedony was fancied by the ancients in any form. Onyx stone has been identified from Mesopotamia in archeological levels dating from about 4000 to 3200 B.C. onward. Chalcedony is also known to occur in the western desert of Arabia, and especially in the Mahd adh Dhahab-Wadi al Aqiq area (Fig 3-1). (That *aqiq* can mean "agate" in Arabic may signify that the onyx mentioned in Genesis 2:12 came from this area.)

There is bdellium. Also mentioned in Genesis 2:12 is bdellium. Bdellium is a fragrant gum resin obtained from plants of the bursera (balsam) family. Frankincense comes from trees of the genus *Boswellia* of the bursera family, while myrrh and bdellium come from trees of the genus *Commiphora*. All of these types of gum-resins (frankincense, myrrh, and bdellium) were used in the ancient Near East for religious (incense), cosmetic (perfume), and medicinal purposes. The trees from which myrrh and bdellium are extracted grew exclusively in southern Arabia (modern-day Yemen) and northern Somalia during ancient times (Fig 3-1). The Old Testament records the Queen of Sheba coming to King Solomon's court by caravan along the great Arabian incense route from Mariaba (Marib), now a part of Yemen, carrying gold, precious stones (like onyx), and spices (1 Kings 10:1-13; Fig 3-2).

The Name of the Second Is Gihon

Proceeding counterclockwise in a continuous sweep around southern Mesopotamia (as Genesis does), the second river is the ancient "Gyudes," which Genesis calls the Gihon. The location of the river Gihon is debatable. The problem revolves around the identity of the "land of Cush," which in the King James Version was translated "Ethiopia." Not only is this translation questionable, it also does not make sense. A river in Ethiopia would flow to the Red Sea or to the Mediterranean or Indian Ocean, not to the confluence of the Euphrates and Tigris Rivers as stated by Genesis. According to biblical scholar Ephraim Speiser, "The land of Cush has been mistakenly identified with Ethiopia, rather than with the land of the Kassites." And, according to biblical archeologist Kenneth Kitchen: "It should be obvious that this Kush cannot have been Upper Nubia in East Africa, most of two thousand miles away to the southwest of the Tigris zone. The land of 'Kush' would be the land of the Kassites (Kashshu)." The Kassites (or *kaššû*) lived to the east of Mesopotamia in the Old Babylonian Period (2000-1600 B.C.; Table 3-1), but before then this area was known as the land of Elam or Susiana, where the inhabitants on the Plain of Susa lived (Fig 3-3). If

Figure 3-2. *Solomon and the Queen of Sheba* by Giovanni Demin (1789-1859). *Wikimedia Commons.*

Cush is the territory of the Kassites, then the river referred to in Genesis 2:13 must have flowed from the east, or from what is today western Iran. In addition, if *kush* denotes the territory of the Kassites, then this would be an indication that Genesis 2:10-14 was written during or just after the Old Babylonian Period, not during the much later time of the eighth to sixth centuries B.C. as maintained by the "critical scholars" (see Chapter 10). In other words, this reference to "the land of Cush" may be yet another of the "historical memories" preserved in Genesis.

It compasseth the whole land. If the Gihon River flowed from the east, or western Iran, then the most likely candidates for the biblical Gihon are the present-day Karun or Karkheh Rivers. The word *compasseth* in Hebrew means "to revolve, surround, or border, or to pursue a roundabout course, to twist and turn," and that is exactly what the Karun River does: it is a meandering river with great bends. Its course is 510 miles long, but its lateral distance across (as the crow flies) is only 175 miles wide. The reason for this zigzagging, "roundabout course." is that the sedimentary rocks of the Zagros Mountains are folded into great anticlinal and synclinal structures, and the Karun winds back and forth along, and then cuts across, this series of folded rocks. The Karkheh also twists and turns on its path westward to the Tigris River. It also flows past the archeological mound of Susa in the land of Elam (Fig 3-3). Susa was the primary city of ancient Persia, and is known by archeologists to have existed as early as around 4000 B.C.

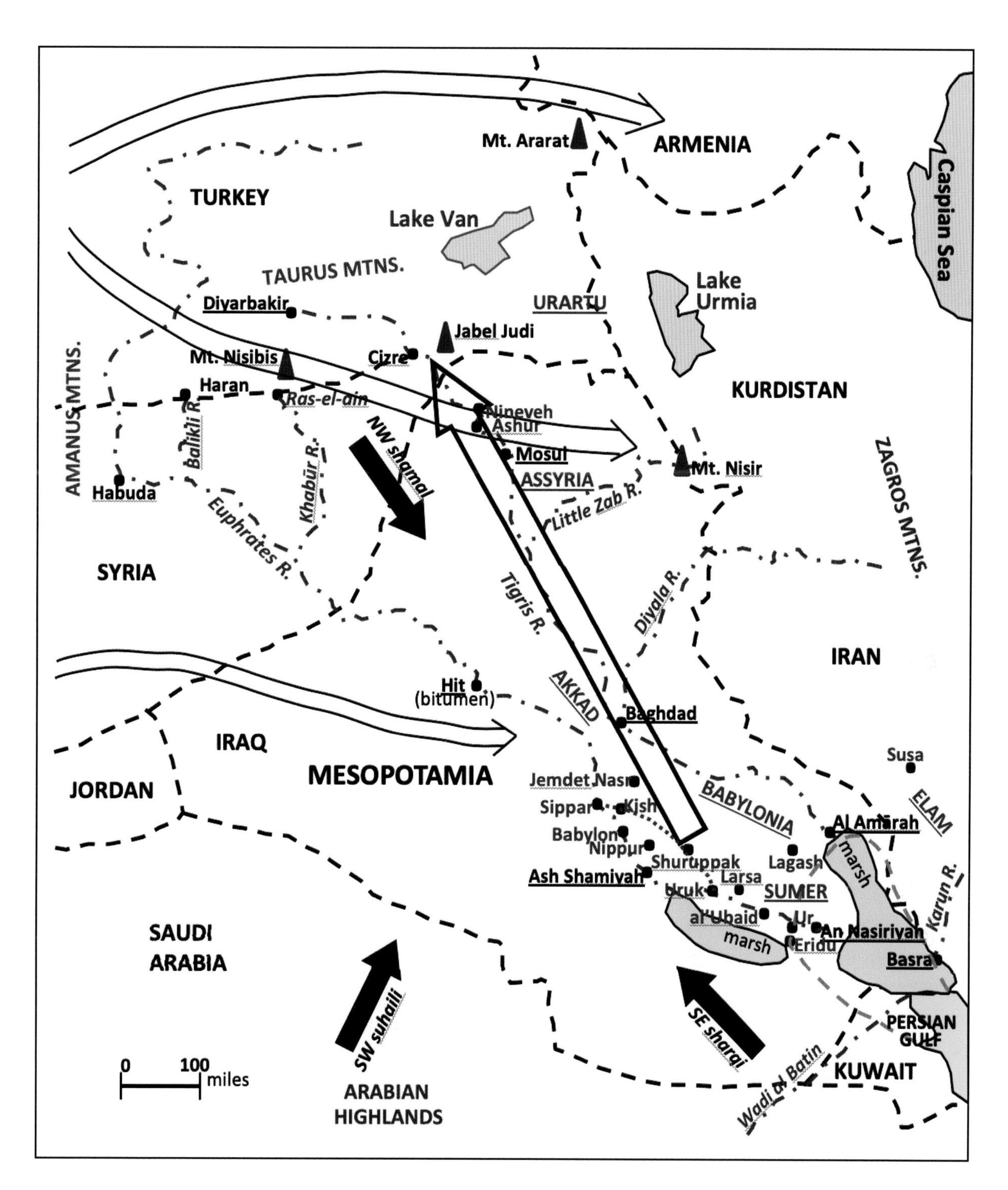

Figure 3-3. The land of Mesopotamia, showing (in red) the ancient regions and cities of Babylonia, Sumer, and Akkad in southern Mesopotamia, Assyria in northern Mesopotamia, Elam in Iran, and Urartu in southern Turkey. The brown triangles represent the four most-frequently cited landing places of Noah's ark. The red dotted line from Sippar to Kish to Shuruppak to Uruk denotes the route of the Euphrates River in Noah's time and was called *Purattu* by DeGraeve. The blue-green dashed line shows the limit of the Persian Gulf in Noah's time. Also shown are the cyclonic weather patterns for the region (curved white arrows), the predominant winds (bold black arrows), the marshlands (green areas), ancient cities (now archeological mounds) mentioned in the text (red), modern countries, and modern cities (underlined). Rivers and river/spring names are in blue. The large white straight arrow shows the proposed direction that Noah's ark could have taken from Shuruppak (Noah's "home town") to the "mountains of Ararat" (Urartu). The "Mesopotamian hydrologic basin" is the whole watershed area surrounded by the Zagros Mountains, Taurus Mountains, Amanus Mountains, and Arabian Highlands.

Archeological Period	Archeological Assigned Age	C-14 Dates (calibrated)	Biblical Person/Event
Ubaid	~5500-3800 B.C.	ca. 6000-4000 B.C.	Eridu, Adam and Eve?
Uruk	~3800-3100 B.C.	ca. 4000-3350 B.C.	Tubal-Cain, Jabal, Jubal?
Jemdet Nasr	~3100-2900 B.C.	3350-2960 B.C.	Shuruppak, Noah and flood?
Early Dynastic I	~2900-2750 B.C.	2960-2760 B.C.	Nimrod?
Early Dynastic II	~2750-2600 B.C.	2760-2655 B.C.	Tower of Babel?
Early Dynastic III	~2600-2350 B.C.	2655-2260 B.C.	
Dynasty of Akkad	~2350-2150 B.C.		
Third Dynasty of Ur	~2150-2000 B.C.		
Old Babylonian	~2000-1600 B.C.		Abraham = ~2000 B.C. Joseph = 1800 B.C.

Table 3-1. Archeological periods of Mesopotamia and their possible correlation with people, places, and events in Genesis. If Adam lived in southern Mesopotamia, where Genesis says the garden of Eden was located, it would have been at the beginning of the Ubaid Period, since that is the earliest archeological period identified for that area. The radiocarbon (calibrated C-14) dates are from a variety of sources.

The Name of the Third Is Hiddekel

The third river of Genesis is the Hiddekel, which is the Hebrew name for the Tigris River. The Tigris River rises on the southern slopes of the Taurus Mountains in eastern Turkey and cuts a bed almost 1,160 miles long on its way to the Persian Gulf. The Tigris was the great river of ancient Assyria. On its banks stood many of the cities mentioned in Genesis (Fig 3-3), including Nineveh, Nimrud, and Ashur (Ashur was one of the children of Shem and built the city of Nineveh; Gen. 10:11, 22). Genesis 2:14 identifies the Hiddekel as "that is it which goeth toward the east of Assyria" (KJV) or the land of Ashur (also sometimes spelled Asshur or Assur). And the Tigris does (and did) flow east of ancient Ashur (now the mound of Ashur), in perfect concordance with Genesis 2:14.

And the Fourth River Is Euphrates

Continuing counterclockwise from the Pishon to the Gihon to the Tigris is the fourth river of Eden, the Euphrates (Gen. 2:14). Draining the western part of ancient Mesopotamia, it starts in the highlands of Turkey, flows southeastward over a limestone hill terrain in northern Iraq, and enters its delta at Hit (about 80 miles west of Baghdad; Fig 3-3). Overall, the river winds its way over a 1,700-mile path on its way to the Persian Gulf. South of Hit, the river has an extremely low gradient. Hit is located more than 500 miles upriver from the gulf, yet is only 175 feet above sea level. At An Nasiriyah, the water level of the Euphrates is only eight feet above sea level, even though the river still has to cover a distance of more than ninety-five miles to Basra. Once Ash Shamiyah is passed, water of the Euphrates is lost in an immense marshland region, where water percolates slowly through the marsh and into the Persian Gulf (Fig 3-3). In ancient times, this entire, very-flat southern Mesopotamian region – from the Euphrates east to the Tigris – could become severely inundated with spring floods.

The flow rate of the Euphrates River exhibits significant seasonal fluctuations, as does the Tigris. The highest flow for both rivers is in April and the lowest is in September. The monthly flow of the Euphrates is between 275 percent (April) and 33 percent (September) of the annual average; the Tigris is between 260 percent (April) and 23 percent (September).

Where the Four Rivers Meet

So exactly where was the garden of Eden located with respect to the four rivers of Genesis 2:10-14? Assuming a date of about 5500 B.C. for Adam and the garden of Eden (Table 3-1), the four rivers would have converged near the Persian Gulf at a position somewhat inland from where the gulf is today. The Pishon River (Wadi al Batin) now enters the Persian Gulf at Umm Qasr, but the cobbles and pebbles from this fossil river system once extended as a fan from southern Kuwait northward to the vicinity of Ur (Fig 3-1). At about 5500 B.C., the sea level began rising and an advancing shoreline transformed the southern part of the Mesopotamian plain into a freshwater marsh and estuarine environment; by around 3500 B.C., this marine transgression had completely inundated the southern plain (Fig 3-3, blue-green dashed line). At this time, Eridu was explicitly described in Sumerian inscriptions as "standing upon the shores of the sea," and Ur (situated about twelve miles from Eridu) was described as "having landing docks where oceangoing vessels changed their cargos."

The location of the Tigris and Karun Rivers at the maximum time of marine transgression is uncertain. Pliny's *Natural History* states that, during the conquest of Alexander the Great (~340 B.C.), the confluence of the Tigris and Karun Rivers was at Charax, at a distance of 1 1/4 miles from the coast at that time, and, afterwards, the Karun appears to have shifted its center of deposition to the southeast. For a short time, Charax represented the location of a temporary seaport on a retreating Gulf.

Despite much speculation concerning the exact location of the garden of Eden, it does seem likely that it was located somewhere about 100 miles northwest of the present-day Basra in Iraq (Figs 3-3 and 3-4). At the latitude of An Nasiriyah, the landscape is dotted with numerous mounds representing ruins of ancient cities such as Ur, but southeast of An Nasiriyah, no mounds exist – presumably because the Persian Gulf waters then extended at least as far inland as Ur, while the land south of these cities was submerged. Of all of these ancient mounds, Eridu is archeologically one of the oldest settlements known in southern Mesopotamia, dating from about 5800 B.C. According to ancient Mesopotamian tradition, Eridu ranks as the oldest city in the world and it was also regarded as a sacred city (Fig 3-5). Cuneiform inscriptions indicate that Eridu was located near a garden, a "holy place" containing a sacred palm tree. On Sumerian tablets found at Nippur (Fig 5-1B) and other sites, a list of ten "pre-flood" kings – starting with Alulem (presumed to be Adam by some; Fig 3-5) and ending in

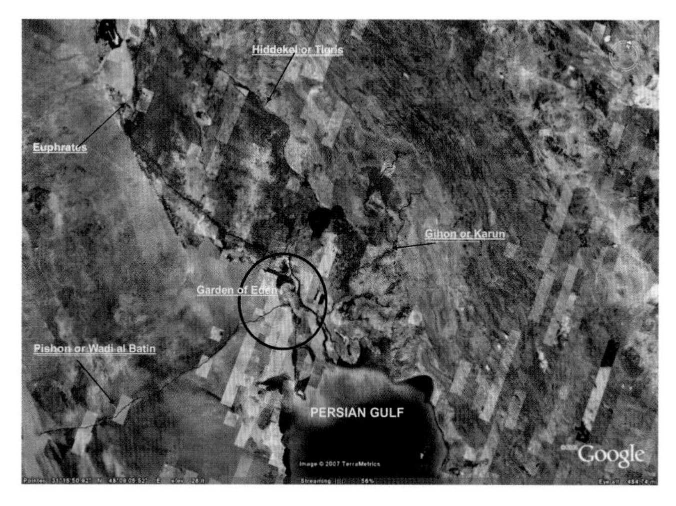

Figure 3-4. Google Earth aerial view of the four rivers of Eden, located in present-day Iraq, Saudi Arabia, Kuwait, and Iran. The circle shows the most likely location for the biblical garden of Eden, as per the description in Genesis 2:10-14. The north-south folded Zagros Mountain belt can be seen on the right.

Figure 3-5. An excavated fifth-millennium B.C. altar at Eridu, one of the oldest settlements in southern Mesopotamia. *Photo supplied by Dick Fischer.*

Ziusudra (Noah's Sumerian name) – displayed an inscribed description of Eridu as:

> When the kingship was lowered from heaven the kingship was in Eridu....In Eridu, Alulem became King....

Dick Fischer, in his book *Historical Genesis: From Adam to Abraham*, presented his view that the mound of Eridu (Fig 3-5) could have actually been the altar of Adam (Adapa or Adamu or Alulim or Atum – thought by Fischer to be possible name variations for Adam in Babylonian, Akkadian, Sumerian, and Egyptian, respectively). Eridu is located very close to where Genesis says that the garden of Eden was located (Figs 3-3, 3-4).

A River Rises *in* Eden

> And a river went out of Eden to water the garden; and from thence it was parted, and became into four heads. (Gen. 2:10, KJV)

This passage has been problematical to biblical scholars for years. One long-established view is that a river originating in Eden went *out* of Eden and then, after leaving the garden, split into four rivers (*heads*), including the Tigris and Euphrates. Therefore, some biblical scholars have interpreted the text to mean that Eden was located somewhere in ancient Armenia near the source of the Tigris and Euphrates (Fig 3-3). However, this locality does not fit with the Pishon River being located in Arabia (as discussed previously). Furthermore, there are other translations of Genesis 2:10 that are viable, one being that of Kenneth Kitchen:

> Very strictly, it is *not* "the Garden of Eden" at all, but "*a* garden *in* Eden." It has to be grasped very clearly that the garden was simply a limited area within a larger area "Eden," and the two are not identical, or of equal area. A realization of this simple but much-neglected fact opens the way to a proper understanding of the geography of Eden and its environment.

Kitchen's take on the rivers of Eden is that out of the "greater Eden," a stream flowed into the "garden" in Eden. Upstream from where this single stream entered the garden, four "head" rivers came together, and these four streams were the Pishon, Gihon, Tigris, and Euphrates.

Another interpretation is that, according to the *Anchor Bible Commentary*, Genesis 2:10 should read: "A river rises *in* Eden to water the garden; outside, it forms four separate branches." A river "rises in" is how the Hebrew should be translated according to Speiser, not the traditional "went out of." A river which "rises in" Eden strongly suggests ground flow or the rise of subterranean waters (i.e., a spring), and the word *outside* (which in the Hebrew literally means "from there") has the sense of being "beyond it" (Eden). The term *heads* also has nothing to do with streams into which the river breaks up *after* it leaves Eden, but designates instead four separate branches which have merged *within* the vicinity of Eden.

A spring rising forth in or near Eden also makes sense hydrologically. All four rivers – the Pishon, Euphrates, Tigris and Gihon – once converged near the (then) head of the Persian Gulf to create a fertile land

A GARDEN *IN* EDEN

Very strictly, it is *not* "the Garden of Eden" at all, but "a garden *in* Eden." It has to be grasped very clearly that the garden was simply a limited area within a larger area "Eden," and the two are not identical, or of equal area. In addition, a river "rises in" is how the Hebrew should be translated, not the traditional "went out of." A river which "rises in" Eden strongly suggests ground flow or the rise of subterranean waters (i.e., a spring), and the word *outside* (which in the Hebrew literally means "from there") has the sense of being "beyond it" (Eden).

fit for a garden. Not only was this garden located near the junction of these four rivers, but, in addition, a spring rose up in or near the garden to water it, and then it (a smaller river created by the spring) flowed out of the garden where it met within the confluence of the four great rivers. But what evidence is there for a spring rising in the vicinity of Eridu, the most likely site for Eden?

There is geologic evidence. The Dammam Formation ofTertiary (Paleogene-Neogene) age (Fig 3-7) is the principle aquifer (water-bearing rock) for all of Kuwait, Saudi Arabia, and Bahrain. Currently, much of the subterranean freshwater upwells from beneath the central gulf through karstic (cave-bearing) limestone – called *khawakb* in local Bahraini dialect – and the Damman is one of those aquifers. The Dammam Formation is composed of sedimentary limestone rock which covers an extensive part of western Iraq, occurring both on the surface and in the subsurface west of the Euphrates River. The formation is known to crop out only a few miles southwest of Eridu. Thus, a spring in the vicinity of Eridu (Eden?) would not be at all surprising, geologically speaking.

A Worldview Approach to a Garden in Eden

A Worldview Approach to the garden of Eden (or rather, *a* garden *in* Eden) is that, while Genesis identifies the garden as being a real place in southern Mesopotamia, the language used to describe the garden must be interpreted from the literary worldview of the ancient Hebrews.

Ziggurat/temple complexes were often featured in Mesopotamian literature as containing gardens (Fig 5-4) with these gardens being watered from springs that rose within and flowed from temples. Typical of some of these accounts was the idea of four streams flowing from the temple to water the garden and from there to water the four corners of the earth. A number of these elements are present in the garden of Eden story. A river (spring) "rises" in Eden, "and from thence it was parted and became into four heads" (Gen. 2:10, KJV). Eden is presented as a real place, but the significance of it is to be found in what it represents theologically. To quote John Walton:

> Eden was the place of God's abode and was the source of life-giving water that flowed through the rivers.... The presence of God was the key to the garden and was understood by author and audience as a given from the ancient worldview.

This theme of waters flowing out from the sanctuary of God is also symbolically portrayed in other parts of the Bible; e.g., the four streams of Ezekiel 47:1-12 and in Revelation 22:1-2, where the river of life flows from the throne of God. Some authors have carried this symbolic analogy even further, such as identifying Jerusalem with Eden. Also, like

Figure 3-6. The *Garden of Eden* by Flemish Baroque painter Jan Breughel the Younger (1601-1678), son of painter Jan Breughel the Elder. The resplendent and exquisite style of this painting is typical of the Renaissance. *Wikimedia Commons, public domain.*

the Mesopotamian concept of ziggurats and their gardens being a link between heaven (the gods) and earth (man), God's presence in the garden theologically makes the garden of Eden God's temple (Fig 3-6). Despite the symbolism involved, the garden of Eden is considered to have been a real place located by the four rivers of Genesis – a place where God sought out and directly encountered Adam and Eve. Subsequently, God's presence became identified with other temples – either ziggurat temples of the Mesopotamians (such as the Tower of Babel; page 130) or Israelite temples (such as that in Jerusalem) when heaven (the gods or God, respectively) and earth (humans) met. Thus, the real and symbolic seem intertwined into a composite historical/figurative account told from the worldview of the ancient authors/scribes who wrote the biblical text.

Garden of Eden on a Modern Landscape

In this chapter, we have shown from the identification of the four rivers of Eden that the garden of Eden was located on a *modern landscape* at the head of the Persian Gulf – possibly at or near Eridu (Fig 3-5), one of the oldest settlements in southern Mesopotamia. This identification has important implications for Young-Earth Creationists' claims of where the garden of Eden was located and for their view that Noah's flood was global.

Implications for Flood Geology

Flood geology is the Young-Earth Creationists' position that Noah's flood was global in extent, and that all (or almost all) of the sedimentary rock on Earth formed as a consequence of this flood. It is also their position that the topography of the Earth was completely changed during the flood due to the global extent of the flood. We will critique flood geology later in Chapter 7, but here we will explain, in the context of this garden of Eden chapter, why the biblical author's placement of the garden of Eden on a *modern landscape* presents a major conflict between Genesis and the flood geology position of a global flood. The reason is this: there are *six miles* of sedimentary rock *beneath* the garden of Eden/Persian Gulf. How could Eden, which existed in *pre-flood* times, be located *over* six miles of sedimentary rock supposedly deposited *during* Noah's flood?

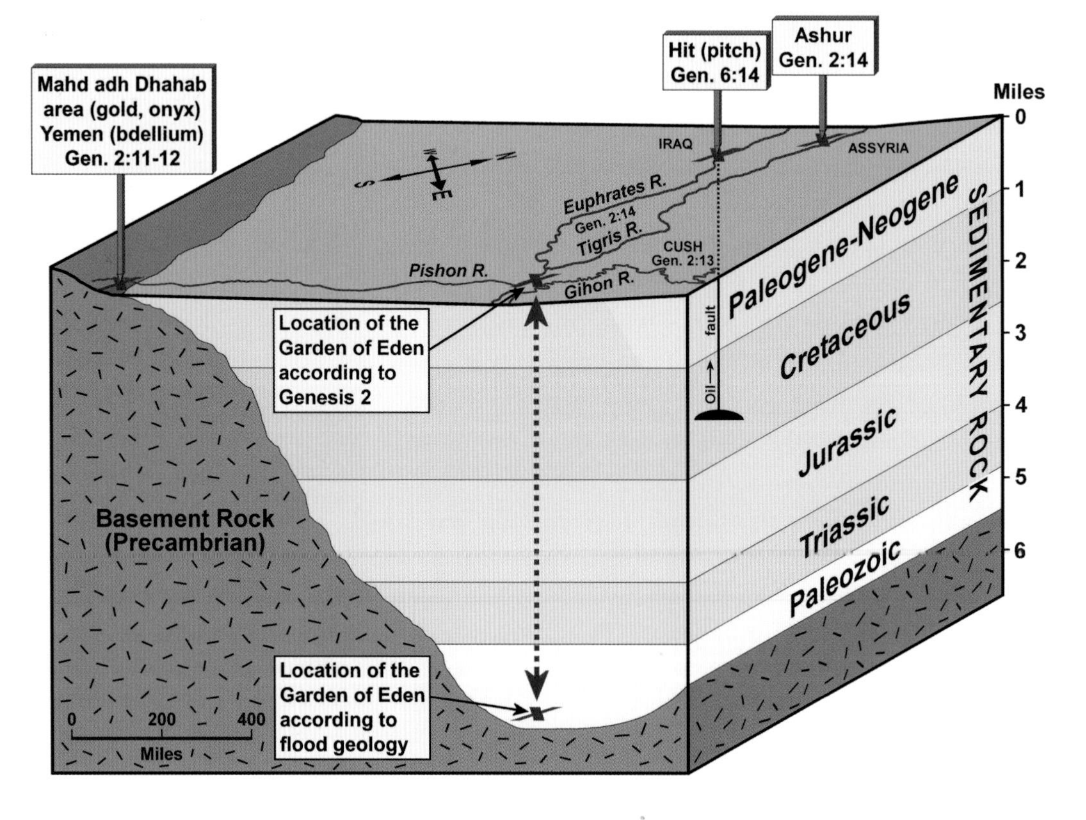

Figure 3-7. Schematic diagram of the subsurface geology beneath the Persian Gulf/garden of Eden area. If all sedimentary rock formed at the time of Noah's flood, as claimed by flood geologists, then Eden would have existed on Precambrian basement rock six miles below where Genesis says it was located. Ashur (now a mound) is mentioned in Gen. 2:14 (NIV) to document for post-flood readers exactly where the Tigris River flowed in pre-flood time.

What flood geologists are implying is that the garden of Eden existed on a Precambrian basement of granite-rock terrain in the southern Iraq (southern Mesopotamia) area (Fig 3-7), and then Noah's flood covered up the garden of Eden with six miles of sedimentary rock. But this is not what Genesis says. It says that Eden was located where the four rivers existed on a modern landscape, *on top of* six miles of sedimentary rock. Thus, this sedimentary rock must have existed in *pre-flood times*.

Geologists know that six miles of sedimentary rock exist beneath the Persian Gulf because this area has been extensively drilled for oil down to the Precambrian basement. The six miles of sedimentary rock below the garden of Eden area include Tertiary, Cretaceous, Jurassic, Triassic, and Paleozoic rock up to a depth of about 32,000 feet before the Precambrian basement is encountered. A cross-section of the sedimentary rock existing below the Persian Gulf/garden of Eden area is shown in the 3-D schematic of Figure 3-7. Precambrian granite rock is exposed at the surface in the western part of Saudi Arabia (geologists call this the *Arabian Shield*), and this rock becomes progressively overlain by a thicker sedimentary rock cover northeastwards, towards Iran. The upper red X in Figure 3-7 indicates the approximate location of the garden of Eden according to Genesis, and the lower red X indicates its approximate location according to flood geologists, since no (or very little) sedimentary rock supposedly existed on planet Earth before Noah's flood.

Pitch for the Ark

Another Genesis passage confirms that the pre-flood world was located on a modern landscape: "Make thee an ark of gopher wood; rooms shalt thou make in the ark, and shalt pitch it within and without with pitch" (Gen. 6:14, KJV). Pitch (or bitumen) is a thick, tarry, oil product composed of a mixture of hydrocarbons of variable color, hardness, and volatility. Bitumen was used extensively by the ancient people of Mesopotamia for every type of adhesive-construction need, including the waterproofing of boats and mortar for buildings (e.g., the "slime" of Gen.11:3). It is known to have been used in the construction of reed boats even as early as the Ubaid Period (Table 3-1). The center of bitumen production in Noah's time was (and still is) at Hit, which is located along the Euphrates River about eighty miles west of Baghdad (Fig 3-3). The Hit bitumen occurs in "lakes" where a line of hot springs are upwelling along deep faults. These faults connect the surface with the source of hydrocarbons at depth – the source being *sedimentary rock* (Fig 3-7). In southern Iraq, oil and gas are produced from the limestone and sandstone sedimentary rocks of the Jurassic Najmah Formation; the Cretaceous Yamama, Zubair, Nahr Umr, Mishrif, and Hartha Formations; and the Miocene (Tertiary) Fars and Ghar Formations.

The essential point of the above discussion is this: How could Noah obtain pitch from sedimentary rock for building his ark if no sedimentary rock existed? Genesis identifies Eden with four rivers that flowed *over* and *cut into* sedimentary rock. The Pishon River incised into Tertiary sedimentary limestone and sandstone rock near the border of Saudi Arabia and Kuwait. The gold of Genesis 2:12 is found in quartz veins that cut across sedimentary-type metamorphic rock. The Karun (Gihon?) River winds around folded sedimentary rock in western Iran. And the Tigris and Euphrates Rivers encounter sedimentary rock throughout their drainage systems from the mountains of Turkey to the Persian Gulf. All of this is *biblical* evidence for sedimentary rock being present on the Earth *before* Noah's flood rather than being formed *during* the flood.

When we get to the three flood chapters (Chapters 5, 6, 7), it will be important to recall this biblical evidence because it undermines the main assumption of flood geology – that is, that all of Earth's sedimentary rock with its contained fossils was formed in Noah's flood. But for now, we will continue on to Genesis 4 where the chronologies of Genesis begin, and we will ask the question: Can the age of the Earth be calculated from the six days of Genesis 1 and the chronologies of Genesis 5 and 11?

Stained glass window in the Canterbury Cathedral, depicting Methuselah, who according to Genesis 5:27 lived to be 969 years old. In this chapter it is argued that the ages of the patriarchs are numerological, not numerical, numbers. *Wikimedia Commons, public domain.*

CHAPTER 4

THE NUMBERS AND CHRONOLOGIES OF GENESIS

Genesis 5:27. And all the days of Methuselah were nine hundred sixty and nine years. (KJV)

In Chapter 2, we discussed the traditional position of the Christian church that creation took six literal days. In this chapter, we will discuss the traditional church view that the Earth is only about 6,000 years old. We will also expand on our discussion in Chapter 2 of the sacred numbers in Genesis, and then proceed to the chronologies of Genesis to see if these numbers and genealogies can be reliably used to determine the age of planet Earth.

How does biblical tradition come up with a 6,000-year age for the Earth? It assumes a six-day creation, and then it tallies up all of the patriarchal generations (since Adam) and reigns of kings recorded in Genesis and the rest of the Bible to arrive at a figure of approximately 6,000 years. Using this literal chronologies method, a number of attempts have been made to determine the date for Adam and Eve and the creation of the world. One of the first attempts was that of Jewish scholar Jose Ben Halafta, who, in the second century A.D., calculated that Adam was created in 3761 B.C., which date of approximately 3760 B.C. has become part of orthodox Jewish tradition and is the basis for the Jewish calendar. Most famous of these "literal" chronologies, and the one most cited, is Bishop Ussher's age for the world published in 1654. Ussher was the Anglican archbishop of Armagh, Ireland, and one of the most formidable biblical scholars of his time. The exact date he calculated for the creation of the Earth and universe was Sunday, October 23, 4004 B.C. The results of these (and other) dates vary, partly because three of the earliest Old Testament manuscripts (Masoretic, Septuagint, and Samaritan; Table 4-4) contain different numbers for the patriarchal ages, and partly because scholars disagree on when to begin counting backwards from the beginning of the Christian era.

There are two main difficulties with a traditional "literal" view of a young creation, both of which have caused major blockages to faith for millions of people:

(1) The incredibly long lifespans of the patriarchs such as Methuselah, who supposedly lived for 969 years (Gen. 5:27), are in direct conflict with the known lifespans of people living today or with

those living in the past according to the archeological record. As stated by Hugh Ross in his book *The Genesis Question:* "When readers encounter the long lifespans in Genesis, they become convinced that the book is fictional, or legendary at best, whether in part or in whole."

(2) The 6,000-year age for the Earth is in direct conflict with the science of geology that emphatically maintains that the Earth is 4.56 billion years old.

We will discuss problem (1) in this chapter because whether the numbers in Genesis are real or symbolic is crucial for testing the claim that the Earth is 6,000 years old. Difficulty (2) will be covered in Chapter 7, where flood geology is compared to modern geology.

Two Christian Views on the Age of Earth

Before we begin, it should be emphasized that not all Christians are Young-Earth Creationists and not all "creationists" are Young-Earth Creationists – a misconception many non-Christians hold. *Creationism* is the theological position that the universe and Earth were deliberately planned and made by a creator. It is the position held by all Christians, regardless of their viewpoints on the age of the Earth. But under this general creationist position, there is a division between Christians as to the age of the Earth. Christians who believe that God formed the universe and Earth approximately 6,000 years ago are called *Young-Earth Creationists.* Their position follows from a belief that the opening chapters of the book of Genesis provide an exact chronology of events in the creation of the cosmos over a period of six solar days and then they add in the genealogies presented in Genesis Chapters 5 and 11. Christians who believe that God formed the universe and Earth over a long period of time are called *Old-Earth Creationists* (or "Evolutionary Creationists" if evolution is accepted). The foundation of Old-Earth Creationism rests on the bedrock of science; that is, on more than four centuries of astronomical and geological observations. The cumulative findings of these studies support a 13.82 billion year old age for the universe and a 4.56 billion year old age for our solar system and planet Earth (Box 4-1).

This chapter is divided into four sections. First, we will discuss the Mesopotamians' concept of numbers in order to provide a worldview background for the next three sections. Second, we will discuss the difficult problem of the long ages of the patriarchs. Third, we will cover the *scriptural* problems involved with biblical genealogies and chronologies. Fourth, we will end this chapter by promoting a Worldview Approach to biblical chronologies.

The Mesopotamians' Worldview of Numbers

The Mesopotamians were the first people to develop writing, astronomy, mathematics (algebra and geometry), a calendar, and a system of weights and measures, accounting, and money. Even as early as the Ubaid Period (~5500-3800 B.C.; Table 3-1), Mesopotamian architects were familiar with geometric principles such as 1:2, 1:4, 3:5, 3:4:5, and 5:12:13 triangles for laying out buildings. By about 3000 B.C., Mesopotamian mathematicians were working with very large and very small numbers, and

Box 4-1. Two Christian Views on the Age of the Earth

Young-Earth Creationist	Old-Earth Creationist
From the six days of Genesis 1 and the chronologies of Genesis 5 & 11, the age of creation and planet Earth can be set at about 6,000 years (10,000 years maximum).	From the scientific evidence of astronomy and geology, the age of the universe is about 13.82 billion years and the age of our solar system and Earth is about 4.56 billion years.

were the first people group to arrive at logarithms and exponents from their calculations of compound interest and the first to solve systems of linear and quadratic equations in two or more unknowns. They also calculated the value of pi (π) to an accuracy of 0.6 percent. The so-called "Pythagorean theorem" was invented by the Mesopotamians more than a thousand years before Pythagoras lived, and was known not only for special cases, but in full generality. In other words, the Mesopotamians were the first accomplished mathematicians and astronomers.

Sexagesimal Numbers

The mathematical texts of the Sumerians or Babylonians (people who lived in southern Mesopotamia; Fig 3-3) show that these people were regularly using a sexagesimal (base 60) numbering system at least by Uruk time (~3100 BC; Table 3-1). There are also influences of a decimal system that operated together with the sexagesimal system. Examples of the Mesopotamian sexagesimal system are still with us today in the form of the 360-degree circle, with 60-minute degrees and 60-second minutes, and the 60-minute hour and 60-second minute. A sexagesimal basis for time is also reflected in a 360-day (60 x 6) year, where seven lunar months needed to be intercalated in every nineteen years to keep to the solar year. The Sumerians wrote their numbers in cuneiform – a series of wedged marks impressed onto clay tablets (Fig 4-1).

Sacred Symbolic Numbers

The Mesopotamians incorporated two concepts of numbers into their worldview: (1) numbers could have real values, and (2) numbers could be symbolic descriptions of the sacred. "Real" numbers were used in the everyday administrative and economic matters of accounting and commerce (receipts, loans, allotment of goods, weights and measures, etc.), construction (architecture), military affairs, and taxation. But certain numbers of the sexagesimal system, such as *sossos* (60), *neros* (600), and *saros* (3600) were considered sacred and, thus, occupied a special place in their religious views. The major gods of Mesopotamia were assigned numbers according to their position in the divine hierarchy: e.g., Anu, the head of the Mesopotamians' pantheon of gods, was assigned the number 60, the most perfect number in the hierarchy. The Mesopotamians also sometimes used numbers cryptographically; e.g., names could have a corresponding numerical value. For example, during the construction of his palace at Khorsubad, Sargon II stated: "I built the circumference of the city wall 16,283 cubits, the number of my name."

The Purpose of Symbolic Numbers

The Mesopotamians (and other ancient peoples) had a totally different concept of numbers than we have today. To us, a number is just a number and one number is no better than any other number. But to many ancient peoples, numbers had intrinsic meaning beyond their numerical

Figure 4-1. A Sumerian cuneiform (clay) tablet from Jemdet Nasr, Iraq, recording the rations allotted to 40 men in the course of a five-day work week. The tablet indicates that each of the men received rations equivalent in value to two units of barley per day, barley being the currency of that time period. The earliest writing denoted commodities, but it later evolved into a narrative form (see Figs 5-1A, 5-1B). *British Museum #00579266001.*

SYMBOLIC NUMBERS

The purpose of numbers in ancient religious texts could be *numerological* rather than *numerical*. Numerologically, a number's *symbolic* value was the basis and purpose for its use, not its secular value in a system of counting, and one of the religious commitments of scribes involved with numbers was to make certain that any numbering scheme used added up to the right numbers *symbolically*. This is distinctly different from a secular use of numbers in which the overriding concern is that numbers add up to the correct total *arithmetically*. Another way of looking at it is that the sacred numbers used by the Mesopotamians had honorific value, which gave a type of religious dignity or respect to important persons or to a literary text.

value; for example, just as a name could hold a special significance for that person (Gen. 5:29), so could a number have significance in and of itself. That is, the purpose of numbers in ancient religious texts could be *numerological* rather than *numerical*. Numerologically, a number's *symbolic* value was the basis and purpose for its use, not its secular value in a system of counting, and one of the religious commitments of scribes involved with numbers was to make certain that any numbering scheme used added up to the right numbers *symbolically*. This is distinctively different from a secular use of numbers in which the overriding concern is that numbers add up to the correct total *arithmetically*. Another way of looking at it is that the sacred numbers used by the Mesopotamians had honorific value, which gave a type of religious dignity or respect to important persons or to a literary text. For example, Genesis 7:6 states that Noah was 600 years old when the flood started. What other "perfect" age could have been assigned to Noah from the numerological worldview of the Mesopotamians, since he was considered a man "perfect in his generations" (Gen. 6:9, KJV)?

Preferred or Figurative Numbers

Besides the perfect, "sacred" exaggerated sexagesimal (base 60) numbers used in the early chapters of Genesis, the rest of the Bible often uses "preferred" numbers. These numbers were preferred by the biblical authors because they had spiritual (figurative) meaning rather than a strict numerical meaning. Even a cursory reading of the Bible will reveal that certain numbers are used over and over again. Among these preferred numbers are three, seven, twelve, and forty.

Three. Three is the number of *emphasis* in the Bible; e.g., "holy, holy, holy" signified that God was being especially hallowed. Jesus often repeated himself three times to emphasize a point, or things were done three times for emphasis (e.g., John 21:15-17). Three as a number also symbolized completeness; for example, when Jesus rose from the dead in 3 days, his mission was complete. Jonah was in the "great fish" for 3 days and 3 nights, in 3 days the temple will be raised, etc. The preferred number three is also used symmetrically in the genealogies of Genesis. For example, three sons are named for Adam, Noah, and Terah, even though these men probably had more than three sons (e.g., Gen. 5:4). And three men, Esau, Jacob, and Ishmael, each had twelve sons.

Seven and seventy. The number seven was especially sacred to the Jews because of the Sabbath, the seventh day of their week. As the last day of the week, it is a recurrent biblical symbol of fullness and perfection. The number seven is used ubiquitously throughout both the Old and New Testaments. Some specific examples are: 7 golden candlesticks, 7 spirits, 7 words of praise, 7 churches, 70 (7 x 10) nations, 70 (7 x 10) elders, Daniel's 70 (7 x 10) weeks, forgive 70 x 7 times, Terah's age of 70 (7 x 10), Lamech's age of 777, and so on. Seven is especially prevalent in Revelation, where it is used over fifty times and where its use is certainly symbolic. The use of the number seven starts in Genesis, but it is carried throughout both the Old and New Testaments in the form of parallelisms, or "seven stanza templates," by which the Hebrews constructed their poetry and some prose (e.g., Isa. 28:8-14; Mark 15:40-47; John 1:1-18).

The number seventy (7 x 10) is also used over and over in the Bible. Seventy may, or may not, have represented an exact number, but this was unimportant to the ancients. The number seventy could symbolize a *numerological ideal,* not a quantitative reality. Thus, in Chapter 10 of Genesis, 70 nations are mentioned – which number was symbolic among the Israelites for any family blessed with fertility (e.g., the 70 descendants of Jacob who came to Egypt in Gen. 46:27). In Genesis 10, the biblical author selectively penetrates into the various descendant lines in order to achieve his final number of 70. The Israelites were in captivity in Babylon for 70 years (Jer. 29:10) – and there are many other examples.

Twelve. Another number that is repeated over and over in the Bible is twelve. There are 12 pillars; 12 wells; 12 springs; 12 precious stones; 12 silver bowls; 12 golden spoons; 12 bullocks, rams, lambs, and goats; 12 cakes; 12 fruits; 12 pearls; 12 tribes of Israel; 12 tribes of Ishamael; 12 tribes of Esau; 12 tribes of Nahor; 12 districts of Solomon; 12 gates of the New Jerusalem; 12 disciples of Jesus; 12,000 horsemen; 144,000 (12 x 12 x 1000) people in the remnant of Israel, and so on. The preferred use of the number twelve may have had its foundation in the Mesopotamian sexagesimal system (6 x 2), but it extended all the way to the time of Christ and is well represented symbolically in the book of Revelation.

Forty. The number 40 also occurs many times in the Bible in different contexts and can be taken either literally or figuratively. Noah's flood lasted 40 days and 40 nights, Moses fasted 40 days and 40 nights, and Jesus fasted 40 days and 40 nights. The Israelites were in the wilderness for 40 years; Jesus was seen by his disciples after his resurrection for 40 days; Jonah preached to Nineveh for 40 days; Solomon, David, and Saul are each credited with a reign of 40 years; Eli judged Israel for 40 years; Goliath presented himself for 40 days; 1 Kings 6:1 describes a time of 480 (12 x 40) years, and so forth. The number forty was also considered to represent "a generation"; e.g., God made Israel wander in the wilderness for "40 years" until that "generation" who had done evil in the sight of the Lord had died (Num. 32:13).

In addition to preferred numbers, figurative ages were used for certain important individuals. An example of this is found in Deuteronomy 34:7: "And Moses was an hundred and twenty years old when he died; his eye was not dim, nor his natural force abated" (KJV). The number 120 (60 x 2, or 40 x 3) is first mentioned in Genesis 6:3: "Yet his days shall be an hundred and twenty years" (KJV). This number has also been mentioned in a similar context in a cuneiform text found at Emar: "One hundred twenty years (are) the years of mankind – verily it is their bane." This is the only known extra-biblical parallel to Genesis 6:3. The figure 120, shared by Genesis 6:3 and the Emar text, is to be regarded as a maximal and *ideal* age, which in the worldview of that time could be reached only by extremely virtuous individuals. Indeed, in the Bible there is only one person to whom this life-span was attributed – namely, Moses. Similarly, Joseph and Joshua were each recorded as dying at age 110, a number considered "perfect" by the Egyptians. In ancient Egyptian doctrine, the phrase "he died aged 110" was actually an epitaph commemorating a life that had been lived selflessly and had resulted in outstanding social and moral benefit for others. And so, for both Joseph and Joshua, who came out of the Egyptian culture, quoting this age was actually a tribute to their character. But it did not necessarily bear an exact relationship to their actual lifespans.

What, then, is to be construed from the preferred or figurative numbers in Genesis? Which numbers are to be taken "literally" (as exact numerical numbers), and which are to be taken symbolically or figuratively? Scholars have been struggling with this question for centuries, but two fairly obvious conclusions can be construed concerning these numbers:

(1) Many preferred numbers were used by the Hebrews *just because* the number was considered sacred. For example, the 7 golden candlesticks, 12 silver bowls and spoons, and so on, were objects

constructed in these configurations or quantities just because these numbers reflected sacredness. Once a number had been established as being sacred, it then was used repeatedly because it conferred sacredness on items, persons, or events.

(2) In some cases the numbers are to be taken literally because the Bible *specifies* a literal interpretation. For example, in the New Testament account of Jesus's death and resurrection, the three days are to be taken literally since the time within these three days is accounted for. Another literal example is in the Old Testament where the Israelites wandered in the wilderness for 40 years. This 40-year period began on the day Israel left Egypt (the day after the first Passover) – the fifteenth day of the first month of the first year (Num. 33:3) – and lasted until the first Passover in the Promised Land, on the fourteenth day of the first month of the forty-first year (Josh. 5:10). On the other hand, the seventy years of captivity for the Israelites, as in Jeremiah 25, may or may not be literal.

Unless we assume that God prefers certain numbers over other numbers and somehow passed that preference down to the biblical authors, we must acknowledge that in many cases where preferred numbers are used, they are to be taken symbolically or figuratively. The symbolic reading of numbers does not mean that the Genesis stories are to be considered as mythological. It just means that the biblical author(s) were trying to impart a spiritual or historical truth to the Genesis text – one that in their worldview surpassed the exact mathematics of rational numbers.

How the Concept of Numbers Has Changed over Time

Today, we don't consider certain numbers to be sacred or special (at least we in the West don't – in some parts of the Near and Far East this worldview is still prevalent). So how did the concept of numbers change over time from numerological to numerical?

Biblical scholars have tried over the years to compare how the common traditions of ancient cultures – including the creation and flood stories and the numbers contained in Genesis – might correspond to one another. The ancient Akkadian (northern Mesopotamia), Sumerian (southern Mesopotamia), and Egyptian cultures all had exaggerated "long reigns" for their gods and kings, and this seems to have been a common religious tradition for peoples of the ancient Near East. A number of scholars have also attempted to mathematically determine a numerical connection between the long lifespans in the Sumerian King List and the long ages of the patriarchs in Genesis. But, despite these attempts, only a superficial similarity has been established.

What has emerged from such comparative studies, however, is that the numbers used in the very early chapters of Genesis are definitely sexagesimal in nature, and thus, point to Mesopotamian roots for these stories. Abraham, when he lived in Ur in southern Mesopotamia at around 2000 B.C. (Table 3-1), most probably would have used a sexagesimal numbering system. However, when Abraham left Ur, he and his descendants came in contact with other Semitic people groups who were using the decimal system. Thus, gradually the decimal system replaced the sexagesimal system in the Israelites' numerical worldview as they moved from Mesopotamia to Haran to Palestine to Egypt and back to Palestine (Table 4-1). Along with slowly acquiring the decimal system came a concept-loss of the sacredness of certain numbers that were formerly attached to the Mesopotamian sexagesimal system, but the use of these and other preferred numbers still persisted. How the concept of numerological numbers may have changed over time is shown in Table 4-1.

It seems certain that a sound and really historical chronology had become established in Israel by the time of David (~900 B.C.), as two hundred or so chronological dates in the books of Samuel, Kings, and Chronicles are, with a few exceptions, of remarkable consistency. But even then, and long after, preferred or figurative numbers were used up to and throughout the New Testament. Before and during the Middle Ages in Europe, the concept of "sacred" numbers was lost – much like the three-tiered cosmology of the ancients that we discussed in Chapter 1 was lost (Fig 1-1) – and it was not until the discovery

>2000 B.C.	~1500 B.C.	~1000 B.C.	1st Century A.D.	Middle Ages	1800s-Present
Mesopotamia	*Egypt*	*Palestine*	*Palestine*	*Europe*	*Western World*
Sexagesimal, exaggerated, *sacred* numbers (60, 7); used in Genesis up to Abraham	Decimal numbers, Joseph-Joshua-Moses; preferred and figurative use of numbers (40, 12, 3)	Decimal numbers, Solomon-David; real numbers but religious use of preferred numbers	Time of Christ, real numbers; preferred numbers (3, 7, 12, 40) still used but waning	Real numbers only; concept of sacred/preferred numbers lost; interpreted as literal numbers	Cuneiform tablets found; concept of sacred numbers rediscovered by archeologists

Table 4-1. How the concept of numbers may have changed over time.

and publication of the Babylonian mathematical cuneiform texts in the late 1800s and early 1900s that the numerological nature of the Genesis numbers was rediscovered.

Long Ages of the Patriarchs

Having discussed the Mesopotamians' concept of sacred and secular numbers with their dual numerological and numerical meanings, and the Hebrews' use of preferred numbers for sacred purposes, we can now tackle the difficult problem of the long lives of the patriarchs.

Long Patriarchal Lifespans

A number of attempts have been made over the centuries to explain the long lifespans of the patriarchs. In this book, the word *patriarch* will refer to the line of Adam down to Abraham, or "the book of the generations of Adam" (KJV) specifically outlined in Chapters 5 and 11 of Genesis. In other words, we will use the term *patriarch* in this book in its general sense of the "fathers" in the line of Adam leading to the Israelites and Christ rather than in the specific sense of Abraham, Isaac, Jacob, and Joseph. Here are four of the most common explanations for the long ages of the patriarchs:

One year equals one month or season hypothesis. This explanation proposes that perhaps a "year" to the people of the ancient Near East had a different meaning than it does today. Instead of being marked by the orbit of the Sun, a "year" then marked the orbit of the Moon (a month) or a season (three months). Among the Greeks, years were sometimes called "seasons" (*horoi*), and so some ancient authors such as Augustine thought that a "year" in Genesis might be equivalent to a one-month or three-month time period.

However, this theory is nonsensical if one looks at the "begetting" ages of the patriarchs (Table 4-2). If the ages of Adam and Enoch are divided by twelve (1 year = 1 month), then Adam would have fathered Seth at age eleven and Enoch would have been only five when he fathered Methuselah. Enoch's age (65; Gen. 5:21) divided by four (1 year = 1 season) would result in an age of sixteen, which is biologically possible. But if the number 500 – Noah's age when his first son(s) were born (Gen. 5:32) – is divided by four, the "begetting" age would have been 125 years, another unlikely possibility.

Astronomical hypothesis. Astronomical explanations have also been proposed to explain the incredibly long patriarchal ages. Perhaps the rotation period of the Earth has changed, so that the days then were not equivalent to those we have now. Or perhaps a supernova could have damaged the Earth's ozone layer, thus increasing ultraviolet radiation and systematically decreasing the age of humans. The problem with such astronomical explanations is the absence of supporting evidence. There have been no known supernova explosions within the last 10,000 years that could account for the long ages of the patriarchs.

Tribal, dynastic, or "clan" hypothesis. Another explanation is that, when the Bible makes a statement like "Adam was the 'father' of Seth," it means that the Adam "clan" had exercised dominion for 130 years (the age of Adam when Seth was born). In this view, Seth would be a direct-line descendent of Adam (grandson, great-grandson, and so on), but not the immediate son of Adam. Then, Seth's "son" descendants would become part of the Seth dynasty or tribe. While this theory has merit in that "father" and "begot" do not necessarily mean a direct father-son relationship, it still does not account for the personal encounters that the "fathers" reportedly had with their "sons"; for example, Noah was 500 years when his son(s) were born (Gen. 5:32), yet he coexisted with them on the ark (Gen. 7:13).

Canopy theory hypothesis. Other people have tried to explain the long ages of the patriarchs by creating a "different world" for pre-flood humans. A still-popular view in the Young-Earth Creationist camp is that before Noah's flood a vapor canopy shielded Earth from harmful radiation so that people lived to a very old age. After the flood, harmful radiation slowly increased so that the patriarchs' ages exhibit a steady decline to the biblical lifespan of 70 (or 80) years mentioned in Psalm 90:10.

The problem with the vapor canopy hypothesis is that there is no physical or geological evidence to support it (refer to Chapter 7). In addition, there is no archeological evidence that substantiates incredibly long lives for people in the past – either in Mesopotamia or anywhere else in the world. It is known that humans living in the Bronze Age (which time span includes most of the patriarchs; Table 9-1) had an average lifespan of about 40 years, based on human skeletons and legal documents of the time. If infants and children are included in this lifespan average, it would be even lower. From the examination of skeletons in a number of graves at al'Ubaid (one of the oldest known archeological sites in Mesopotamia; Table 3-1), it has been determined that some people did live to be over sixty – a great age at that time.

How then can the long lives of the patriarchs and other problematic numbers of Genesis be explained? Does one have to construct a fantastical world based on incredible ages in order to come up with an adequate explanation? The answer is actually quite simple – *if* one considers the worldview of the people living during the time of the patriarchs; that is, the Mesopotamians (the people who lived in what is now mostly Iraq) and the Israelites in Palestine who lived in Mesopotamia up until Abraham moved to Haran (Fig 3-3). This worldview includes both the religious ideas of these people and their dual numerical-numerological numbering system.

A list of the patriarchs from Adam to Abraham, containing their ages when their first son was born, their remaining years of life, and total years, is shown in Table 4-2. These ages are then "deciphered" into their common components as to the sacred sexagesimal numbers (7, 60) or the preferred number 40.

The first thing to notice in Table 4-2 is that most of the numbers listed in the Genesis chronologies are based on the sexagesimal (60) system and can be placed into one of two groups: (1) multiples of *five*; that is, numbers exactly divisible by five, whose last digit is 5 or 0; and (2) multiples of *five* with the addition of *seven* (or two sevens). The significance of the number five is that 5 years = 60 months, and combinations or multiples of 60 years + 5 years (60 months) are basic to the patriarchal ages. Note that for the 30 (10 rows, 3 columns = triplet) numbers listed for the antediluvian patriarchs up to the flood (from Adam to Noah), *all* of the ages end in 0, 5, 7, or 2 (5 + 7 = 1$\underline{2}$), with only one number ending in 9 (5 + 7 + 7 = 1$\underline{9}$). However, since the third number of each triplet is entirely determined by the sum of the first two numbers, it cannot be treated

Table 4-2. *(facing page)* The chronologies of Genesis: age of patriarchs and corresponding sexagesimal and preferred numbers. All age-numbers (30 in all) from Adam to Noah are some combination of the sacred numbers 60 (years and months) and 7. No numbers end in 1, 3, 4, 6, or 8. Thirteen numbers end in 0 (some multiple or combination of 60), 8 numbers end in 5 (5 years = 60 months), 3 numbers end in 7, 5 numbers end in 2 (5 yrs + 7 yrs = 12), and 1 number ends in 9 (5 yrs + 7 yrs + 7 yrs = 19). *All of this cannot be coincidental.* The Mesopotamians were using numerological, *not* numerical, numbers for the patriarchal ages. Therefore, these numbers should not be interpreted as real numbers.

Patriarch	Age (yrs) when first son born	Sexagesimal and Preferred Numbers	Remaining years of life	Sexagesimal and Preferred Numbers	Total years	Sexagesimal and Preferred Numbers
Adam	130	60x2yrs + 60x2mos	800	60x10x10mos + 60x60mos	930	60x3x5yrs (60mos) + 6x5yrs (60mos)
Seth	105	60x10x2mos + 60mos	807	60x10x10mos + 60x60mos + 7yrs	912	60x3x5yrs (60mos) + 5yrs (60mos) + 7yrs
Enosh	90	(6+6+6) x 60mos	815	60x10x10mos + 60x60mos + 60x3mos	905	60x3x5yrs (60mos) + 5yrs (60mos)
Kenan	70	7x2x5yrs (60mos)	840	60x10x10mos + 60x60mos + 60x8mos	910	60x3x5yrs (60mos) + 2x5yrs (60mos)
Mahalalel	65	60yrs + 5yrs(60mos)	830	60x10x10mos + 60x60mos + 60x6mos	895	60x3x5yrs(60mos) – 5yrs(60mos)
Jared	162	60x6x5mos + 5yrs (60mos) + 7yrs	800	60x10x10mos + 60x60mos	962	(60+60+60+6+6)x60mos – 5yrs (60mos) + 7yrs
Enoch	65	60yrs + 5yrs (60mos)	300	60x5yrs(60mos)	365	60x6yrs + 5yrs (60mos) = 1 solar year
Methuselah	187	60x3yrs + 7yrs	782	60x10x10mos + 60x60mos – 6x3yrs	969	(60+60+60+6+6)x60mos – 5yrs (60mos) + 7yrs + 7yrs
Lamech	182	60x7x5mos + 7yrs	595	60x10yrs – 5yrs(60mos)	777	7x10x10 + 7x10 +7yrs
Noah	500	60x10x10mos	450	40x2x5yrs (60mos) + 10x5yrs (60mos)	950	60x3x5yrs (60mos) + 10x5yrs (60mos)
Flood						
Shem	100	60x10x2mos	500	60x10x10mos	600	60x10yrs
Arphaxad	35	7x5yrs(60mos)	403	40x2x5yrs(60mos) + 3yrs (6x6mos)	438	40x2x5yrs(60mos) + 60x6 + 60 + 6x6mos
Shelah	30	60x6mos	403	40x2x5yrs(60mos) + 3yrs (6x6mos)	433	40x2x5yrs(60mos) + 6x(60+6)mos
Eber	34	60x6mos + 6x8mos	430	40x2x5yrs(60mos) + 6x60mos	464	40x2x5yrs(60mos) + 60yrs + 6x8mos
Peleg	30	60x6mos	209	40x5yrs(60mos) + 5yrs(60mos) + 6x8mos	239	40x5yrs(60mos) + 6x6yrs + 6x6mos
Reu	32	60x6mos + 6x4mos	207	40x5yrs(60mos) + 5yrs(60mos) + 6x4mos	239	40x5yrs(60mos) + 6x6yrs + 6x6mos
Serug	30	60x6mos	200	40x5yrs (60mos)	230	40x5yrs(60mos) + 60x6mos
Nahor	29	60x6mos – 6x2mos	119	60x2yrs – 6x2mos	148	60x10x2mos + 6x8yrs
Terah	70	7x2x5yrs(60mos)	135	60x2yrs + 60x2mos + 5yrs(60mos)	205	40x5yrs(60mos) + 5yrs(60mos)
Abraham	100	60x10x2mos	75	5yrs(60mos) x 3x5yrs(60mos)	175	60x10x2mos + 15x5yrs(60mos)

as independent. The truly independent calculation has 20 numbers that end in 0, 2, 5, and 7 – *a chance probability of about one in ninety million!* For the entire 60-number list (antediluvian and postdiluvian), none of the ages end in 1 or 6 – a chance probability of about one in a half million.

Such mathematical improbabilities continue for the 26 generations between Adam and Moses, as shown in Table 4-3. If one includes the six generations from Abraham to Moses (Isaac, Jacob, Levi, Kohath, Amram, Moses), then the total number of years for these men becomes 12,600 (70 x 180), which total reflects both the sexagesimal (base 60; 60 x 3 = 180) system of the Mesopotamians and the Mesopotamian-Hebrew preferred number 7 (or 70). The ages in the first, Genesis 5, column add up to 8,575 (25 x 7 x 7 x 7) and the 7 ages in the third column add up to 1,029 (3 x 7 x 7 x 7). The 17 ages of the first and third column combined add up to 9,604 (4 x 7 x 7 x 7 x 7); the middle age for these two columns is that of Lamech (777), and the 7 ages on either side of Lamech add up to a total of 7,777!

It is inconceivable that all of this should be accidental! Surely, if all of the ages listed in Tables 4-2 and 4-3 are statistically random numbers, such numerical improbabilities should not exist. The conclusion must be that these patriarchal numbers were *purposely* contrived by scribes knowing both mathematics and the religious importance of these numbers. The question that then must be asked is: What could have been the significance of these numbers? Could these ages somehow be mathematically connected to the real ages of the patriarchs? Could they be cryptographic (gematria) numbers, where numerical values were assigned to different letters of the patriarchs' names? Why do the "begetting" ages of the patriarchs decrease over time (Table 4-2)? Is it because successive biblical authors gradually lost their concept of sacred exaggerated sexagesimal numbers? Were these numbers "assigned" to the patriarchs on the basis of their character, accomplishments, or relationship with God? For example, in the generally decreasing age trend, there is an enormous jump in the "begetting" age of Noah, which may signify an attempt by the biblical author to favor the more righteous or those who "stand out" from the rest due to their prominence in the unfolding story (i.e., Noah, the righteous hero of the flood). Specifically for Genesis and the patriarchal ages, how these numbers were meant at the time of writing is something that we may only guess at today, and if a mathematical or theological principle underlies such numbers, it is no longer readily apparent. What does seem apparent, however, is that the specific intent of the biblical author(s) for the patriarchal ages was to preserve numerical symmetry and harmony in the written text, one commensurate with their mathematical worldview.

Genesis 5		Genesis 11		Abraham to Moses	
Adam	930	Shem	600	Abraham (Gen. 25:7)	175
Seth	912	Arphaxad	438	Isaac (Gen. 35:28)	180
Enosh	905	Shelah	433	Jacob (Gen. 47:28)	147
Kenan	910	Eber	464	Levi (Exod. 6:16)	137
Mahalalel	895	Peleg	239	Kohath (Exod. 6:18)	133
Jared	962	Reu	239	Amram (Exod. 6:20)	137
Enoch	365	Serug	230	Moses (Deut. 34:7)	120
Methuselah	969	Nahor	148		
Lamech	777	Terah	205		
Noah	950				

Table 4-3. The 26 generations between Adam and Moses and total years of each person's lifespan.

Numerical Symmetry and Harmony

As discussed in Chapter 2, there is a symmetry and regularity to the early chapters of Genesis that cannot be accidental. Rather, there seems to have been an intentional impartation of religious harmony and prosaic beauty to the text corresponding with the style of literature of that time. Love for symmetry and order also applies to numbers and to early Mesopotamian numbering on tablets. For example, each genealogy presented in Chapters 5 and 11 of Genesis includes ten names. Adam to Noah contains ten names and Shem to Abraham contains ten names (Table 4-2). To break a text into a ten-generational pattern was common for many Near Eastern people groups of that time, and doing so reflected an overall sense of numerical importance and harmony (ten was the base of the decimal numbering system for most of these peoples, including the Egyptians and Hebrews). In addition, the description of each of these ten generations ends with a father having three sons; for example, in Genesis 5:32, Noah begot Shem, Ham, and Japeth; and in Genesis 11:26, Terah begot Abram, Nahor, and Haran. This is likewise the case for the Cananite genealogy with respect to Jabal, Jubal, and Zillah (Gen. 4:20-22). By ending each of these sections with three sons, an overall symmetry was established using the preferred number three for emphasis. Thus, it appears that the symmetry of the Genesis text was *artificially* constructed by the authors/scribes/mathematicians who wrote, copied, or worked on the biblical text.

Scriptural Problems with Biblical Genealogies and Chronologies

Besides its numerological construct, a literal interpretation of the Genesis chronologies has a number of serious scriptural problems. We will discuss the two major problems: (1) condensed genealogies and gaps in chronology, and (2) the three different genealogical numbering systems in the Masoretic, Samaritan, and Septuagint versions of the Old Testament.

Condensed Genealogies and "Gaps" in Chronology

The matter of obtaining a creation date based on a literal addition of the patriarchal ages is not that plausible if the whole genealogical record of the Bible is examined. Genealogies are frequently abbreviated by the omission of unimportant names. In fact, *abridgment* was the general rule for biblical authors who did not want to encumber their texts with more names than necessary for their intended purpose. Numerous examples of abridgment exist, the most notable example being the genealogy of Jesus in Matthew 1:8 where Uzziah was not the son but the great-great grandson of Joram (three names have been dropped). The tradition of breaking down long genealogical lists into a ten-generational pattern also suggests that only the ten *most important persons* were retained. Since a number of generations could have been omitted from the biblical genealogies, it is reasonable to conclude that these genealogies should be used in a broad sense to indicate overall descent ("X fathered the line culminating in Y") rather than a direct father-to-son relationship ("X fathered Y"). Biblical father-to-son sequences can represent the actual genealogical line, or they can be a condensation from an originally longer series of generations. The term *son* was also often used to mean "beings belonging to class X;" for example, "the sons of the prophets" does not necessarily mean those whose fathers were prophets.

Furthermore, the fact that each member of a series is said to "beget" the next succeeding member is not evidence in itself that some genealogical links have not been omitted. For example, in Genesis 46:16-18, the list of children born to Jacob by Zilpah actually includes not only their sons (Gad, Asher), but their grandsons and great-grandsons. Sometimes "beget" does not even apply to people. It can also refer to geography (e.g., Elishah, Tarshish, Gen. 10:4 and 1 Chron. 1:2), to cities (e.g., Sidon, Gen. 10:15), to people groups or tribes (e.g., Kittim and Dodanim, Gen. 10:4 and 1 Chron. 1:17), and even to nations (e.g., Canaan, the grandson of Noah is said to have begotten the Jebusites and Amorites; Gen. 10:16-18). The flexibility of the word "beget" must therefore be considered in any interpretation of the stated ages of the patriarchs. When it is said, for example, in Genesis 5:9: "And Enosh lived ninety years, and begot Cainan [Kenan] (KJV), how do we know that "begot" means that Kenan was the immediate son of Enosh or if he was in the descendent line of Enosh? Perhaps Enosh was ninety years old when his grandson Kenan, or great-grandson Kenan, was born.

An additional piece of evidence suggesting that there could be gaps in the Genesis chronologies is the "overlap" of patriarchal lifespans. If the genealogies in Genesis 5 and 11 are both literal and complete, then the death of Adam has to be dated to the generation of Noah's father, Lamech, and Methuselah would have died in the year of the flood. Shem, Arphaxad, Shelah, and Eber would have outlived all of the generations following as far as, and including, Terah, the father of Abraham. Noah would have been the contemporary of Abraham for 58 years and Shem (Noah's son) would have survived Abraham by 35 years. But where does the Bible indicate that any of these men were coeval? They are spoken of as respected ancestors, not as contemporaries that interacted with them or who were to be cared for in their old age. The whole impression of the biblical narrative in Abraham's day is that the flood was an event long-since past, and that the actors in it had passed away generations ago. Attempting to maintain that the patriarchal ages are to be taken as literally precise thus seems contrary to the biblical record.

Three Different Genealogical Numbering Systems: Masoretic, Septuagint, and Samaritan

There is another serious problem with assuming absolute literal ages for the Genesis chronologies: the ages ascribed to people in the biblical texts differ significantly between the Masoretic, Septuagint, and Samaritan Pentateuch versions of the Old Testament. The antediluvian ages before the birth of the first sons from Adam to Noah is 1,656 years in the Hebrew Masoretic text, 2,262

Person	Masoretic	Septuagint	Samaritan
Adam	130	230	130
Seth	105	205	105
Enos (Enosh)	90	190	90
Cainan (Kenan)	70	170	70
Mahalalel	65	165	65
Jared	162	162	65
Enoch	65	165	62
Methuselah	187	187	67
Lamech	182	188	53
Noah	500	500	500
Years to flood	100	100	100
TOTAL	**1,656 years**	**2,262 years**	**1,307 years**

Table 4-4. Three biblical versions of the patriarchal numbers of Genesis; taken from J.A. Borland. There is no way for us to now determine which, if any, of these three chronologies is the correct one. Most scholars favor the Masoretic text since it is in the Bible we now have. However, the Septuagint came before the Masoretic and was in circulation at the time of Christ.

years in the Greek Septuagint text, and 1,307 years in the Samaritan text (Table 4-4). The postdiluvian ages before the birth of the first son in the interval between the flood and Abraham is 292 years in the Masoretic text, 1,172 years in the Septuagint text, and 942 years in the Samaritan text. If the Bible is literally correct with respect to patriarchal ages, *which* Bible version is correct?

The Septuagint was the first translation of the Old Testament into another language, having been translated from Greek in the third century B.C. In the centuries following the time of the Dead Sea Scrolls (150 B.C. to A.D. 70), Jewish scribes became increasingly focused on copying manuscripts as they received them, leading up to the work of the Masoretes (A.D. 600 to 900). The Masoretes were the most careful of any ancient copyists in transmitting Scripture, and for that reason, later Hebrew scholars, such as Cassuto, argued that the Masoretic text was the autograph copy of the Old Testament (from which the others were modified) and thus the most reliable. And, since this is the version that has made it into our Bible, it is the one that has been used in this discussion and in Table 4-4. But the discrepancies between the three accounts logically entail that they cannot all be true, and thus cannot be considered inviolate from an absolutely literal point of view.

A Worldview Approach to Biblical Chronologies and Age of Planet Earth

What then is to be made of the Genesis chronologies as they apply to the age of the Earth? The most important premise of this discussion is that the patriarchal ages do *not* represent real numbers – they represent *numerological* numbers. Therefore, the genealogies of Genesis cannot be used for the construction of a chronology on an *absolute* time scale. It is forcing a purpose for the biblical chronologies that was never intended. Biblical chronologies were intended by the biblical author(s) to confirm a specific "line of Adam" descent for the Jews in the Old Testament, and for Jesus in the New Testament. They were not meant to set an absolute date for the creation of the universe or Earth. A Worldview Approach to the numbers and chronologies of Genesis is important because it proposes that since the biblical author(s) held a dual and pre-scientific worldview of numbers, the Genesis chronologies cannot imply a 6,000-year-old age for planet Earth or universe. Therefore, the scientific evidence for a 4.56 billion-year-old Earth is *not* contrary to the Bible! We will continue our discussion of the age of planet Earth in Chapter 7 when we compare flood geology with modern geology.

OLD EARTH OR YOUNG EARTH?

The most important conclusion of this chapter is that the patriarchal ages do *not* represent real numbers - they represent *numerological* numbers. Therefore, the genealogies of Genesis cannot be used for the construction of a chronology on an *absolute* time scale. It is forcing a purpose for the biblical chronologies that was never intended. A Worldview Approach to the numbers and chronologies of Genesis is important because it proposes that since the biblical author(s) held a pre-scientific worldview of numbers, the Genesis chronologies cannot imply a 6,000 year-old age for planet Earth, and, therefore, the scientific evidence for a 4.56-billion year-old Earth is not contrary to the Bible.

The Deluge by Francis Danby, 1840.
Tate Gallery.

CHAPTER 5

NOAH'S FLOOD: HISTORICAL OR MYTHOLOGICAL?

Genesis 7:19. And the waters prevailed exceedingly upon the earth; and all the high hills, that were under the whole heaven, were covered. (KJV)

Of all the stories in Genesis, there is none more fascinating to biblical scholars and lay persons alike than Noah's flood. Chapters 6-8 of Genesis have been used by some Evolutionary Creationists to "prove" the non-historical nature of the first eleven chapters of Genesis, and out of Young-Earth Creationism a new subdivision (called *flood geology*) has emerged to promote a global (planet-wide) view of the flood. Significantly, the subject of Noah's flood has emerged in the last few decades as perhaps the most contentious subject in the science-Scripture debate, since flood geologists promote not only a 6,000-year-old Earth but also the view that almost all of Earth's sedimentary rock with its contained fossils formed in the one-year-long Noah's flood. The question rightly asked by the Christian community subjected to this debate is: How does modern geology fit with the Genesis text on Noah's flood? This is the topic of the next three chapters. Chapter 5 (this chapter) presents the archeological and geographical evidence for a historical flood, Chapter 6 discusses the evidence related to a global versus local flood, and Chapter 7 critiques the credibility of flood geology when compared with the science of modern geology.

To begin, let's examine the flood events of Genesis 7-8 so we can ascertain what the Genesis text actually says about Noah's flood (Table 5-1). This table will be referred to in this and the following two chapters.

A Worldview Approach to Noah's Flood

As stated in Chapter 1, the basic premise of a Worldview Approach is that the Bible in its *original context* records *historical* events if considered from the *worldview* of the biblical authors who wrote it. Based on this premise, and from various pieces of evidence presented in this chapter, a Worldview Approach considers Noah to have been a real historical person and the flood to have been a real historical event – even though many aspects of the flood story appear to be *mythological* (as the term is popularly understood). This is the objective of a Worldview Approach: to blend science, ancient history and literature, and worldview together to arrive at the most reasonable interpretation of Scripture. Or, to quote biblical archeologist Kenneth Kitchen, in his book *On the Reliability of the Old Testament*:

> By and large, the ancients did not *invent* spurious history, but normally were content to *interpret* real history in accord with their views.…Once detected, this viewpoint can be "peeled back" if need be and the basic history made clear (italics in original).

Genesis 7:11, the beginning of the flood account (Table 5-1), is a prime example of how a Bible verse can describe real historical events, yet at the same time contain the worldview of the ancient author(s). Genesis 7:11 states that: "In the six hundredth year of Noah's life, in second month, the seventeenth day of the month, the same day were all the fountains of the great deep broken up, and the windows of heaven were opened" (KJV). First, Noah is said to have been 600 (60 x 10) years old when the flood started. As discussed in the last chapter, this was probably not his actual age but a sacred, symbolic age – one of the most "perfect" numbers in the sexagesimal numbering system, according to the worldview of the Mesopotamians. And, while Noah's 600-year-old age is immense compared to human lifespans today, it is not nearly as exaggerated as in the Sumerian account, where Ziusudra (Noah) is said to have been 36,000 (60 x 60 x 10) years old. Next comes the "second month, seventeenth day," which was intended by the biblical author(s) to pinpoint the exact time of the year when the flood started. This date corresponds to the beginning of the actual rainy season in Mesopotamia (middle March) because it records a real historical memory. Next, the "fountains of the great deep" could also be a historically accurate statement because a rainfall of long duration (figuratively *40 days and 40 nights*) would have caused springs to immediately start flowing from the shallow water table of the Mesopotamian alluvial plain. However, from the worldview of the Mesopotamian biblical author(s) this phrase would have meant waters emerging from a primordial deep/netherworld (Fig 1-1 [9]). Then the "windows of heavens were opened," which meant that the sluice gates of the solid firmament had been opened so that rain began pouring down from the immense ocean of water held above the firmament (Fig 1-1 [7]).

A Worldview Approach to Noah's flood also applies to other numbers besides Noah's age in the Genesis flood account. The size of the ark is said to have been 300 cubits long, 50 cubits wide, and 30 cubits high (Gen. 6:15). These are also probably figurative numbers rather than real numbers. Three hundred is exactly one-half of the sacred number 600, and 30 is one-tenth of 300. Three hundred divided by six (6, a bi-sexagesimal number) is 50, or the width of the ark. Genesis 7:24 says that the flood water prevailed upon the earth for 150 days, or one-half of 300. The ark rested in the seventh month (the preferred number 7) on the seventeenth day (10 + 7, two preferred numbers) of the month, upon the mountains of Ararat (Gen. 8:4). After 40 days (again, a figuratively long period of time), Noah opened the window of the ark (Gen. 8:6); Noah stayed yet another seven (7) days, twice (Gen. 8:10, 12). The flood lasted *exactly* 365 days, or one solar year, if the first and last days of the duration of the flood event are included (Table 5-1). All of this is not a *numerically* accurate account of a real flood event, but it is a *numerologically* sacred account of what could have been a record of a very large historical flood. In other words, the literal meaning of the account is what was literal from the religious point of view of the biblical authors, not what would be considered literal today. Thus, a perfectly "literal" interpretation of the Genesis flood story combines a probable historical event with the worldview of the biblical author(s), and one must filter out the worldview aspects to get to the historical core of the text. If this is not done, the numbers make the text seem mythological and at odds with science and reason (*our* reasoning, not *theirs*).

Day	Genesis	Account
Year of flood	Gen. 7:6	Noah was in his 600th year when the flood was upon the earth.
7 days pre-flood	Gen. 7:7, 10	Noah went into the ark and waited 7 days for the flood to start.
Day 1 of flood	Gen. 7:11	Fountains of great deep broken up and windows of heaven opened. Rain started and springs began to flow. (2nd month, 17th day)
Day 1 to Day 40	Gen. 7:4, 12, 17	It rained for 40 days and 40 nights. (3rd month, 27th day)
	Gen. 7:17	The waters increased and bore up the ark and it was lifted above the ground (earth).
	Gen. 7:18	The waters prevailed and increased greatly upon the earth, and the ark was lifted up on the face of the waters. Water still rose and the ark floated upon the surface of the water.
	Gen. 7:19	And the waters prevailed exceedingly upon the earth, and all the high hills were covered. Water still rose, so that all surrounding hills were covered.
	Gen. 7:20	Fifteen cubits upward did the water prevail and the high hills (mountains) were covered. The depth of the water above ground level was 15 cubits (cubit-units vary from about 21 ft to 26 ft).
Day 40 to Day 150	Gen. 7:24	And the waters prevailed upon the earth 150 days (Gen. 7:24) and then were abated (Gen. 8:3). The land was flooded for 150 days or 5 months. (7th month, 17th day)
	Gen. 8:1	God made a wind to pass over the earth, and the waters subsided. Wind caused the water to evaporate.
	Gen. 8:2	The fountains of the deep and the windows of heaven were stopped, and the rain from heaven was restrained. Springs stopped flowing at 150 days and no more rain fell after 150 days.
	Gen. 8:3	And the waters returned off the earth continually. Waters drained away from the land into rivers and toward sea level (Persian Gulf).
Day 150	Gen. 8:4	And the ark rested in the 7th month, 17th day on the mountains of Ararat (Urartu or ancient Armenia).
+73 more days = Day 223	Gen. 8:5	And the waters decreased continually until the 10th month. On the first day of the month, the tops of the mountains were seen by Noah. (10th month, 1st day)
+40 more days = Day 263	Gen. 8:6-7	At the end of 40 more days, Noah opened the window of the ark and sent forth a raven that circled the sky until the waters dried up. (11th month, 10th day)
+ 7 days = Day 270	Gen.8:8-9	And Noah sent forth a dove, but the dove found no resting place for the sole of her foot so she came back to the ark and Noah took her in. (11th month, 17th day)
+ 7 days = Day 277	Gen. 8:10-11	And yet another 7 days, the dove was sent out again and came back with an olive leaf, so Noah knew that the waters were abated from off the earth. (11th month, 24th day)
+ 7 days = Day 284	Gen. 8:12	7 more days, dove is sent out 3rd time but doesn't come back. Water now abated enough for the dove to survive on dry land. (12th month, 1st day)
+30 more days = Day 314	Gen. 8:13	Waters were still drying up from off the earth, but not yet completely dry. Noah removes the ark's waterproof covering and sees that the ground was becoming dry enough to leave the ark.
+ 50 more days = Day 364	Gen. 8:14-16	Ground now completely dry, and God tells Noah to leave the ark. If the first and last days are included, the total time of the flood account is 365 days, or the length of 1 solar year.

Table 5-1. Sequence of flood events according to Genesis 7-8.

A Time and Place for Noah

In order to understand the worldview of the Mesopotamians, we must transport ourselves in time and place back into the ancient world in which Noah lived. To do this, we will examine both the archeological record and the biblical text – or to again quote Kenneth Kitchen, we must "peel back" the worldview to get at the real history.

Who was Noah? Was he a mythological or historical person? In the New Testament, Jesus considered Noah to be a literal person and the flood to be an actual event: "As it was in the days of Noah, so it will be at the coming of the Son of Man. For in the days before the flood people were eating and drinking, marrying and giving in marriage, up to the day Noah entered the ark... the flood came and took them all away" (Matt. 24:37-39; see also Luke 17: 26-27; NIV). Even though in this verse, taken in context, Jesus is comparing the story of Noah with the end of the age and his return, it still implies that Jesus believed Noah to be a real person, just as in Romans 5 it implies that Paul believed Adam to be a real person (see discussion in Chapter 10).

If Noah was a historical person, then when and where does the Bible place him in time and place, and what was his cultural background? To answer these questions, we must go to Genesis, since this is the primary document we have concerning Noah and the flood. However, it is not the only document. Cuneiform tablets have been discovered that refer to a great flood that inundated the Mesopotamian region (Fig 5-1A) and to kings that lived before and after this great flood (Fig 5-1B). Other ancient cuneiform tablets have revealed to scholars what life was like back in Noah's time. Since the massive archeological undertakings in Iraq in the middle of the nineteenth century, more than one million cuneiform texts have been excavated that disclose the literature by which scholars have gained insight into this ancient world.

NOAH'S FLOOD: A HISTORICAL EVENT?

Noah's flood, as described by the authors/scribes of Genesis, is not a *numerically* accurate account of a real flood event, but it is a *numerologically* sacred account of what could have been a record of a very large historical flood. In other words, the literal meaning of the account is what was literal from the religious point of view of the biblical authors, not what would be considered literal today. Thus, a perfectly "literal" interpretation of the Genesis flood story combines a probable historical event with the worldview of the biblical author(s), and one must filter out the worldview aspects to get to the historical core of the text. If this is not done, the numbers make the text seem mythological and at odds with science and reason (*our* reasoning, not *theirs*).

When Did Noah Live?

Biblical and archeological evidence constrain the time when Noah could have lived. The Bible dates Noah because it traces the genealogy of Adam to Abraham through the line of Seth (Genesis 5). Abraham is known to have lived about 2000-1900 B.C., and since Genesis chronologically puts the birth of Noah approximately 900-1,000 years before Abraham, it thus places Noah in a time frame of ca. 3000-2900 B.C. While a "generation" in the Bible does not necessarily mean a direct father-to-son descent (as was discussed in Chapter 4), it is still notable that Genesis lists nine "generations" between Adam and Noah and seven "generations" between Adam (through Cain) and Jabal, Jubal, and Tubal-cain (Genesis 4). Assuming the average length of a generation was the same for both lines (a 40-year "full generation"), this places Jabal, Jubal, and Tubal-cain sometime before Noah – perhaps a hundred or so years before. The importance of Jabal, Jubal, and Tubal-cain to this discussion is that Genesis

Figure 5-1A. 3D image of cuneiform tablet containing the Sumerian flood story: Larsa Babylonia, 19th-18th centuries B.C. In this account of the great flood, the hero is Ziusudra. In the Assyrian version of the flood, the equivalent of the biblical Noah is Ut-napišhtim, and in the Akkadian version it is Atra-hasis. *The Schøyen Collection MS3026, Oslo and London.*

Figure 5-1B. The Sumerian King List; Babylonia, 1813-1812 B.C. The Sumerian King List displays striking similarities to the genealogies of Genesis, although the patriarchs of Genesis have much shorter "reigns" than the more exaggerated ages of kings in the King List. It begins with: "When the kingship was lowered from heaven the kingship was in Eridu." It also talks about "pre-flood" and "post-flood" kings. *The Schøyen Collection MS2855, Oslo and London.*

mentions their occupations, which can be linked to archeological and technological evidence.

> [19]And Lamech took unto him two wives: the name of the one was Adah, and the name of the other, Zillah.
> [20]And Adah bore Jabal: he was the father of such as dwell in tents, and of such as have cattle.
> [21]And his brother's name was Jubal: he was the father of all such as handle the harp and organ.
> [22]And Zillah, she also bare Tubal-cain, an instructor of every artificer in brass and iron: and the sister of Tubal-cain was Naamah. (Genesis 4:19-22, KJV)

The word *father* in verses 20 and 21 suggests that Jabal and Jubal may have been the first to ever practice these occupations. However, the word can also be translated as "ancestor" or "instructor," which alternately suggests that these men might have been only the first of their lineage (Adam's line) to be instructors of these professions. Cuneiform texts from Uruk, one of the earliest cities to exist in southern Mesopotamia, mention the following occupations as being already established by 3100-3000 B.C.: plowman (farmer), shepherd (sheep and goats), cowherd (cattle), fisherman (fish), smith (worker in metal), weaver (of textiles), and potter (maker of pottery). Since these occupations were already in existence by 3100 B.C., it implies that both Jabal and Tubal-cain lived around or before this time (if they were the first).

Jabal, the "father of such as dwell in tents and have cattle," is difficult to place timewise in an archeological context. It is known that the domestication of cattle, sheep, and goats occurred around 10,000 YBP (years before present) in areas surrounding Mesopotamia. It is also known that

Figure 5-2. *Queen's lyre* found by Leonard Woolley in the Royal Cemetery of Ur and dated to around 2600 B.C., or just after Noah lived. This artifact shows the sophistication of technology around Noah's time, when gold and precious gems were being imported from around Mesopotamia and then fashioned into works of art. *British Museum #00832465001.*

nomads occupied the Negev (west of southern Mesopotamia) by the Early Bronze Age (~3000 BC) and in Arabia long before this (see Ch. 9). These ancient nomads appear to have had a similar lifestyle to that of modern-day Bedouin, who pitch their tents and move seasonally in order to provide grazing for their domesticated animals.

The musician occupation of Jubal also fits within an approximate 3000 B.C. time frame. Sumerian characters representing harps have been found on stone tablets from late in the Uruk Period (~3100 B.C.; Table 3-1), and pieces of several harps, including the famous Queen's lyre (Fig 5-2), found by archeologist Leonard Woolley in 1929 in the Royal Cemetery of Ur, were made almost 5000 years ago.

The most pertinent of the three professions named in Genesis 4 is that of Tubal-cain: "he was an instructor of every artificer [craftsman] in brass [bronze] and iron." Bronze is a metal consisting of copper alloyed with arsenic, antimony, or tin, but the word can also mean "copper" in Hebrew. The generally recognized date for the beginning of the Bronze Age in the Near East is around 3200 B.C. (Table 9-1). Thus, Tubal-cain could have lived as early as 3200 B.C. if he was one of the first in Mesopotamia to craft bronze objects. Again, all of these dates (~3200-3000 B.C.) for Jabel, Jubal, and Tubal-cain place Noah in the approximate time frame of 3000-2900 B.C.

Additional archeological evidence exists for the time of Noah and the flood from two cuneiform texts: the Gilgamesh Epic and Sumerian King List (Figs 5-1A, B). Both of these documents attest to a great flood survived by Ziusudra, the Sumerian Noah (or Ut-napišhtim or Atra-hasis, alternate names for Noah), who was "king" of the ancient city of Shuruppak (Fig 3-3). Gilgamesh was listed as the fifth king of the first dynasty of Uruk following the great flood, and he supposedly reigned around 2650 B.C. Therefore, it is reasonable to assume that the flood happened sometime before 2650 B.C. – and perhaps at least two hundred years before.

The Sumerian King List mentions ten antediluvian kings, with Ziusudra (Noah) being the "king" who lived in Shuruppak just before the flood (Fig 5-1B). The mention of Shuruppak is important because the ancient ruins of this city still exist today as the archeological mound of Fara, which has been partially excavated in modern times. Based on pottery types, cylinder seals, and proto-cuneiform tablets found at Fara, archeologists have determined that this city was founded in the Jemdet Nasr Period (3100-2900 B.C.; Table 3-1), and that it was a significant urban center during this time. There is no archeological evidence for a settlement at Shuruppak earlier than the Jemdet Nasr Period; therefore, if Noah did live at Shuruppak, he could not have lived there much before 3100 B.C. If the end of the Jemdet Nasr Period is represented by the flood, then that would put Noah's flood at about 2960 B.C. From the archeological and radiocarbon-dating evidence, it can thus be estimated that Noah most likely lived somewhere around 2900 B.C. (± 200 years).

Where Did Noah Live?

The Sumerian King List (Fig 5-1B) mentions Noah (Ziusudra) as being the "king" of the city-state of Shuruppak in Babylonia (Fig 3-3). Genesis also attests to Noah having lived in the area of Mesopotamia. The garden of Eden was located in the land of the four rivers of Mesopotamia: Euphrates, Tigris, Pishon, and Gihon (Gen. 2:10-14; Fig 3-4). The ark landed upon the mountains of Ararat (Urartu), located just north of Mesopotamia (Gen. 8:4). The names of some of Noah's descendants mentioned in Genesis 10 (e.g., Ophir, Havilah, Ashur) represent places in or bordering Mesopotamia, and Noah's descendant Nimrod (Gen. 10:8-12) was the founder of Babel, Uruk (Erech), and Calneh, all Mesopotamian cities in the land of Shinar, the biblical counterpart of the cuneiform "Sumer." The "tower of Babel" was located on the plain of Shinar and probably refers to the ziggurat (high temple) of ancient Babylon (Fig 3-3; page 130).

Both the "when" (ca. 2900 B.C.) and the "where" (Mesopotamia) of the flood does not support the popular hypothesis of Ryan and Pittman that Noah's flood was an actual deluge that took place around 5600 B.C. in the area of the Black Sea. These authors have proposed that this Black Sea inundation may have been the source of the ancient Sumerian (Gilgamesh Epic) and biblical (Genesis) flood stories. However, this hypothesis is contradicted by the archeological, geological, and biblical evidence. Archeological discoveries by undersea explorers of a Neolithic site (containing stone tools) that once existed by a freshwater lake before the Black Sea rose, raise the question: How could Neolithic humans (or any ancient culture) have constructed a boat the size of the ark (or even a much smaller boat if the claimed size is numerological, not numerical)? Also, there is some geologic evidence that does not support a catastrophic flooding of the Black Sea; rather, it supports a complex and progressive reconnection of the Black and Mediterranean Seas over the past 12,000 years. Finally, the Genesis text does not support the Ryan and Pittman hypothesis: the Black Sea never dried up as the Genesis text claims for Noah's flood, and its flooding was much longer than one solar year. So, if Noah's flood is historical and not mythological, as maintained by a Worldview Approach, it is *not* the same event as proposed by Ryan and Pittman. This is another example of how *both* scientific and biblical scholarship is needed when unraveling science-Scripture controversies.

Ancient Mesopotamia

What would it have been like to live in the ancient world of Noah? What geographical, climatological, political, technological, and cultural forces would have shaped the worldview of Noah and the people of his time and place? In order for the Noah story to be considered historical, we need to examine its archeological and geographical setting.

The land where Noah lived was called Mesopotamia, which literally means "the land between the rivers." The rivers are the Euphrates and Tigris, the main waterways of ancient Mesopotamia and what is now modern Iraq. The Euphrates and Tigris Rivers receive their waters from the mountains of Iran, Turkey, Syria, and Saudi Arabia, which surround the Mesopotamian alluvial plain (Fig 3-3). Southern Mesopotamia included the provinces of Sumer and Akkad, which together were referred to as Babylonia. Northern Mesopotamia was referred to as Assyria. Southern Mesopotamia is one of the flattest places on Earth, the only elevated areas being occasional mounds on the plain that are the archeological remains of ancient cities such as Shuruppak, Uruk, and Ur. The very low gradient of these rivers between Baghdad and the Persian Gulf causes the flow to sometimes back up and breach the levees.

Before modern dams were built, many hundreds to thousands of square miles of the Mesopotamian alluvial plain could become inundated with an almost continuous sheet of water when the Euphrates and Tigris Rivers were in flood from March to July. These two rivers have continually changed their courses over time. During Noah's time (~3000-2900 B.C.), the Euphrates flowed from Sippar to Kish to Nippur to Shuruppak to Uruk (Fig 3-3), and the mounds that still exist of these cities are found along this once-present river course. Even as early as the Ubaid Period (Table 3-1), these two rivers were diverted into canals to water the fertile alluvial soil left behind by seasonal floods. This canal system consisted of an intricate network of dikes, reservoirs, and small dams constructed to store water and release it at the proper time. These canal systems fed into a giant reed-marshland over the lowest-elevation regions of southern Mesopotamia (Fig 3-3).

The climate of southern Iraq (Mesopotamia) is extremely hot and arid and it almost never rains. Temperatures can reach 120°F in July and August, and the average rainfall amounts to only a few inches per year – mainly in March through April from intermittent storms brought in from the Mediterranean (Fig 3-3). The prevailing wind is the *shamal*, or north wind, which sweeps almost continuously down the valleys of the Tigris and Euphrates from June to September and which dries up the flooded ground. During spring, the southerly *sharqi* or *suhaili* winds can bring violent rainstorms to the region from the Persian Gulf.

The natural resources of ancient Babylonia consisted of water from the two rivers, a very rich and fertile soil, clay derived from the alluvial soil, bitumen derived from hydrocarbons coming up faults at Hit (Fig 3-7), and reeds from the marshlands (Fig 3-3). The region was almost completely lacking in metals, gemstones, and high-quality wood, and these commodities had to be imported from the surrounding highlands. Wool (from sheep) and linen (from flax) were common textile export items, as was barley, the main constituent of bread and beer. Clay was used for practically everything: for the making of clay bricks, pottery, sickles (Fig 9-5), tablets, and even nails. Bitumen was used for cementing bricks and for caulking boats. Reeds were used to strengthen mud bricks, to build small boats, to build and roof houses, and to make baskets and mats. Agriculture based on canal irrigation and animal husbandry (sheep, goats, cattle, pigs, fowl) were the primary sources of food. The fertile river lands provided the Mesopotamians with a varied and nutritious diet.

By Noah's time, land trade routes had become established with neighboring regions, especially during Late Uruk expansion in the latter part of the fourth millennium B.C. Lapis lazuli was brought to Mesopotamia along overland trade routes from Afghanistan and carnelian along sea routes from India (Fig 5-3). Tin came from Iran; copper and obsidian from Turkey; and gold, silver, and onyx from Arabia. The timber used by Noah to build the ark probably came from the Amanus Mountains (Fig 3-3) because southern Mesopotamia is almost lacking in timber. Timber, if destined for Babylonia (and Shuruppak, Noah's home town), was cut in the cedar forests of the Amanus, and then dragged or carted down roads 120 miles eastward to the Euphrates River to the port city of Habuda (Fig 3-3), where the logs were tied into rafts and floated downstream. Bitumen for calking the ark would have come from Hit, which was also located upstream from ancient Shuruppak along the Euphrates.

Sea routes were established between Mesopotamia and India and Egypt by the end of the fourth millennium B.C., but it was not until the third millenium B.C. – or after Noah's time – that these routes became commercially important. Trade centers, such as those established in ancient Urartu (Ararat), mainly surrounded the margins of Mesopotamia. Thus, for the great majority of people living in Mesopotamia at that time, knowledge of the world was limited to the Mesopotamian alluvial plain and surrounding mountainous areas.

Cities arose in southern Mesopotamia around 3400 B.C. in the Uruk Period (Table 3-1), or approximately when Cain lived and supposedly established one of the first cities (Gen. 4:17). Polit-

Figure 5-3. Jewelry from the Royal Cemetery of Ur, around 2600 B.C. The blue beads in the necklaces are lapis lazuli (lazurite, the most prized stone in Sumer); the red beads are carnelian (a red form of chalcedony); and the golden beads and pendants are made of gold. The gold probably came from the Mahd adh Dhahab gold mines in Arabia (Fig 3-1) and then transported to Ur to be fashioned into jewelry. That is, by this time the Mesopotamians had already mastered the craft of metal- and jewelry-working. *British Museum #00032463001.*

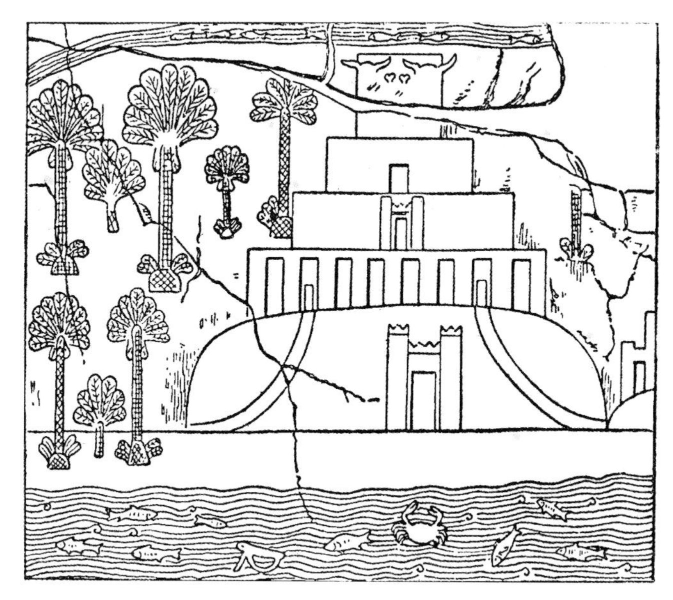

Figure 5-4. Representation of a ziggurat on an Assyrian bas-relief found at Nineveh. The water represents the Tigris River when it, or one of its canals, flowed past ancient Nineveh. The lines crossing the drawing are cracks in the relief. Note also the garden, typical for temples (ziggurats) of that time. The bullheads on the top temple alter-building were possibly made of bronze, as bronze bullheads have been reported for Susa (Fig 3-3). *From A. Parrot, The Tower of Babel (1955). Copyright permission granted by SCM Press.*

ically, ancient Mesopotamia consisted of a number of city-states that competed with each other for power. Wars and power struggles between these city-states were constant and violent. The buildings of these ancient cities consisted of houses, public buildings, and temples. Simple huts were built of reeds plastered with mud, but some houses were built with clay bricks. Temples were built on the tops of ziggurats, which were essentially artificial "mountains" of sun-baked brick and bitumen mortar, constructed in a stair-step, pyramidal manner (Fig 5-4). The purpose of the ziggurat and temple tower was to "reach unto heaven" (Gen. 11:4, KJV), so as to provide a link between heaven and earth and a sanctuary for the pantheon of gods that the polytheistic Mesopotamians worshipped. In a practical sense, high ziggurats were constructed in order to escape the annual floods that inundated the Mesopotamian plain.

The Mesopotamians were the technological leaders of the ancient world. They developed astronomy and mathematics. They invented the wheel and potter's wheel; they discovered how to make glass. In architecture, they developed the arch, dome, and vault; they laid out the plans for cities, temples, and canals. They invented writing and a numbering system, and they set up a legal system and compiled collections of laws. Their literature included epic texts, ritual texts, chronicles, prayers, hymns, proverbs, love poems, laments, and myths. They invented games to play (Fig 5-5). The Mesopotamians had what is considered to have been one of the first technologically based civilizations.

Pictographic writing arose in Mesopotamia around the end of the fourth millennium (or about the same time as it arose independently in Egypt), as did the establishment of a well-developed system

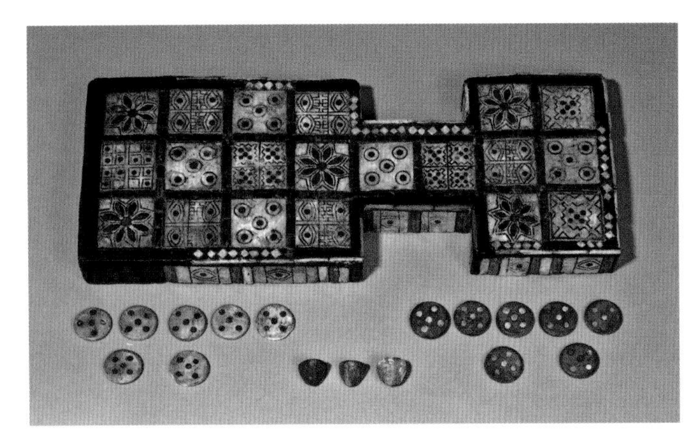

Figure 5-5. Board game played at Ur, around 2600 B.C., from the Royal Cemetary of Ur. *British Museum #00098326001.*

of numbers and measures. (The dual numerical and numerological sexagesimal numbering system used by the Mesopotamians has already been discussed in Chapter 4.) Writing evolved from clay tokens (Fig 5-6A), to markings on envelopes enclosing these tokens, to impressed signs on tablets, to pictographic script (Fig 5-6B). By Jemdet Nasr time (ca. 3000-2900 B.C., or when Noah lived; Table 3-1) the tablet-pictographic stage had been reached, but it wasn't until about 2500-2400 B.C. that narrative and religious writings were being recorded on clay (cuneiform) tablets. Thus, ca. 2500 B.C. would have been about the earliest that the story of Noah's flood could have been written down and copied by scribes. Before that time, it would have been passed down orally (see Chapter 10).

Noah's Worldview

From this cultural context, let us now try to place ourselves into Noah's worldview, or into the worldview of the ancient biblical authors/scribes who first wrote down the story of the flood from oral accounts. If Noah was the Sumerian "king" of Shuruppak, as stated in the King List (Fig 5-1B), he would probably have been well-educated and rich. He would have spoken the archaic Sumerian language, which is known from during or just before the Jemdet Nasr Period (Table 3-1). He could have acquired a knowledge of mathematics and also of how to build ships, such as plied the Euphrates River next to his city. He, or his extended clan, may have been involved in canal building and maintenance since this activity was basic to every citizen's livelihood. Noah could have been engaged in importing cedar wood from the Amanus Mountains and bitumen from Hit, both necessary items for building the ark. Noah would have been surrounded by a polytheistic and violent culture. However, the Bible says that Noah

Figure 5-6A. Tokens from Susa (in present-day Iran; Fig 3-3), around 3300 B.C., used throughout the Near East for counting commodities. *Musée du Louvre, Département des Antiquités Orientales, Paris; image sent by Denise Schmandt-Bessart with her permission.*

Figure 5-6B. Early pictographic script from Uruk III; Sumer, around 3100 B.C. Pictographs were made using a reed stylus with a prismatic tip. The script denotes beer production at the Inanna Temple in Uruk. *The Schøyen Collection MS1717, Oslo and London.*

NOAH'S WORLD

If the patriarch Noah was the Sumerian "king" of Shuruppak, as stated in the King List (Fig 5-1B), he probably would have been well-educated and rich. He would have spoken the archaic Sumerian language, which is known from during or just before the Jemdet Nasr period (Table 3-1). He could have been educated in geography, agriculture, husbandry, canal building, commodities, mathematics, and pictographic script. He could have also acquired a knowledge of how to build ships, such as plied the Euphrates River next to his city, and he could have been engaged in importing cedar wood from the Amanus Mountains and bitumen from Hit, both necessary items for building the ark. Noah would have been surrounded by a polytheistic and violent culture. However, the Bible says that Noah was "a preacher of righteousness" (2 Peter 2:5), a "just man and perfect in his generations, and Noah walked with God" (Gen. 6:9, KJV).

was "a preacher of righteousness" (2 Peter 2-5), a "just man and perfect in his generations, and Noah walked with God" (Gen. 6:9, KJV).

Despite being a man of God, Noah's scientific knowledge would have been limited to what he could observe. That is, he would have had the same cosmological worldview as the ancient Mesopotamians, which worldview consisted of a flat alluvial plain "held up" (surrounded) by mountains (Fig 1-1). The "earth" was the dry land, a flat disc extending to the horizon upon which people walked and worked. "Sluice gates" in the firmament (sky) could open up and pour forth rain, analogous to the opening of canal sluice gates, allowing water to access fields. This cosmology makes perfect sense from the pre-scientific knowledge and everyday experience of the ancient Mesopotamians.

Most important, the *known world* of Noah would have consisted basically of the Mesopotamian alluvial plain and surrounding areas. People living at this time would have had scant knowledge of civilizations outside this Near Eastern boundary, and certainly no knowledge of other distant civilizations then existing over planet Earth (see Chapters 6 and 9). It is into this limited, non-global world that the Genesis flood occurs, and it is within the constraints of this pre-scientific and limited geographical worldview that the story of the flood unfolds and should be interpreted.

Meteorology and Hydrology of the Flood

Besides the numbers problem of the Noah story, there are other factors that make the Genesis text seem mythological or non-historical. We will ask two pertinent questions involving the meteorology and hydrology of the flood that seem to be myth-based, and we will attempt to answer these questions by elaborating on some of the things we have already learned about the ancient Mesopotamia of Noah's time.

How Could It Have Rained for Forty Days and Forty Nights?

While the phrase "forty days and forty nights" in the Bible can be used figuratively for a long, but undisclosed, period of time (see Chapter 4), let's try to determine how "long" that time could have been. To realistically determine this length of time, it is necessary to first understand the weather patterns (meteorology) of the Mesopotamian region and surrounding mountainous terrain. Then these patterns can be compared with the Genesis account of weather associated with Noah's flood.

Cyclonic Storms. The "land of the five seas" refers to the lands encompassed by the Mediterranean, Black, Caspian, Red, and Arabian Seas. This entire region is, and has been for thousands of years, controlled by the Asiatic pressure system (Fig 3-3). In winter, storms originating over the Atlantic Ocean sweep eastward along a low-pressure trough that exists over the Mediterranean Sea. During temporary breakdowns of this system, cyclonic storms travel along the low-pressure Mediterranean trough to the region of the Aegean, and then – still traveling eastward – these storm tracks bifurcate either north to the Black and Caspian Sea areas and the mountains of Turkey, Armenia, and Iran, or south to the Palestine, Syria, Iraq, and Persian Gulf areas (Fig 3-3). For each of these winter tracks, there are about three storms a month that move across the Mesopotamian region, with the peak of rainstorm activity occurring in March and April. During the summer, the low-pressure system over the Mediterranean is replaced by high pressure, and the paths of resulting storms are northward of the "five seas" area. This pervasive situation has caused both northern and southern Mesopotamia to experience nearly rainless conditions in the summer months for millennia.

In addition to the general weather pattern just described, when low-pressure centers exist both in the Mediterranean and over the Persian Gulf and Arabian Sea, Iraq (ancient Mesopotamia) becomes susceptible to the influence of colliding maritime air masses. The eastern tropical maritime air masses originate in the Indian Ocean and can travel northwestward via the Arabian Sea and Persian Gulf to as far as the Mosul area (Fig 3-3). The lower of these two maritime air masses is usually warm and moist, while the upper layers are cool – conditions that favor instability. This results in heavy rainfall in the mountainous parts of the country and also considerable rain in the lowlands in the winter-spring. Continuous downpours that last for days are characteristic of this type of maritime condition, and rains are often accompanied by strong winds.

Long-duration downpours are caused by the stalling or blocking of a Mediterranean frontal system, and depending on how long the system stalls, a "100-year" or "1,000-year" precipitation (flood) event can result. Rare occurrences of extremely high precipitation events such as these are referred to as the "Noah effect" by meteorologists and hydrologists. When circulation patterns persist, then high amounts of rain and snow can also precede or follow a cyclonic event. An example of this happening was in 1969 over the Jordan basin, when cyclonic circulation patterns persisted for twenty-four days, and rain and snow fell during a period of almost two months. When this front stalled over the area, it brought an average of seventy-five inches of rain to the basin – the highest amount for a 150-year period – and caused considerable flooding.

Precipitation. As mentioned previously, southern Iraq (southern Mesopotamia) has an average annual rainfall of less than four inches. The Mediterranean storms (called "westerlies") that pass through Iraq in winter-spring provide the only significant rain of the year, and even this meager rain can be fickle, with some years having no rain at all and with other years having substantial amounts of rain. The alluvial plain of this area (ancient Mesopotamia) is surrounded on the east by the Zagros Mountains, on the north and northeast by the Taurus Mountains, on the northwest by the Amanus Mountains, and on southwest by the Arabian highlands (Fig 3-3). These mountains receive rain and snow that feed the Euphrates and Tigris Rivers in the spring. The mountains of Armenia and Kurdistan in the northeastern Taurus range experience especially severe winters of up to six to eight months' duration, and snow in these mountains frequently reaches depths of six feet. Before modern dams were built, the Tigris River overflowed its banks practically every spring from snow melting in the Zagros Mountains, reaching its highest flood level in late spring.

Wind. The predominant wind in Iraq (Mesopotamia) is the northwest *shamal* (Fig 3-3). The shamal is the more-or-less constant flow of air down the Tigris-Euphrates river valley, following topography and gradient from the Taurus Mountains in Turkey southward to the Persian Gulf. The shamal wind operates all year long, but is especially prevalent in summer-fall from June to October when the wind direction is not interrupted by the passage of cyclonic storms. These are dry, warm, persistent winds, signifying clear skies and stable weather. The very dry air brought by the shamal permits intense heating and evaporation of the land surface.

In winter and spring, the regularity of the shamal decreases and the *sharqi* becomes the predominant winds up to an elevation of approximately 7,000 feet. These winds coming from the Persian Gulf are relatively cool and damp and may bring clouds and rain to the entire region as they develop in front of advancing cyclonic depressions. Sometimes in the Persian Gulf region, these southeasterly sharqi winds are followed by southwesterly *suhaili* winds after the passage of a trough. The suhaili are often strong winds that can pose a danger for ships in the gulf. Only with the passing of a cyclonic storm are pressure-gradients steep enough for violent winds to result.

Accordance with the biblical account. If the "second month, the seventeenth day of the month" of Genesis 7:11 is interpreted as denoting the season of the year when the flood started, rather than a month-day extension of Noah's age, then Genesis is in remarkable accordance with the weather patterns that actually exist and have existed in the Mesopotamian area for millennia. When the ancient Mesopotamian sidereal calendar is coordinated with today's tropical calendar, this puts the "second month, seventeenth day" in about the middle of March when meteorological conditions bring the most abundant rain to the Mesopotamian region. Genesis 7:12 implies that it was a "heavy" rain which fell upon the earth (land) for forty days and forty nights, and this is the type of continuous downpour that can result from the activity of maritime air masses characteristic of this season. The duration of rain (up until 150 days; Gen. 8:2; Fig 5-1) could have been caused by the stalling of a Mediterranean cyclonic front over the Mesopotamian area in combination with maritime air masses moving up from the Persian Gulf and Arabian Sea/ Indian Ocean. This stalled storm would have been associated with southerly winds (the sharqi and/ or suhaili), but not with the northwesterly shamal wind. These winds could have been very intense, both in strength and duration.

Genesis 7:24 and 8:1 records that five months after the flood began – or about in the middle of August assuming a middle-of-March starting-date for the flood – a wind passed over the Earth causing the waters to subside (Table 5-1). This wind could correspond with the northwest shamal wind that blows almost continuously during the summer months. In spring, the melting of snow and steady rain in the mountains of northern Iraq produces flooding in the valleys of the south. Then in summer, the wind howls southward along the narrow fertile strip between the Euphrates and Tigris Rivers, and the drying process begins.

A REASONABLE METEOROLOGY SCENARIO?

The purpose of this chapter is to show what the real world of Genesis was like in Noah's time. Genesis chapters 6, 7, and 8 collectively show that the Genesis account of when the rains began (March-April), duration of the rain (forty days and forty nights), length of flooding (150 days), flow duration of springs (5 months), severity of the winds (sharqi, suhaili winds), mechanism (shamal wind) for the flood water to completely dry, and the total time of the flood (365 days), are not unreasonable considering the meteorological and hydrological conditions known to have existed in Mesopotamia for millennia. This accordance supports a historically based flood, rather than a mythological (non-historical) flood, and is remarkable considering that the Noah story had to have been passed down orally for hundreds of years before it could have been written down as a narrative text after around 2500 B.C. (see Chapter 10).

How Did Prolonged Flooding Occur?

A number of hydrologic factors could have been responsible for the prolonged (150 days) flooding of the Mesopotamian region, as claimed by the Genesis story of the flood (Gen. 7:24; Table 5-1). ("Hydrology" is the science that deals with all aspects of water and its cycling from the oceans to land and back to the oceans again. The "Mesopotamian hydrologic basin" is the area surrounded by mountains in Figure 3-3, where the watershed from these mountains supplies water to the four rivers of Mesopotamia, which then empty into the Persian Gulf.)

Rain. Not only does Genesis 7:12 say that it rained heavily and continuously for forty days and forty nights, but after this time it could have also rained intermittently until day 150, when Genesis 8:2 says the rain finally stopped. The exact location and duration of rain is unclear. If the cyclonic storm was regional, it could have rained over all of Mesopotamia and the surrounding highlands for an extended period of time. To quote archeologist André Parrot, who excavated in the region:

> I believe also that this inundation was occasioned primarily by more than usual violent flooding of the Euphrates and the Tigris, which could easily sweep away human habitations in an area that is quite flat. The destructive action of the flooding was probably increased by torrential rains – a Westerner can have no idea of the intensity of which they are capable.

Snow. While the Bible doesn't specifically mention the involvement of snow in the Genesis flood, melting of mountain snows by the rains of Genesis 7:17 could also have been an important factor contributing to prolonged flooding. When deep snow is melted by heavy rains, water is released quickly and can produce immediate flooding, but if the snow is deep and not subject to quick melting, water will be released over a long period of time. If the snow had been exceptionally deep during the winter of Noah's flood, this snow could have added a great amount of runoff water to the Mesopotamian hydrologic basin for a period of months. In particular, it could have been responsible for prolonged flooding in the upper parts of the basin in the northern Iraq (ancient Urartu) region. Such an occurrence is recorded as having happened during the 1954 flood along the upper Tigris River.

Springs. The Bible mentions the "fountains of the great deep" (springs) twice in its narrative – once when the springs started flowing (Gen. 7:11) and again when they stopped (Gen. 8:2) (Table 5-1). Springs could have been a prime factor causing prolonged flooding of the Mesopotamian hydrologic basin. When it rains or when snow melts, water doesn't only flow over the ground as stream runoff; it can also travel underground as "groundwater," finally emerging at the surface as springs. Genesis 7:11 records that the "fountains of the great deep" (subterranean water or groundwater) were "broken up," which phrase derives from the Hebrew word *bâqa*, which means to "break forth" or be "ready to burst," and so the literal meaning of the text is that these springs began gushing water.

Springs exist all over Iraq and surrounding highlands, and many of these are limestone (karst) springs. Ras-el-ain near the border of Syria and Turkey is one of the largest karst springs in the world and is the effective head of the Khabūr River, a major tributary to the Euphrates (Fig 3-3). Water from this complex of springs comes from maximum winter snow melt and rain in the Taurus Mountains in January to February, but this water does not discharge at Ras-el-ain until the following July or August. This type of delay is typical of springs, where recharge may be distant or convoluted from the discharge point, and where a continuous supply of water may be supplied for many months after a heavy rain storm. Genesis seems to indicate that at least some springs began gushing water immediately after the flood started (Gen. 7:11), but that others continued for up to 5 months (Gen. 8:2).

Specific springs (among many) that could have contributed water to the Mesopotamian hydrologic basin during Noah's flood are those located near ancient Sippar, Babylon, and Kish; those in the vicinity of Hit; and those in the Jezira desert region between Baghdad and Mosul (Fig 3-3).

Tributaries to the Tigris River also emerge from large caves along the foothills of the Zagros Mountains. When severe rains occur in the Zagros, these springs respond with a strong outflow, causing the rivers to spill and overflow onto the plains. In antiquity, one of the most important of these springs emerged from Shalmaneser's Cave, which was thought to be the "source" of the Tigris River when Shalmaneser III visited the cave in 852 B.C. It is also recorded that Sargon II had learned the secret of tapping water from subterranean strata (groundwater) during his campaign against Ulhu and Urartu (the land of Ararat).

Numerous springs also exist in the deep canyons of the Cudi Dag (Jabel Judi), Cizre region of southeastern Turkey (Fig 3-3). Various karst features such as springs, sinks, and caves occur in the Jurassic-Cretaceous Cudi Limestone of these mountains. The best known of these springs is located west of Beytişebab, and smaller ones occur further south. Runoff from these springs helps prolong flooding in the upper Tigris River valley-Cizre Plain region.

Storm surge. In addition to rain, snow melt, and spring activity, there is a possibility that a storm surge may have prolonged flooding in southern Mesopotamia during Noah's flood. Storm surges occur where a low-pressure meteorological system causes high winds and tides, which can drive seawater inland for hundreds of miles, especially if the terrain is very flat, as it is in southern Iraq. According to archeologist A. Parrot, the possibility of a tidal wave cannot be ruled out. This hypothesis is also supported by written cuneiform texts. The technical word for flood or deluge is *amaru* in Sumerian or *abubu* in Akkadian; similarly, the Hebrew word *mabbūl* for "flood" used in the Genesis text is applicable to both an inundation or an "over-flooding" caused by a sweeping, wind-driven rainstorm. Specifically, *abubu* indicates moving water caused by a rainstorm or a storm that drives seawater landward. In the Sumerian Gilgamesh epic, it is said that a "hurricane raged" and after the flood "the sea became quiet, the storm was still, and the *abubu* ceased." Thus the term *abubu* not only depicts an inundation, but it can also describe the destructive winds and gales that accompany the rainstorm. In the Sumerian cuneiform tablets found at Nippur, the Noachian deluge is described as: "the mighty winds blew violently... and the ship moved along over the face of the great waters, driven by the wind" (page 54). In the Akkadian Atra-hasis epic, the text speaks of thunder and savage winds. In the Gilgamesh epic, the flood of Ziusudra (Noah) is recorded as having been a "south storm" accompanied by wind and thunder, where the flood-winds blew over the land and the south-wind tempest swept over it. This "south wind" could have been either the sharqi or the suhaili, both of which can be violent.

Accordance with biblical account. The point of this long discourse on the meteorology and hydrology of Mesopotamia should now be clear: the biblical verses discussed above are examples of "historical memories," such as mentioned at the beginning of this chapter. These verses collectively show that the Genesis account of when the rains began (March-April); duration of the rain (forty days and forty nights); length of flooding (150 days); flow duration (5 months) of springs (fountains of the deep); and the time (365 days) and mechanism (prevalent wind patterns) for the flood to completely dry (Table 5-1), are not unreasonable considering the meteorological and hydrological conditions known to have existed in Mesopotamia for millennia. This accordance supports a historically based flood, rather than a mythological (non-historical) flood, and is remarkable considering that the Noah story had to have been passed down orally for hundreds of years before it could have been written down as a narrative text around 2500 B.C. (see Chapter 10).

The purpose of this chapter is to show what the real world of Genesis was like in Noah's time; that is, to show the historical background of when Noah lived and how different ancient texts describe the flood. All of this information relates to the topic of the next chapter: Was Noah's flood global, as maintained by Young-Earth Creationists/flood geologists, or was it local and confined to the Mesopotamian hydrologic basin? If it was local, then there are other critical hydrological questions that need to be addressed.

Noah's Ark riding on a swell after the Great Flood. 3D computer illustration by James Steidl.

CHAPTER 6

NOAH'S FLOOD: GLOBAL OR LOCAL?

Genesis 8:4. And the ark rested in the seventh month, on the seventeenth day of the month, upon the mountains of Ararat. (KJV)

In this chapter, we will cover another Young-Earth/Old-Earth Creationist controversy: Was Noah's flood global or local? By "global," Young-Earth Creationists mean that Noah's flood extended over the entire planet Earth, covering even the highest mountains. And, since Genesis 7:21 states that "all flesh" died in the flood, by extension it means that almost all of Earth's sedimentary rocks containing fossils were deposited in Noah's flood. By "local," Old-Earth Creationists mean that Noah's flood was local to the Mesopotamian region – an area that for millennia has been inundated by floods.

Box 6-1. Two Christian Views on Extent of Flood

Young-Earth Creationist	Old-Earth Creationist
Global	Local

"Universal" Language of Genesis 6-8

We are going to start our global-local flood discussion with a hermeneutical word study, where we compare words and passages contained in the flood account with other passages in the Bible. This is because the main support for a global flood is the seemingly "universal" language of Genesis 6-8, which is no doubt the primary reason why people in centuries past have believed that Genesis was talking about planet Earth, and why this traditional interpretation has continued until today in the form of Young-Earth Creationism and flood geology. In Genesis 6-8, *earth* is used 42 times, *all* is used 20 times, *every* is used 23 times, and *under heaven* (literally, *under the sky*) is used 2 times. However, such language is not necessarily all-inclusive when understood from the cultural and linguistic context of Genesis and the rest of Scripture.

The Hebrew Word "Earth"

The Hebrew word for "earth" used in Genesis 6-8 is *eretz* or *adâmâh*, both literally meaning "earth, ground, land, dirt, soil, or country." The word *earth* (uncapitalized) cannot be taken to mean planet Earth (capitalized), since the Genesis authors/scribes had no concept of Earth as a planet and thus had no word for it. According to a Worldview Approach, the biblical account should be interpreted within the narrow limit of what was known about the extent of the world at *that time*, not what is known about its extent today.

Biblical context also makes it clear that *earth* does not mean the whole planet Earth. For example, the "face of the ground," as used in Genesis 7:23 and Genesis 8:8 in place of *earth*, does not imply planet Earth. "Land" is a better translation than "earth" for *eretz* because it extends to the "face of the ground" we can see around us; that is, the land that is within view of the horizon. It can also refer to a local geographic or political landscape. For example, when Zechariah 5:6 says "all the earth" the verse is literally talking about Palestine – a tract of land or country. Similarly, the Mesopotamian concept of the land (*kalam* in Sumerian) seems to have meant the entire alluvial plain. The clincher to the word *earth* meaning ground or land (and not the planet Earth) is Genesis 1:10: "God called the dry land earth" (*eretz*). If God defined *earth* as "dry land," then so should we.

PLANET OR DRY LAND?

Biblical context makes it clear that earth does not mean the whole planet Earth. For example, the "face of the ground," as used in Genesis 7:23 and Genesis 8:8 in place of earth, does not imply planet Earth. "Land" is a better translation than "earth" for *eretz* because it extends to the "face of the ground" we can see around us; that is, the land that is within view of the horizon. It can also refer to a local geographic or political landscape. For example, when Zechariah 5:6 says "all the earth" the verse is literally talking about Palestine – a tract of land or country. Similarly, the Mesopotamian concept of "the land" (*kalam* in Sumerian) seems to have meant the entire alluvial plain. The clincher to the word *earth* meaning ground or land (and not the planet Earth) is Genesis 1:10: "God called the dry land earth" (*eretz*). If God defined earth as "dry land," then so should we.

All, Every, under Heaven

While these three terms appear to confer a global meaning to the flood event, all three are used elsewhere in the Bible for local events. For example, Acts 2:5 states: "And there were dwelling at Jerusalem Jews, devout men out of every nation under heaven" (KJV). Does this passage include every nation on Earth? Or does it specifically refer to the nations mentioned in Acts 2:9-11 that Luke, the writer of Acts, knew about? Certainly it didn't include North America, South America, or Australia, as these continents were unknown to Luke in the first century A.D. Such "universal" language is simply the way people expressed themselves in those days to emphasize a level of inclusiveness – a type of "Bible-speak" that is not supposed to be taken absolutely literally, but in the context of what the biblical author was trying to emphasize. This passage in Acts simply means that devout men (Jews) of many nations from some extended region of the then-known world were dwelling at Jerusalem.

Another example of how such "Bible-speak" is used in Genesis to describe a regional (non-global) event is Genesis 41:56: "And the famine was over all the face of the earth" (KJV). This is the exact language used in Genesis 6:7; 7:3, 4; 8:9; and elsewhere when describing the Genesis flood. So was the biblical author claiming that the whole planet Earth (North America, Australia, and so on) was experiencing famine? No, the universality of this verse applies only to the lands of the Near East (Egypt, Palestine, Mesopotamia), and perhaps even to the Mediterranean area; that is, the whole world that had become known by Joseph's time (about 1800 B.C.; Table 3-1). Another Old Testament example of such universal language can be found in Daniel 6:25 where "King Darius wrote unto all people, nations, and languages that dwell in all the earth..." (KJV). The phrase "in all the earth" in this case would have referred to the whole world that had become known by Daniel's time (~600 B.C.). In other words, as the knowledge of geography expanded over time, the word *earth* took on a more global meaning rather than a localized meaning, and only since the fifteenth century A.D. has the word *earth* been assumed to mean planet Earth as we know it today.

The same principle of a limited universality commensurate with the knowledge of the biblical author(s) also applies to the story of Noah's flood. From Noah's perspective, the "earth" included just the land (ground) in the visible horizon, as Noah saw it "under heaven." The language used in the scriptural narrative is thus simply what would have been observable to an eyewitness (Noah). Then this eyewitness account was passed down orally to biblical scribes, who ultimately wrote the story down from the worldview perspective of *their* culture at *that* time. Biblical archeologist Leonard Woolley aptly described the situation thusly: "It was not a universal deluge; it was a vast flood in the valley of the Tigris and the Euphrates which drowned the whole of the habitable land...; for the people who lived there that *was all the world*" (italics mine).

Thus, the "universal language" of the Genesis flood account does not support a global flood if one considers how the ancient biblical authors/scribes wrote down the story of Noah from *their* limited geographical worldview. (The subject of how the Genesis stories were passed down will be discussed in Chapter 10.)

Depth of the Flood

Another key verse as to whether the Noachian flood should be interpreted as being global or local is Genesis 7:20: "Fifteen cubits upward did the waters prevail; and the mountains were covered" (KJV). Young-Earth Creationists assume this passage means that the floodwater rose at least fifteen cubits above Mount Ararat, their preferred landing place for the ark because of their *presumption* of a global flood. But there is a major difficulty with this interpretation that involves the translation of the Hebrew word *har* as "mountain" in Genesis 7:20. This word can also be translated as "a range of hills" or "hill country," implying with Genesis 7:19 that it was "all the high hills" (also *har*) that were covered with floodwater rather than the high mountains. The Young-Earth Creationist rationale for the ark having landed atop the highest mountain in the area (Ararat) seems to stem from Genesis 8:5 where the tops of "mountains" were not seen for some time after the ark landed. Yet if the ark had grounded in the lower foothills, the water still would have had to recede before the tops of the "hills" where the ark rested could have come into view. The ark did not need to be on a mountain top for this description to hold true.

To make matters more complicated, the Sumerians considered their temples (ziggurats) to be "mountains," calling them *É. kur*, which in Sumerian means "house of the mountain" or "mountain house." Also, the specific Mesopotamian word for "mountain" (*šadû*) is derived from "mounds," and may indicate that the Mesopotamians actually thought of their high temple mounds on the very flat alluvial plain as hills or mountains. So to which of these alternatives was the biblical author referring to in Genesis 7:20? Were the flood waters fifteen cubits above the highest mountains of planet Earth? Were they fifteen cubits above the "hill country" of northern Mesopotamia? Were they fifteen cubits above the Mesopotamian alluvial plain? Or, were they fifteen cubits above the tops of ziggurat temple mounds ("mountains") in southern Mesopotamia, thus dooming all the people who ran to the high temples for safety from the floodwaters? Could the "fifteen cubits upward" even refer to the draft (draught) of the ark; that is, how deep its 30-cubit depth (Gen. 6:15) was submerged in the water when the ark was loaded? David Snoke, in his book *A Biblical Case for an Old Earth*, put it this way:

> The passage [Gen. 7:20] says that the water rose only twenty feet, not six miles. For no reason other than to make sure there would be enough water for a global flood, this verse is frequently altered to "the water rose to twenty feet higher than the highest mountains." This latter reading is *not* the "literal" reading; it is interpolated, that is, read into the text.

Another difficulty with Genesis 7:20 is: How did Noah measure the depth of the floodwaters at fifteen cubits? In riverboats of that day, people used rods or poles to measure water depth. But how could Noah have possibly taken a pole measurement on top of a 17,000-foot-high mountain like Ararat in the midst of a tempestuous global ocean? Rather, the biblical account (Gen. 7:17-20) seems to suggest that the waters increased continuously until the ark was lifted up above the earth (land), and in this context one can imagine Noah measuring the depth of water from where the ark began floating on the alluvial plain. In any case, the phrase "fifteen cubits upward" does not necessarily imply a global flood; if anything, it favors a local flood where the depth from the deck of the ark to the ground surface could be easily measured.

Continuing our investigation of the biblical text, we will next examine what Genesis has to say about Noah's ark and where it supposedly landed. We will discuss this topic in order to try to ascertain how realistic the story of the ark is from a global versus local, and historical versus mythological, perspective. Finally, we will try to answer the important hydrological questions related to Noah's flood, assuming that it was a local flood: Where is all of the sediment left by Noah's flood?; and, most difficult of all, how could the ark have made it from southern Mesopotamia where Noah lived all the way to the mountains of Ararat in northern Mesopotamia since drainage would have been southward towards and into the Persian Gulf?

Noah's Ark

An Ark Encounter

Noah's ark has been portrayed in many ways, mostly in children's books as a small boat containing happy people and cute animals or as an artwork theme by major artists (page 70 and Fig 6-1). A supposed replica of the ark (called the "Ark Encounter") has also been built by the Young-Earth Creationist organization, *Answers in Genesis*, at its theme park in Kentucky. But what did Noah's ark really look like? In this section we will discuss not only what the Genesis text says about the ark, but also what knowledge biblical scholars and nautical historians have contributed to understanding these passages.

There are two descriptions in Genesis of Noah's ark – Genesis 6 and Genesis 8. We will discuss both of these descriptions.

> **Genesis 6:**
> [14] Make thee an ark of gopher wood; rooms shalt thou make in the ark, and shalt pitch it within and without with pitch.
> [15] And this is the fashion which thou shalt make it of: The length of the ark shalt be three hundred cubits, the breadth of it fifty cubits, and the height of it thirty cubits.
> [16] A window shalt thou make to the ark, and in a cubit shalt thou finish it above; and the door of the ark shalt thou set in the side thereof; with the lower, second, and third stories shalt thou make it. (Gen. 6:14-16, KJV)

> **Genesis 8:**
> [6] Noah opened the window of the ark which he had made…
> [13]…and Noah removed the covering of the ark, and looked, and behold the face of the ground was dry. (Gen. 8:6, 13, KJV)

Wood for the ark (Gen. 6:14). As was mentioned in Chapter 5, southern Mesopotamia where Noah lived was almost completely lacking in high-quality wood, and wood for constructing

Figure 6-1. *Noah's Ark* by Edward Hicks, 1846. *Philadelphia Museum of Art, Bequest of Lisa Norris Elkins, 1950-92-7.*

the ark would have had to have been imported from the surrounding highlands. The identity of "gopher wood" is lost in antiquity, but most likely it was cedar, which is strong, durable, and ideal in the making of large ships. It has been documented that a lumber trade had already become established between southern Mesopotamia and the Amanus Mountains (Fig 3-3) by Jemdet Nasr (Noah's) time (Table 3-1), and that cedar trunks up to one hundred feet long were being imported southward during this period. To make the journey to southern Mesopotamia from the Amanus Mountains, the timber was floated down the Euphrates River on rafts, usually during spring flooding. Lumber and other goods were also hauled upriver throughout Mesopotamia in antiquity, either by rowing or towing (Fig 6-2).

Figure 6-2. Timber transported by rowboats for the construction of Sargon II's palace at Khorsabad, Assyria. Bas-relief, eighth century B.C. *Google Image.*

Rooms in the ark (Gen. 6:14). According to the Hebrew scholar Cassuto, the word *qinnnīm* for "rooms" literally means "nests" in the sense of compartments or dwellings. Speiser also translated Genesis 6:14 as "make it an ark with compartments [cells]," but Yahuda thought the correct translation should be "fiber-tight shalt thou make the ark" – in the sense that reeds may have been used in the ark's interior construction. Compartments (rooms or cells) would have been needed to house and separate different animals, and these could have been made out of marsh reeds, a common building material in southern Mesopotamia at that time.

Pitch for the ark (Gen. 6:14). The bitumen (pitch) for the ark would not have been difficult to obtain. As discussed in Chapter 3, and as shown in Figures 3-3 and 3-7, the center of bitumen production in Mesopotamia was (and still is) at Hit, located along the Euphrates River about 80 miles west of Baghdad and 200 miles north of the mound of Shuruppak (Noah's "home town"). Even today, bitumen is packaged into reed baskets at Hit and floated down the Euphrates in boats. Bitumen for the caulking of boats has been documented since the Ubaid Period (Table 3-1), but the bitumen industry only became well established by around 3000 B.C., or about the time that Noah lived (see Chapter 5). Boats were the most general means of transportation in southern Mesopotamia, and large quantities of bitumen were used for the coating and caulking of boats, large or small. Since Genesis 6:14 says that the ark was pitched both "within and without," it implies that the ark may have been grayish to blackish in color.

Size of the ark (Gen. 6:15). A major archeological and technological difficulty involves the stated size of the ark (300 cubits long, 50 cubits wide, and 30 cubits high). If one supposes a 20-inch Mesopotamian cubit in proto-Sumerian time (~3000 B.C.), then the stated dimensions of the ark would have been about 500 feet long, 85 feet wide, and 50 feet high – or about the size of a modern ocean liner, which far exceeds the known capacity for boats of around 3000 B.C. In addition, the size of wooden ships is limited to about 300 feet due to their inherent-strength instability above this size, so Genesis's claimed dimensions for the ark seem excessive.

A possible explanation for the "technology gap" between the ark's stated size and the ship-building techniques of Noah's day may involve the Mesopotamians' use of the sexagesimal-decimal numbering system (see Chapter 4). As pointed out by Cassuto, the dimensions of the ark are multiples of 60 or 10, the fundamental numbers of the sexagesimal system. Thus, like Noah's age of 600, the stated dimensions of the ark may be numerological numbers with some sacred meaning that is lost to us today. The stated length/width ratio of the ark (300:50, or 6:1) is particularly interesting because this ratio is important for the stability (the control of pitching and rolling) of ships. Thus, this 6:1 ratio may imply an ancient understanding of ship dynamics that extends as far back as Noah's time or before.

A smaller size for the ark also makes sense for the number of animals that Noah could have gathered onto the ark. Our previous discussion on the limited meaning of *all*, *every*, and *under heaven* also applies to Genesis 6:19, where it says "And of every living thing of all flesh, two of every sort shalt thou bring into the ark, to keep them alive with thee; they shall be male and female" (KJV). This limited interpretation would *not* mean animals that occupied the entire planet Earth, but *only* the animals in the Mesopotamian area, and primarily domesticated animals and birds that could reproduce after the flood or be eaten on the ark's one-year journey; that is, dietary "clean" animals and birds should be gathered by "sevens, the male and his female" (Gen. 7:2-3, KJV). "Thus did Noah; according to all that God commanded him, so did he" (Gen. 6:22, KJV). That is, Noah gathered all of the animals *himself*, so again this implies animals local to his area and an ark size that could accommodate these animals.

Shape of the ark (Gen. 6:15). While Genesis 6:15 gives the ark's dimensions, it does not indicate its actual shape – whether it was oblong or rectangular. Biblical scholars Cassuto and Yahuda were both of the opinion that the ark's architecture was Egyptian in origin, as the Hebrew word for "ark" (*têbâh*) comes from the Egyptian word for "box" (*têbâh*). Also, the fact that the ark is never described by the Hebrew word for boat indicates that it was not boat-shaped, but simply a large box that would float with its base on the surface of the water and not need to be steered. This description suggests a barge-like vessel, such as is depicted in Figure 6-3. Barges of that time typically had slightly upturned prows, straight gunwales, and almost square sterns. Barges made of wood and square at each end are documented by around 3000 B.C. in Egypt so they could have

Figure 6-3. Noah's ark resting in the Mountains of Ararat, from Johann Scheuchzer's *Physica Sacra* (1731). *Courtesy of the David Mongomery Collection.*

also been plying the Euphrates and Tigris Rivers in Mesopotamia by that time.

"Window" to the ark (Gen. 6:16). The word *window* used in the King James Version is put in quotes because the New International Version translates the word as "roof." Biblical scholar Speiser translated it yet in another way: "Make a *skylight* for the ark, terminating it within a cubit of the top." *Tsōar* (or *zōhar*) is the Hebrew word used for "window" in Genesis 6:16, but it actually denotes a skylight-like feature characteristic of Egyptian houses and temples in ancient times. Such skylights consisted of square or semi-circular openings, situated over the door or in a corner fairly near the roof, which admitted light and air from above when windows and doors had to be shut against overbearing heat or driving rain. A skylight interpretation for Genesis 6:16 makes sense because the text states that the *zōhar* should not be more than a cubit in height, and that it should be fixed high up; that is to say, beneath the roof, in order for the ark to be lighted and ventilated from a spot no water could penetrate. This provision was necessary because the window proper (*challōwn* or *hallōn*) was to remain closed during a storm so that water could not enter the vessel.

Door of the ark (Gen. 6:16). Genesis 6:16 also states that a "door" (*pethach*) shall be put in the "side" (*tsad*) of the ark, with its obvious function being for entry and exit. But does this verse mean a door in the side of the ship's frame, or a door into a cabin or deck that then led down into the insides of the ark? A door into a ship's frame seems unlikely in that it would make the ship un-seaworthy. In addition, boats by the fourth millennium B.C. are known to have had a cabin (or two cabins) placed amidship, so perhaps entry into the interior of the ship was via a ramp covered by a cabin? A cabin on deck would also make sense as living quarters for Noah and his family.

Decks or stories to the ark (Gen. 6:16). The description of the ark specifies that it contained three decks or stories; in addition, the plural form of the Hebrew specifies that there were many rooms on each deck. Accounts of decked boats in Mesopotamia are uncommon, and only one account of a cargo boat used for horse transport and having a deck "half-way the bilge" has been documented. But the Egyptian Cheops ship (~2650 B.C.) is known to have had decks, which suggests that decked ships may well go back in time as the product of a long-established Near Eastern tradition.

Window of the ark (Gen. 8:6). It appears likely that the "window" of Genesis 8:6 is not the same "window" of Genesis 6:16 because the Hebrew word used in Genesis 8:6 is *challōwn* (or *hallōn*), not the *tsōar* (or *zōhar*) of Genesis 6:16. This difference in wording implies that there could have been two types of "windows" in the ark – skylights to let in light and air, even in the presence of pounding rain, and windows that could be opened or shut depending on the weather. Evidently, the "window" of Genesis 8:6 was one that could be alternately opened or shut (as opposed to a skylight), as Noah first sent forth a raven from a window (Gen. 8:7), and then a dove (Gen. 8:8) (Fig 6-5), whereas a skylight would have been too high near the roof.

Covering for the ark (Gen. 8:13). The Hebrew word for "covering" is *mikceh*, which literally means "weather-boarding." Besides having a skylight to keep out the rain, the ark evidently had some type of weather-proofing over its top deck to help keep the rain from entering its inner parts. The word *covering* is exactly the same as used in Exodus 26:14 (and elsewhere in Exodus and Numbers) for the outermost covering placed over the ark of the covenant. The King James Version specifies that the Exodus covering was made of "badger's skins," but the New International Version identifies the skin as that of "dugongs;" i.e., sea cows which are native to the shores of the Indian Ocean. This latter translation makes sense for a covering which was supposedly water repellent, and such a covering over a top deck makes even more sense if the ark was significantly smaller than the stated dimensions of Genesis 6:15. Or, in other words, Noah built the ark to be "environment-friendly" and storm-worthy, having a skylight for light and ventilation even under storm conditions, and also a deck covering to help protect the ark from leaking.

Summary of Noah's ark. What does all of this biblically- and scholarly-based information tell us about what the ark looked like, how it functioned, and its ability to withstand a violent storm? It was probably a barge made of cedar wood exported from the surrounding highlands, with a size commensurate with other boats of that time, assuming that its stated dimensions are numerological rather than numerical. And, if of a smaller size, it might have even had sails, as it is known that sails were used on both sea and river craft from the fourth millennium B.C. onwards. Direct evidence for the use of sails dating to as early as the Ubaid Period is a small clay replica of a reed-bundle boat, and a painted clay disc depicting a boat with masts. Genesis doesn't mention sails, but a summary of the Gilgamesh epic by Berossus in the third century B.C. does mention a "sail" and "sailing" with regards to the flood ship.

All of this information implies that Noah was familiar with boat construction and that he built the ark using the nautical technology available at his time (around 3000 B.C.). That is, Noah's flood occurred in the *Neolithic* and could *not* have happened before this time because the technology was not yet available for the building of large boats. This practical knowledge of ship building helps make the Genesis account of Noah and the flood *historical* rather than fictional or mythological. It also impacts a global-flood interpretation in that Noah was evidently building a boat out of local materials to house his family, and he was gathering local animals in order to ride out what he anticipated would be a local storm. Noah could have also been familiar with the trade region of Urartu ("mountains of Ararat"; Fig 3-3) along the Tigris River north of Assyria, as this area was known by that time.

Landing Place of the Ark

The landing place of Noah's ark is important to a global-local discussion because *where* the ark landed is crucial to the credibility of the flood account. Where the ark landed is actually one of the most controversial aspects of the flood-story, with Young-Earth Creationists insisting that Genesis identifies the site as Mount Ararat – the huge volcanic construct Agri Dag in northeastern Turkey (Fig 6-4). Why this insistence? It is because if the ark had landed on the top of this 17,000 foot-high mountain, it would imply that floodwater had reached this elevation and must be planet-wide.

The ark has been assigned to at least eight different landing places over the centuries, including Saudi Arabia, India, and even the mythical Atlantis. One reason for this ambiguity is that Genesis does not actually pinpoint the exact place where the ark landed; it merely alludes to a region or range of mountains where the ark came to rest: "the mountains of Ararat" (Gen. 8:4). Although many sites have been proposed for the landing place of the ark, only four appear to meet the requirement of being located within the

SHIP BUILDING IN NOAH'S TIME

All of the information supplied by Genesis for the building of the ark implies that Noah was familiar with boat construction and that he built the ark using the nautical technology available by his time (around 3000 B.C.). That is, Noah's flood occurred in the *Neolithic* (Table 9-1) and could *not* have happened before this time because the technology was not yet available for the building of large boats. This practical knowledge of ship building helps make the Genesis account of Noah and the flood *historical* rather than fictional or mythological. It also impacts a global-flood interpretation in that Noah was evidently building a boat out of local materials to house his family, and he was gathering local animals in order to ride out what he anticipated would be a local storm. Noah could have also been familiar with the trade region of Urartu ("mountains of Ararat," Fig 3-3) along the Tigris River north of Assyria, as this area was known by Noah's time.

boundaries of ancient Ararat: Mount Nisir, Mount Nisibis, Mount Ararat, and Jabel Judi (Fig 3-3). The most often cited and most likely contenders of these four sites are Mount Ararat (Fig 6-4) and Jabel Judi.

Mount Ararat

"Ararat" is the biblical name for the mountainous region of Urartu, as this area was known to the ancient Assyrians. Urartu was geographically centered between Lake Van and Lake Urmia (Fig 3-3) and was part of the ancient region of "Armenia" (not limited to the country of Armenia today). The Hebrew word for mountain in Genesis 8:4 is plural. Therefore, the Bible does not specify that the ark landed on the highest peak of the region (Mount Ararat), only that the ark landed somewhere on the mountains or highlands of Armenia (both *Ararat* and *Urartu* can be translated as "highlands"). In biblical times, *Ararat* was actually the name of a province (not a mountain), as can be seen from its usage in 2 Kings 19:37 and Isaiah 37:38: "And they escaped to the land of Ararat" (NIV), and Jeremiah 51:27: "Summon against her [Israel] these kingdoms: Ararat, Minni, and Ashkenaz..." (NIV).

While Mount Ararat (Agri Dag) appears to have been part of the province of Urartu in the seventh

Figure 6-4. Mount Ararat in present-day Turkey; Turkey's border with Armenia is along its eastern flank (Fig 7-3). *Creative Commons; photo by Andrew Behesnillan.*

and eighth centuries B.C., it is probable that it was not part of this province in Noah's time. At its zenith, the region of Urartu stretched from the eastern bank of the upper Euphrates River to the western shore of Lake Urmia, and from the mountain passes of northern Iraq to the Caucasus Mountains, thus including the western part of Mount Ararat in what is now the region of the Republic of Armenia (Figs 3-3, 7-3). However, this northern Mount Ararat section was added to the Urartu region only in the eighth century B.C. during a time of major Urartian expansion. By contrast, it is known that the Urartian language was present in the northern fringes of Mesopotamia (that is, the Jabel Judi area) at least sometime by the third millennium B.C. After the eighth to seventh centuries B.C., the name *Urartu* faded from view and was transformed into *Ararat* by later vocalizations imposed on the Hebrew Bible.

Search for Noah's Ark on Mount Ararat

If Mount Ararat is not the landing site of Noah's ark, then what about all of the books, movies, and television shows that claim the ark has actually been found on Mount Ararat? None of these "ark fever" accounts have been verified, some have been shown to be hoaxes, and all have been demonstrated to be scientifically unfounded. Since the early 1800s, there have been more than a dozen expeditions to Mount Ararat, none of which have proved successful in finding the ark.

The first popularized modern search for Noah's ark on Mount Ararat was by Fernand Navarra in 1955 and then again in 1969. On the northwest side of Mount Ararat, Navarra collected sections of worked timber from beneath a glacier at about 14,000 feet elevation. These specimens were identified as *Quercus* (oak) and were radiocarbon dated by six different dating labs at A.D. 720-790 for wood collected by Navarra in 1955 and at A.D. 620-640 for wood collected in 1969. These dates suggest that the wood may have been part of a Byzantine or Armenian shrine commemorating what was believed by the people of that region to have been the landing site of Noah's ark.

In 1993, CBS aired a two-hour television special entitled "The Incredible Discovery of Noah's Ark," which was reportedly seen by an estimated 20 million viewers. In this case, an actual hoax was involved in that a piece of modern pine wood was made to look ancient and was claimed to be a piece of the ark.

Noah's ark was again reported by the popular press in the early 1990s to have been found near Doğubayazit, Turkey, about 12 miles southwest of Mount Ararat. Supposedly, a "boat" having the dimensions of the ark had been found – a boat made out of petrified gopher wood and containing ribs, iron rivets, and stone anchors (Fig 6-5). In reality, the "boat" turns out to be a natural rock formation that mimics the shape of a boat due to resistant bedrock being carved by debris from a huge landslide coming down from distant mountains. The supposed fossilized "gopher wood bark" is metamorphic rock, the "iron rivets" naturally occurring concentrations of limonite and magnetite, and the "anchor stones" pieces of a local volcanic rock type. In short, the scientific evidence demonstrates that the "boat" found near Doğubayazit is nothing but a "phantom ark."

Jabel Judi

Jabel Judi (Cudi Dag) is a mountain range located just east of Cizre, Turkey near the border of Iraq and just within the northern boundary of the Mesopotamian hydrologic basin (Fig 3-3). The range rises above the Cizre Plain, which is surrounded by low hills in the north, gently sloping ridges in the south, and hilly land in the west. All of the streams within the Cizre Plain are tributaries to the Tigris River.

Jabel Judi has been another favored landing place for the ark, being the most widely accepted site among Christians, Jews, and Muslims during the latter centuries of the first millennium A.D. It was only in the eleventh and twelfth centuries A.D. that the focus of investigators began to shift toward Mount

Figure 6-5. The so-called "Noah's ark" structure near Doğubayazit, eastern Turkey. In reality, this is a natural geologic formation. *Google Image.*

Ararat as the ark's final resting place, and only by the end of the fourteenth century A.D. does it seem to have become a fairly well-established tradition. The Arab geographer al-Masudi (A.D. 956) stated that the ark "stood on el-Judi...a mountain in the country of Masur... eight farsangs [about 30 miles] from the Tigris River," and said that where it landed could still be seen in his time. In its principal reference to the flood, the Koran (Surah 11, Aya 44) states that the ark eventually came to rest on Mount Djudi (Jabel Judi), which is not far from a region of vineyards and olive trees. Why are vineyards and olive trees important? Because they are both mentioned in the Genesis account of the flood and because both suggest that the flood was local rather than global:

> And the dove came in to him in the evening; and, lo, in her mouth was an olive leaf pluckt off: so Noah knew that the waters were abated from off the earth....And Noah began to be a husbandman [farmer] and he planted a vineyard. (Gen.8:11; 9:20, KJV)

Vineyards. The wine grape of antiquity, *Vitis vinifera*, is what is referred to in both the Old and New Testaments of the Bible. *Vitis vinifera* has been cultivated for thousands of years, probably originating as a wild plant in the Transcaucasian area and then becoming domesticated in the area between the Black and Caspian Seas, eastern Turkey, and the Zagros range sometime around 6000-5000 B.C. It is certain that viticulture was practiced and wine was made in (northern) Mesopotamia sometime before 3000 B.C., so Noah and his family would *not* have been the first people to ever taste wine, as is a popular Christian misconception.

Cultivation restrictions of *Vitis vinifera* limit the location of where Noah could have landed and planted his vineyard. *Vitis vinifera* can only be cultivated at an elevation where the average temperature is at least 60°F in the warmest summer months (for the fruit to ripen), where the winters are not too severe (frost can kill young grapevines), and where the climate is not too hot and dry (grapevines need at least a moderate rainfall). So Noah could not have landed anywhere in southern Mesopotamia because it is too hot and dry there for viticulture to flourish, nor could he have remained in the high regions of Mount Ararat because the severe winters would have killed his vineyard. The region of Mesopotamia where grapevines flourished in ancient times – as they still do today – was Assyria (northern Iraq), which has a moderate rainfall and abundant streams that irrigate orchards and vineyards. The area north and east of Nineveh was especially renowned in antiquity for its wine, corn, and olive oil. In 2 Kings 18:32, King Sennacherib boasted that Assyria was "...a land of grain and new wine, a land of bread and vineyards, a land of olive trees and honey" (NIV).

Olive trees. Olive trees (*Olea europea*) are even more sensitive to climate than grapevines. Olive trees cannot tolerate hot and cold extremes, and they are especially susceptible to flooding. Olive trees need an elevated, well-drained soil to survive: in a waterlogged soil they drown because of their shallow root systems. This fact makes the mention of an olive leaf in Genesis 8:11, supportive of a local flood. If a raging flood had covered the entire planet Earth to 17,000 plus feet (the height of Mount Ararat) with seawater for a whole year, an olive tree (or even its seeds) could not have survived. The return of the olive leaf by the dove (Gen. 8:11) suggests the survival of relatively unharmed trees outside the flooded area. It also implies that the ark may have landed near the same area as suggested for vineyards: that is, north of Assyria in the foothills of the Jabel Judi mountain range.

Doves. Doves are birds that were favored by the ancient Mesopotamians – in fact, they were part of their diet. Noah's dove was probably a rock dove (*Columba livia*), which is native to the Middle East and ancestor to all of the various pigeon breeds that we have today. Doves were kept by Mesopotamians from at least 5,000 to 4,500 years ago onward, and evidence that they may have been at least partially domesticated in Mesopotamia by Noah's time comes from the archeological site al'Ubaid, where a row of sitting pigeons is pictured on the limestone frieze of a temple façade dating from around 3000 B.C. The dove's homing instinct has been recognized and exploited since early times, when the birds were used as carriers of messages in wartime. Noah evidently had knowledge of this homing instinct when he sent forth a female dove from the ark (Gen. 8:8-12; Fig 6-6), and Noah's action in Genesis 8:9 affirms that his dove was a domesticated pigeon: "He put forth his hand, and took her, and pulled her in unto him into the ark" (KJV). Exactly how far an ancient breed of dove like Noah's could have flown from the ark to search for dry land is not known, but it was probably less than one hundred miles. Noah sent out his dove, presumably in the morning, and it came back to him in the evening (Gen. 8:11). Thus, within a one-day's flight from and back to the ark, the dove found an olive tree or sprout growing, picked off a leaf, and returned to the ark. This means that wherever the ark landed, it had to be less than about fifty miles from a region where it was suitable for olive trees to grow in a non-waterlogged soil.

The lower-elevation, Jabel Judi (Cudi Dag) region thus has the following advantages over the higher-elevation Mount Ararat region for being the landing place of Noah's ark:

(1) Jabel Judi is located within the borders of ancient Urartu during Noah's time (~3000 B.C.); Mount Ararat is not.

(2) Jabel Judi is located within the foothills of the Taurus Mountains where optimal conditions exist for the growing of both grapes and olives; Mount Ararat is not.

(3) Jabel Judi is located only about 80 miles from Nineveh (Fig 3-3), a garden region that was renowned in ancient times for both its grapevines and olive trees. Since the northern part of the Nineveh region is within a fifty mile distance of Jabel Judi, it is feasible that a dove could have flown to this area and back to the ark with an olive leaf in one day, as required by the Genesis account.

Figure 6-6. *Noah releasing the dove.* Mosaic by an unknown artist, 12th-13th century. *Wikimedia Commons.*

(4) If the ark did land in the Cizre, Turkey area, then it means that the flood stayed within the (northern) boundary of the Mesopotamian hydrologic basin. This, in turn, implies a flood local to Mesopotamia because if the flood was global, why wouldn't the ark have floated to somewhere outside of the boundaries of Mesopotamia – a place like Europe or Asia?

Because of all the information and evidence presented so far in this chapter and in the previous one, a Worldview Approach interprets Noah to have been a real person and the flood to have been a real event – but only *if* the flood was *local* to Mesopotamia. However, as in Chapter 5 where we discussed meteorological and hydrological problems, there are difficulties with a local-flood stance that must be explained.

Feasibility of a Local Flood

Young-Earth Creationists raise some important objections against a local flood, the most formidable argument being: "How could Noah's ark have floated up-gradient to the mountains of Ararat if the flood was local? Why didn't it instead float downstream to the Persian Gulf? The only way the ark could have gone north to the mountains of Ararat, instead of south to the Persian Gulf, would be for the waters of a global flood to take it there." This argument must be addressed if a local-flood model is to be considered reasonable. Another important question is: If Noah's flood did not deposit all of the sedimentary rock on Earth in a global flood, then where is the flood sediment left by Noah's flood if it was local to Mesopotamia? We will start with the sediment question first.

Where Is the Sediment Left by Noah's Flood, If It Was Local?

As the waters of Noah's flood abated (Gen. 8:3), fine-grained mud particles in the water would have settled slowly out of suspension to form sediments until such time as these sediments dried on a hardening ground (Gen. 8:13-14). Because of the traditional assumption that the flood was global, many Christians up to about A.D. 1750 accepted the idea that sedimentary rock (with its contained fossils) represented the remains of Noah's flood – or roughly the same position held by flood geologists today. Then, starting at the end of the eighteenth century, an agonizing battle over the history of the Earth began between scriptural chronology and the newly founded science of geology. Not only was it evident to geologists that sediments take a long time to be deposited, but it also became clear that the transformation of sediments into sedimentary rock involves an even longer span of time. In addition, it was ascertained that most sedimentary rocks are *not* composed of flood-type sediments. Rather, there are marine sedimentary deposits (the majority) that alternate with eolian (wind), lacustrine (lake), and evaporite (halite and gypsum) deposits. Thus, the sedimentary record of planet Earth does *not* document one catastrophic flood event, but a series of many different sedimentary environments that overlap with each other in time and space. (We shall discuss this record in more detail in Chapter 7 when we talk about the sedimentary rock sequence exposed in the Grand Canyon.)

Before 1850, most geologists had abandoned the idea that all sedimentary rock had been formed at the time of Noah's flood, but many still believed in the former existence of extremely violent floods that had swept over the Earth (with Noah's flood being the last) – floods that had even submerged some of the highest mountain summits and which had created great valleys, gorges, and ravines. The evidence for this belief, called the "diluvialist school of thought" after Noah's deluge, was that many parts of the Earth (especially northern Europe and the Alps) were mantled by a chaotic assemblage of sediments ranging from mud to silt to sand to gravel to huge "erratic" boulders weighing many tons that were left by Noah's flood (Fig 6-7). These deposits led some geologists to propose that the older diluvial deposits (left by the biblical deluge) were overlain by younger alluvial deposits that contained fossils of a recognizable modern type. Also, fossils such as great mammoths trapped in glacial ice and other fossil deposits found in caves were attributed to changes in climate brought about by the Noachian flood.

This was the setting for the emergence of the glacial theory, which rudely shocked the geological community in the late 1830s and early 1840s. It proposed that the action of glaciers accounted for the strangely striated "erratic" boulders and poorly-sorted rock debris (called "till") present in many parts of the world. This "Ice-Ages" revelation then left *no* deposits that could be attributed to Noah's flood. So

Figure 6-7. Erratic boulders in Yorkshire, England called "diluvium" by William Buckland and other early geologists because they were thought to have been left by the Deluge (Noah's flood). These boulders were actually carried south to Yorkshire by a glacier during the last Ice Age. *Photo by Joan Martin, Photo North.*

where, then, are the sediments left by Noah's flood? They are in Mesopotamia, along the extent of its once-flooded hydrologic basin.

Flood sediments in the Mesopotamian hydrologic basin. Flood sediment layers have been found in Mesopotamia at a number of places such as Kish, Shuruppak, Ur, Uruk, Lagash, and Nineveh (Fig 3-3). This is because, before modern-day dams were built, floods were endemic to the region, occurring practically every year somewhere in ancient Mesopotamia. As explained in Chapter 5, the term *hydrologic basin* refers to a water-catchment area surrounded by a drainage divide, where water on one side of the divide flows downhill in one direction, and water on the other side of the divide flows in the opposite direction (the way it does along the Continental Divide in the western United States). The Mesopotamian hydrologic basin includes the entire region drained by the four rivers of Eden (the Pishon, Gihon, Tigris, and Euphrates), all of which flow – or in the case of the ancient Pishon (the now-dry Wadi Batin) did flow – to the Persian Gulf (Fig 3-3).

Some flood deposits in Mesopotamia are from normal-size floods, while others are from larger-magnitude floods, so it is difficult to determine which might be from Noah's flood. The most famous of these flood deposits was found in the late 1920s by archeologist Leonard Woolley who reported eight to eleven feet of "clean water-laid mud" in the Royal Cemetery of Ur and pronounced it the result of "Noah's flood." As it turns out, however, this particular flood deposit seems too early to be a record of the Noachian flood, as it belongs in the Ubaid Period (~4000 B.C.), not to the end of the Jemdet Nasr Period (~2900 B.C.; Table 6-1).

The flood deposits at Nineveh also seem too early to correlate with Noah's flood. At Shuruppak, Kish, and Uruk, the last Jemdet Nasr remains are separated from the subsequent Early Dynastic I Period by clean, water-lain clay deposited by a flood. This clay is nearly five feet thick at Uruk and two feet thick at Shuruppak (Fara). At Shuruppak, the flood stratum was found directly above a polychrome jar, seal cylinders, and stamp seals from the Jedmet Nasr Period and directly below plano-convex bricks from the Early Dynastic I Period, without much time

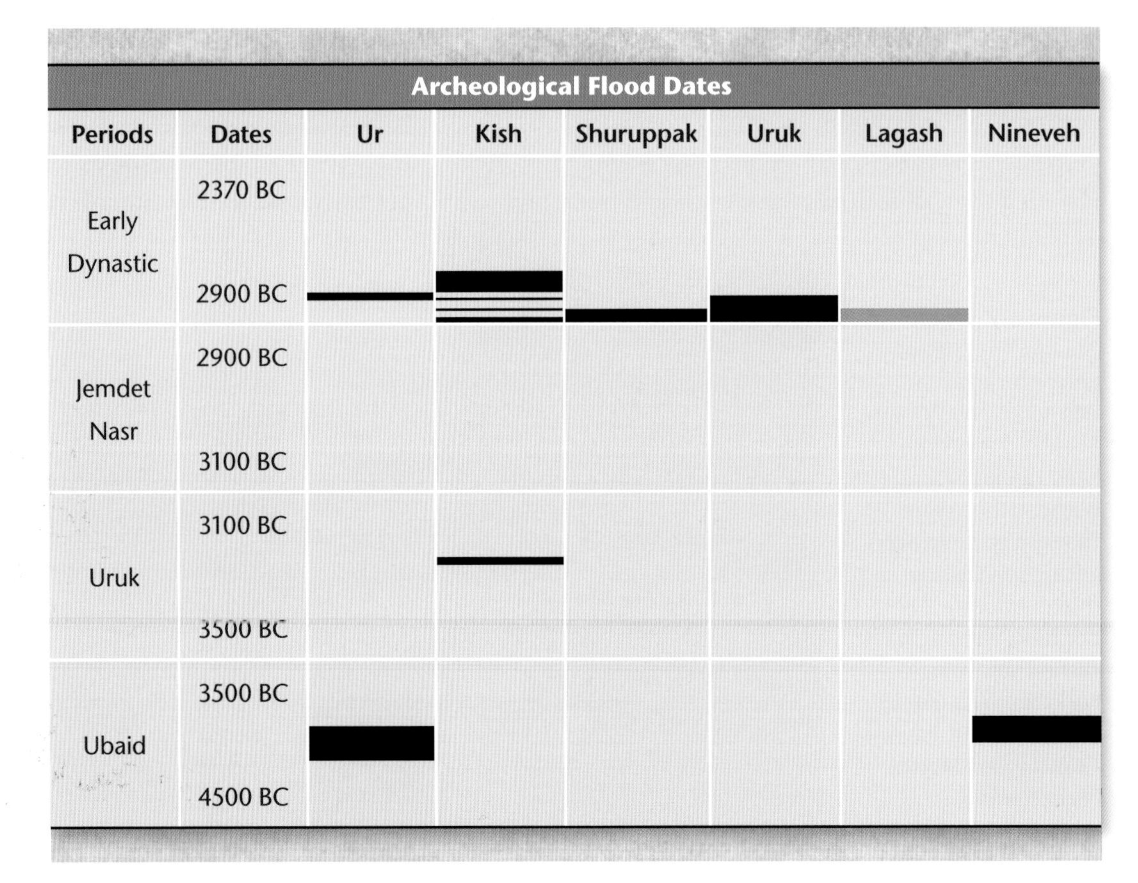

Table 6-1. Record of flood deposits for the most important archeological sites in Mesopotamia. The flood event that occurred at about 2900 to 2800 B.C. is the one most likely to have been Noah's flood. Original table by A. Parrot; table modified by Dick Fischer. *Courtesy of Dick Fischer.*

between the two periods, suggesting that people outside of the flooded area occupied the region soon after the flood event (yet another reason for thinking that Noah's flood was local to Mesopotamia). The Ur, Kish, Shuruppak, Uruk, and Lagash deposits date from around the right time to be attributed to Noah's flood (Fig 6-8). But if these sediments were left by Noah's flood, why are they so limited in extent within and between sites?

Flood deposition and erosion. A popular misconception is that a great inundation such as Noah's flood should have left a widespread layer of sediment all over Mesopotamia. If flood deposits occur at Shuruppak, then why do they not occur at nearby sites? Why in the city-mound of Ur do some pits contain thick flood deposits while other pits nearby contain no flood deposits?

This presumed conflict is completely understood by hydrologists – in fact, it is what they expect. Floods *erode* sediment as well as *deposit* sediment. A large, high, temple-mound ziggurat in the path of a raging flood might be eroded on its side facing the torrent of water, while on its backwater side sediment might be deposited in low areas. Rivers may scour and incise sediment along steep gradients, whereas they may deposit sediment in shallow-gradient locations. In addition, the Euphrates and Tigris river channels have changed their courses over the millennia (Fig 3-3). If a city or village was close to one of these former river channels, then it might be washed away rather than be covered over with river silt. Or sediment left from one flood event might be removed by erosion in a later flood event. Thus, the depth of flood deposits does *not* automatically indicate the depth of a past flood, nor does the lack of flood deposits mean the non-existence of a flood. The only absolute way of knowing when a flood occurred is to dig a series of trenches and date the archeological and/or organic remains both above and below a flood sediment horizon. Dating of archeological material is done on organic material associated with pottery types of known ages, but the dating of flood deposits without pottery clues is harder to determine. Calibrated radiocarbon dating has identified the approximate ages of the archeological periods in Mesopotamia (Table 3-1), but the absolute ages of the flood deposits themselves – and which of these deposits can be correlated with Noah's flood (Fig 6-8) – is less certain.

It is important to this discussion to understand the magnitude of sediment build-up that can occur in a major flooding event. The Mississippi River flood of 1973 was out of its banks for two to three months in some locations ,and the average sediment thickness left by that flood was twenty-one inches along the natural levee and twelve feet in back-swamp areas. Considering that Noah's flood supposedly lasted four to six times as long as the Mississippi River flood, one can roughly estimate that a maximum of about fifty to seventy feet of sediment might have accumulated in an ideal backwater location. This is *nowhere* equivalent to the *miles* of sedimentary rock proposed by Young-Earth Creationists as having formed during the Noachian flood over parts of planet Earth, such as the miles-deep sedimentary rock layers in the Grand Canyon (see the next chapter). But it does fit with a "1000-year" or "5000-year" *local* flood model.

How Could the Ark Have Floated Up-Gradient to the Mountains of Ararat?

The most difficult question of all to answer is: If Noah's flood was local, then how could the ark have gone against the current toward the mountains of Ararat instead of floating down gradient to the Persian Gulf? In order to answer the question of how the ark went "uphill," physicist Alan Hill set up a mathematical model demonstrating how a strong wind could have blown the ark northward from southern Mesopotamia to the "mountains of Ararat." Incorporated into Hill's model were the physical parameters given by the Genesis account: size of the ark (300 x 50 x 30 cubits; Gen. 6:15), period of intense rainfall (40 days and night; Gen. 7:12), and total period of extended rainfall and spring flow (150 days; Gen. 7:24).

Genesis does not give quantitative information on the magnitude of rainfall or spring flow rates, but it does give the initial water depth at the point of the ark's departure (15 cubits), the ark's landing position (mountains of Ararat), and the point in time when Noah disembarked from the ark after the mud had hardened (exactly one year, or 365 days, after the flood started, Table 5-1). Genesis also does not give the exact departure and landing sites of the ark, but the departure point is assumed to have been Noah's "home-town" of Shuruppak on the southern Mesopotamian plain, and the landing point the Cizre Plain in the Jabel Judi area (Fig 3-3). The distance from Shuruppak to Cizre is 440 miles over an elevation change of 2,100 feet. The pre-flood terrain was assumed to have been a modern landscape similar to that existing today; that is, a flat plain rising northward into the foothills of the Jabel Judi Mountains.

It is beyond the scope of this chapter to go into the mathematics and computer analysis of Hill's local-flood model and the interested reader is referred to Hill's paper. Suffice it to say that Hill's calculations demonstrate the feasibility of storm winds blowing the ark upstream, against the hydrologic gradient, to the foothills of Ararat (ancient Urartu). Calculated wind velocities of thirty to seventy miles per hour are considered well within reason for a large, stalled cyclonic storm over the Mesopotamian region (e.g., winds up to 155 miles/hr were recorded in the June 2007 cyclonic storm over Oman and southern Iraq). Furthermore, if the ark's dimensions given in Genesis 6:15 (300 x 50 x 30 cubits, as used by Hill) are considered to be *numerological* numbers instead of numerical numbers (see Chapter 4), then a much smaller ark (ship) – perhaps even one with a sail – could have been driven further northward into the foothills of Ararat by even lower average wind velocities.

Local Flood Model and Route of Noah's Ark

I will now summarize the feasibility of a local flood by proposing a meteorological model for such a flood. A large cyclonic storm stalled over Mesopotamia and provided heavy rainfall for "forty days and forty nights" (figurative for a long period of time) to the lowland regions and snow/rain-melt to the highland regions (see Chapter 5). This rainstorm was accompanied by an intense sharqi and/or suhaili south wind, which blew the ark northward towards the mountains of Ararat (Urartu). The entire Tigris River part of the Mesopotamian basin was inundated up to the area of Cizre because springs and snow melt kept the water flooded in the upper Tigris River valley as well as in the lower Tigris River valley. In the downstream end of the converged Tigris/Euphrates Rivers, water would have discharged extremely slowly through the marshlands near the Persian Gulf (Fig 3-3), thus helping to keep floodwater backed up in the Mesopotamian hydrologic basin for a long time. In addition, the marine transgression mentioned in Chapter 3 (Fig 3-3, blue-green dashed line) could have helped to back up river flow and prevent river-floods, such as Noah's flood, from rapidly draining away.

Powered by strong winds, the ark could have been blown northward from Shuruppak in southern Mesopotamia along the inundated flood plain between the Euphrates and Tigris Rivers up to the area of present-day Baghdad (Fig 3-3). From there, it could have followed the flooded Tigris River channel up to the area of Mosul, where the Tigris still remains a wide, stately river. Northward from Mosul, the terrain becomes hillier, but there is still a wide valley up to Cizre. The ark could have continued to follow the Tigris River channel northward to land somewhere south of Jabel Judi, or it could have made it to the Cizre Plain and landed in the foothills of the Jabel Judi Mountains where the mountain tops could be seen (Gen. 8:5), but where the valleys were still flooded (Gen. 8:9). This location is part of the "mountains of Ararat" of Genesis 8:4 and was known in antiquity for both its olive trees (Gen. 8:11) and vineyards (Gen. 9:20).

The Nature of "Nature Miracles"

From the above discussion it may seem like a completely naturalistic explanation for the flood is being offered. Such is not the case! The Bible claims that Noah's flood was under God's control and direction:

(1) It was God who purposely sent the flood to judge an evil, corrupt, and violent world

(Gen. 6:7; Gen. 6:11-13). But Noah "walked with God" (Gen. 6:9) and "found grace in the eyes of the Lord" (Gen. 6:8). Noah had a personal relationship with God and thus he and his family were spared.

(2) It was God ("I, even I"; Gen. 6:17) who exercised absolute control over the forces of nature that caused the flood.

(3) It was God who commanded Noah to build the ark (Gen. 6:14), to gather the animals onto the ark (Gen. 6:19), and it was the Lord God who shut Noah and his family into the ark (Gen. 7:16).

(4) It was God who restrained the floodwaters (Gen. 8:1-3) and brought the ark safely to the mountains of Ararat (Gen. 8:4).

(5) It was God who established a covenant with Noah and his descendants (Gen. 6:18) and who made the rainbow a sign of that covenant (Gen. 9:13).

Noah's flood was a "nature miracle" because God *intervened* into his physical laws. In nature miracles, one does not have to invoke the notion of the suspension or violation of natural laws. Divine action can simply be understood as higher-order laws (God's ultimate purpose) working seamlessly with lower-order laws (God's physical laws). Is it any less a miracle because it can be explained by natural processes? This is the nature of nature miracles: to have the *timely* action of God into natural processes.

One of the best examples of a "nature miracle" that comes to mind is Jesus rebuking the winds and sea (Matt. 8:23-26). In Matthew 8:26, the calming of the winds and sea could be explained by a sudden change of barometric pressure – which was probably the case. But it was God who caused this change to take place *exactly* when Christ commanded the waves and wind to be still.

Another example is that of the Israelites crossing the Jordan River, where the stoppage of water lasted long enough for the Hebrews to get across the river in one day (Josh. 3:1-4:18). That this type of blockage has happened historically is an established fact: in 1267, 1906, and 1927, landslides upstream from Jericho dammed the river for up to twenty-one hours. The miracle of the Jordan is that God caused the blockage to happen *exactly* when the Israelites needed to cross the river.

One other excellent example of a nature miracle is Joshua's long day that was mentioned in Chapter 1 (Fig 1-3). Not only was God directly involved with this battle (and other battles in the book of Joshua), but the remarkable correspondence of a known solar eclipse in 1207 B.C. with Joshua's victory at Gibeon could be considered as a grand-nature miracle of timing.

Four further points should be made at this time about miracles:

(1) "Nature miracles" – where miracles can be explained by natural processes – are not the only kind of miracle claimed by the Bible. Jesus's walking on water, the virgin birth, and Jesus's resurrection: all of these "non-nature miracles" cannot be explained by natural processes (as we know them). It is not to be implied that "non-nature" miracles cannot or have not occurred.

(2) Just because God can perform "nature miracles" does not mean that all natural disasters are judgments of God, as was Noah's flood. Most natural disasters are due to normal causes where God has allowed the physical processes that he set up to operate naturally.

(3) In order to explain a "nature miracle" like Noah's flood, a slew of miracles that are not recorded in the Bible should not be assumed; that is, miracles should not be "pulled out of a hat" anytime one feels like it. Any theory, no matter how feeble, can be "proved" by recourse to the miraculous or God's omnipotence. It is a weak interpretation that has to invent all sorts of miracles that the text says nothing about in order to compensate for logistical problems. The miracles that the Bible actually claims should be considered to be miracles, and those it does not claim should not be manufactured.

(4) It is inconsistent to question God's ability to perform simple miracles, such as those performed to manage a local flood, yet allow for God to manage a giant-scale miracle (actually a whole set of miracles) necessarily related to a global flood. A "God can do anything" argument is not a defense for a global flood and for the myriad of fantastical explanations required by "flood geology" – the subject of the next chapter.

Aerial photo of the Marble Canyon section of Grand Canyon near the Little Colorado River.
Photo by Wayne Ranney.

CHAPTER 7

FLOOD GEOLOGY

Genesis 2:5, 6. For the Lord God had not caused it to rain upon the earth... but there went up a mist from the earth, and watered the whole face of the ground. (KJV)

This chapter logically follows from the last chapter where we covered the question of whether Noah's flood was global or local. "Flood geology" is the sub-division of Young-Earth Creationism that claims the flood was global and that all (or almost all) of the Earth's sedimentary rock with its contained fossils was deposited during the one-year duration of Noah's flood (Table 5-1). This position is the antithesis of modern geology, which is based on over 400 years of investigation of Earth's rocks, fossils, and landforms. Flood geology not only embraces a 6,000-year-old Earth (the subject of Chapter 4) and a global flood (the subject of Chapter 6), it also makes other far-reaching geologic claims based on a supposed "literal" reading of Genesis. So popular has this position become in the last few decades in the conservative Christian church, that it is a position held by almost half of the American public, despite the fact that it is contrary to all of the evidence provided by modern science.

The main purpose of this chapter is to compare the science of modern geology with the claims of flood geology (Box 7-1), but underlying this comparison is the premise of a Worldview Approach

Box 7-1. Young-Earth Creationist vs. Old-Earth Creationist Positions

Young-Earth Creationist/Flood Geology	Old-Earth Creationist/Modern Geology
• The Earth and universe are 6,000 years old (10,000 years max)	• The universe is 13.82 billion years old and the Earth is 4.56 billion years old
• It never rained on Earth before Noah's flood (most flood geologists favor this; some don't)	• Abundant evidence exists for it having rained throughout Earth's geologic history
• A vapor canopy & "fountains of the great deep" (Gen. 7:11) supplied water for a global flood	• Earth's atmosphere & groundwater can hold only a fraction of the water needed for a global flood
• No death of animals before Adam sinned; all animals were herbivores	• Death of animals has occurred over millions of years and is represented by the fossil record
• All (or almost all) sedimentary rock formed in Noah's flood in one year's time	• Sedimentary rock has formed over millions of years by the process of sedimentation and compaction
• Fossils in sedimentary rocks represent the "all flesh" of Gen. 7:21	• Fossils are plant and animal remains that became buried and preserved over millions of years
• The rainbow of Gen. 9:13 was the first to ever appear in the atmosphere of planet Earth	• Rainbows have refracted light since Earth developed an atmosphere billions of years ago

that since the ancients had a limited scientific and geographic knowledge of their world, their naïvety about the science of geology is present in the Genesis text on Noah's flood.

Flood Geology Theology

The following describes the position and logic of many (or most) flood geologists regarding the history of the Earth based on their so-called "literal" interpretation of Genesis.

Genesis 1. The Earth was created approximately 6,000 years ago based on a 24-hour day, six days of creation of Genesis 1, plus the chronologies of Genesis 5 and 11.

Genesis 2. It never rained upon the earth (planet Earth) before Noah's flood (Gen. 2:5). Rather, there was a *mist* (interpreted by some flood geologists to be a kind of "vapor canopy") that watered the whole face of the ground (of planet Earth) from the time of creation (Genesis 1) to the time of Noah's flood (Genesis 7). Since it never rained on planet Earth before the flood no (or very little) sedimentary rock could have formed before this time, and pre-flood locations like the garden of Eden existed on a crystalline basement devoid of sedimentary rock (Fig 3-7). Therefore, the pre-flood topography of planet Earth bore no resemblance to today's post-flood topography.

Genesis 3 and 4. Before Adam sinned and ate of the fruit of the tree (Gen. 3:6), a world of perfect harmony existed on planet Earth. In this perfect world, there was no death, not even the death of animals. Since no animals died by being eaten by other animals, all animals (created as individual species in Genesis 1) had to have been plant eaters (herbivores) before Adam's fall. Adam's "original sin" brought about a violent imperfect world where both humans and animals died and where some animals became carnivores. This violence is illustrated by the vengeful line of Cain (Gen. 4:23-24).

Genesis 5 and 6. The long ages of the patriarchs before the flood signify decay from a state of perfection in the garden of Eden to a maximum 120-year lifespan for humans after the flood (Gen. 6:3). The violence had become so pervasive by Noah's time that only one man was considered "perfect" by God, and that man was Noah (Gen. 6:9). Consequently, God instructed Noah to build an ark and prepare for a flood, wherein all humans and animals (except for those on the ark) would be destroyed from off "the face of the earth [planet Earth]" (Gen. 6:7).

Genesis 7. Genesis 7:11 (KJV) states that the "windows of heaven" were opened and all the "fountains of the great deep" were broken up. From the perspective of some flood geologists, this verse is interpreted to mean that all of the water in the vapor canopy fell as rain and that a great amount of water in the Earth's crust was expelled along faults and volcanos. And, since the Bible says that all the earth was covered with water including the mountains (Gen. 7:19-20), and that all flesh died (Gen. 7:21), this must mean that Noah's flood left an immense record of itself in the form of sedimentary rock containing fossils. In addition to being subjected to a global deluge, Earth's tectonic forces caused continents to break and spread apart and mountains to heave upwards because sedimentary rock is found today on the highest mountain peaks. (This position is referred to as "catastrophic plate tectonics" by flood geologists.) This tectonic catastrophe has been construed by a few flood geologists to even include the tilting of Earth's axis during Noah's flood.

Genesis 8. Since a global flood covered the highest mountains, the ark would have landed on the highest peak of the region; that is, the almost 17,000-foot-high Mount Ararat (Gen. 8:4) (Fig 6-4). After landing on Mount Ararat, the floodwaters decreased rapidly due to evaporation (Gen. 8:1), and also because they "returned from off the earth continually" (Gen. 8:3, KJV) to low elevations relative to mountains uplifted during the flood. Exactly one year (365 days) after the flood started, the post-flood landscape where Noah landed was completely dry (Gen. 8:14), and the topography of planet Earth was drastically changed from its pre-flood topography.

Genesis 9. Since supposedly it had never rained before Noah's flood, and since *earth* means planet

Earth, the rainbow of Genesis 9:13 was the first ever to refract light in Earth's atmosphere. And, since numerous local floods still occur on Earth, God's promise in Genesis 9:11 (KJV) that "neither shall there anymore be a flood to destroy the earth" implies that the flood had to be global because if it was just a local event, this would mean that God breaks his promise every time a local flood occurs somewhere on Earth today.

Worldview of the Biblical Author(s)

To recap what we have already discussed in previous chapters, a different "literal" picture emerges if the early chapters of Genesis are interpreted from the worldview of the ancient Near Eastern biblical authors/scribes.

Six Days of Creation

It was customary, following the convention and style of literary works prevalent in the ancient Near East of about 4,000 years ago, to divide a narrative work into six days, with the conclusion reached on the seventh day (Chapter 2). It was also customary to divide the six days of work into three pairs. From the worldview of the biblical author(s) of Genesis 1, the six days of Genesis 1 should be taken topically and not chronologically as consecutive 24-hour days.

Chronologies of Genesis

The Mesopotamians' worldview on numbers was both numerical and numerological (Chapter 4). In writing a sacred narrative, such as the early chapters of Genesis, symbolic numbers were *intentionally* used to reflect a symmetry and harmony that gave religious dignity to important persons or to a literary text. Since the patriarchal ages represent sacred symbolic numbers rather than real numbers, the chronologies of Genesis were *not* intended to set an absolute date for the creation of the world, and, therefore, the scientific evidence for a 4.56-billion-year-old Earth is *not* contrary to the Bible.

Earth

As argued in Chapter 6, the use of the term *eretz* to mean planet Earth, rather than a specific piece of land, is the root cause of misinterpreting the flood story. The biblical author(s) of Genesis had no knowledge of Earth as a planet. Their concept of *earth* corresponded to a very local section of real estate (Mesopotamia), and they had a very naïve scientific view of that real estate (Fig 1-1).

Rain and Mist

The misuse of *eretz* to mean planet Earth also leads to an erroneous global interpretation of Genesis 2:5 (KJV): "...for the Lord God had not caused it to rain upon the earth." Does this verse mean that it had never rained over the entire planet Earth before Noah's flood, as claimed by flood geologists? Or does it simply mean that it had not rained over a specific parcel of land in southern Mesopotamia (Eden), which is one of the driest places on Earth? A local interpretation of *eretz* also applies to Genesis 2:6 (KJV): "But there went up a mist from the earth [land or ground around Eden] and watered the whole face of the ground." The key word of this verse is mist (*'ed*), which many traditional flood geologists have envisioned as a thick vapor canopy enshrouding planet Earth before the flood. However, *'ed* (from *edû*, a Sumerian loan word) can also be translated as "flow" in the sense of an underground swell or spring. Therefore, a better translation of Genesis 2:6 might be: "an underground flow went up from the earth [land] and watered the whole face of the ground."

Upper and Lower Waters

For the biblical authors' perspective on the waters of Genesis 7:11, refer again to Figure 1-1. Can the upper waters ("windows of heaven") be interpreted to mean a vapor canopy or were these authors referring to waters above the firmament – the region where an ocean of rain was held above a solid dome? Similarly, can the lower waters ("fountains of the great deep") mean water coming from the Earth's crust via catastrophic plate tectonics, since in the worldview of the ancient Mesopotamians, the upper waters above the firmament (rain) would have joined with the lower waters of the primordial deep (springs) to create the flood? What did the biblical author(s) know about the tectonic forces within the Earth that cause continents to split apart and mountains to rise? Absolutely *nothing*! This is a matter of flood geologists taking a geologic concept that has emerged only in the last fifty years or so and force-fitting it into the Genesis text when no such meaning was intended – or even conceived of – by the biblical authors.

Rainbow

Did the author(s) of Genesis mean that the rainbow of Genesis 9:13 was the first occurrence of refracted light in the atmosphere of planet Earth? Or was it just a sign (a "token," Gen. 9:13, KJV) of the Noachian covenant – a righteous seal of faith given to Noah by God, just like the sign of circumcision was the righteous seal of faith given to Abraham and his descendents (Rom. 4:11)? The subject of the rainbow will be discussed again in Chapter 10 when God's covenants with Adam's line are covered.

Worldviews in Conflict

As has already been emphasized in this book, the main problem with Young-Earth Creationism – besides their faulty assumptions of scriptural interpretation – is that it is in direct conflict with the findings of modern science: with physics, chemistry, astronomy, geology, biology, anthropology, and archeology. In this chapter, we will mainly concentrate on how flood geology is in conflict with modern geology, but first we will briefly present biblical and archeological evidence against a Young-Earth position.

FOSSILS: THE "ALL FLESH" OF GENESIS 7:21?

Young-Earth Creationists (flood geologists) claim that fossils found in sedimentary rock worldwide represent animals and plants that died in a global flood. The Bible itself speaks against Noah's flood as an explanation for the progression of life shown in Table 7-1. Genesis 1 lists major taxonomic groups of plants and animals in the pre-flood creation that match those alive today: seed-bearing plants and trees that bear fruit with seed in it (v. 11), birds (v. 20), and livestock such as cattle and wild animals (v. 24). Yet, *not a single one* of these fossil types exist in the mile-deep sedimentary rock layers of the Grand Canyon, which should be a record of life that existed just *before* the flood if these layers formed *in* Noah's flood. Surely, in a global-flood situation *some* humans should have been deposited chaotically along with dinosaurs if both lived at the same time before the flood, as claimed by flood geologists. Hydrologists (geologists who study floods) will tell you that one of the most diagnostic criteria for recognizing flood deposits is that different kinds of plants, pollen, animal bones, and sediments get all mixed up together. The most logical explanation for Genesis 1 plants and animals not being encased in any ancient sedimentary rock worldwide is that they *did not exist* until recent times. That is, Noah's flood happened recently, while Earth's sedimentary rock layers formed over millions of years.

Biblical Evidence

Contrary to the flood geology position of a young earth, certain verses in the Old Testament seem to suggest that the Earth is old. Habakkuk 3:6 talks about the "everlasting mountains" and "perpetual hills." The Hebrew word for "everlasting" is *'ad*. This is exactly the same word as used in Isaiah 9:6: "The everlasting Father, The Prince of Peace." The word means: "of advanced duration, in perpetuity, old, perpetually, or without end." The Hebrew word for "perpetual" is *olâm* meaning "time out of mind (either past or future), ancient, lasting, of old." Another Old Testament verse that implies the Earth is old is Deuteronomy 33:15: "And for the chief things of the ancient mountains, and for the precious things of the lasting hills" (KJV). The Hebrew word for "ancient" is *gêdmâh*, meaning "antiquity, aforetime, old." "Lasting" is the same as perpetual (*olâm*). All of these words seem to indicate that the biblical authors/scribes thought the mountains (and thus the rocks that make them up) were ancient – albeit from their worldview, the ancients would not have had any conception that they could be millions or billions of years old.

The most important biblical evidence against the flood geology position of Earth's sedimentary rock layers having formed in Noah's flood is that Genesis describes pre-flood Eden as a modern landscape overlying six miles of sedimentary rock – a topic we have already covered in Chapter 3 on the garden of Eden (Fig 3-7). In other words, Genesis never claims that all sedimentary rock on planet Earth was formed in Noah's flood, so how can flood geologists claim it if all of their explanations are supposedly based on a "literal" interpretation of the Bible?

Archeological Evidence

The archeological evidence for a global flood is nonexistent. No flood deposits of recent age correlative with archeological remains have been found in Egypt, Syria, or Palestine, let alone in other parts of the world. Archeological mounds in Syria and Palestine (such as Jericho), which exhibit fairly continuous occupation for the last 10,000 years or so (Fig 9-3), show no signs of a great flood. That the flood did not extend even to the land of Israel is alluded to in Ezekiel 22:24 (KJV): "Thou art the land [Israel] that is not cleansed, nor rained upon in the day of indignation [day of God's judgment by the flood]." Flood legends from around the world – cited by many as evidence for a global flood – exist simply because flooding has occurred in most parts of the world at one time or another. All of these flood stories – except for those from within and surrounding Mesopotamia – are essentially different from the biblical narrative and have only a few indeterminate elements in common with it.

Geologic Evidence

The main evidence against flood geology comes from the science of geology. Modern geologists – which includes hydrologists, geomorphologists, geochronologists, sedimentologists, paleontologists, and geophysicists – know exactly how the different types of sedimentary rocks form, how plants and animals become fossilized and the past ecosystems they represent, how fast continents move, and what flood deposits look like and how they form. They also know how old the Earth is! There is no "conspiracy" between modern-day geologists on any of these subjects: the principles of geology have been *tested* and established for over 400 years, and geologists have now field-surveyed practically the entire planet Earth. If the reader is interested in the history of geology and how the controversies regarding Noah's flood and the age of the Earth were resolved in the 1800s and 1900s, he/she is referred to the book *Great Geological Controversies* by A. Hallam.

Age of the Earth

As already mentioned in Chapter 2, the latest findings of astronomy date the universe to 13.82 billion years ago (Fig 2-1), and the latest findings of geology date planet Earth to 4.56 billion years ago. The age of the Earth is directly based on the *relative dating* of its rock layers and on the *radiometric*

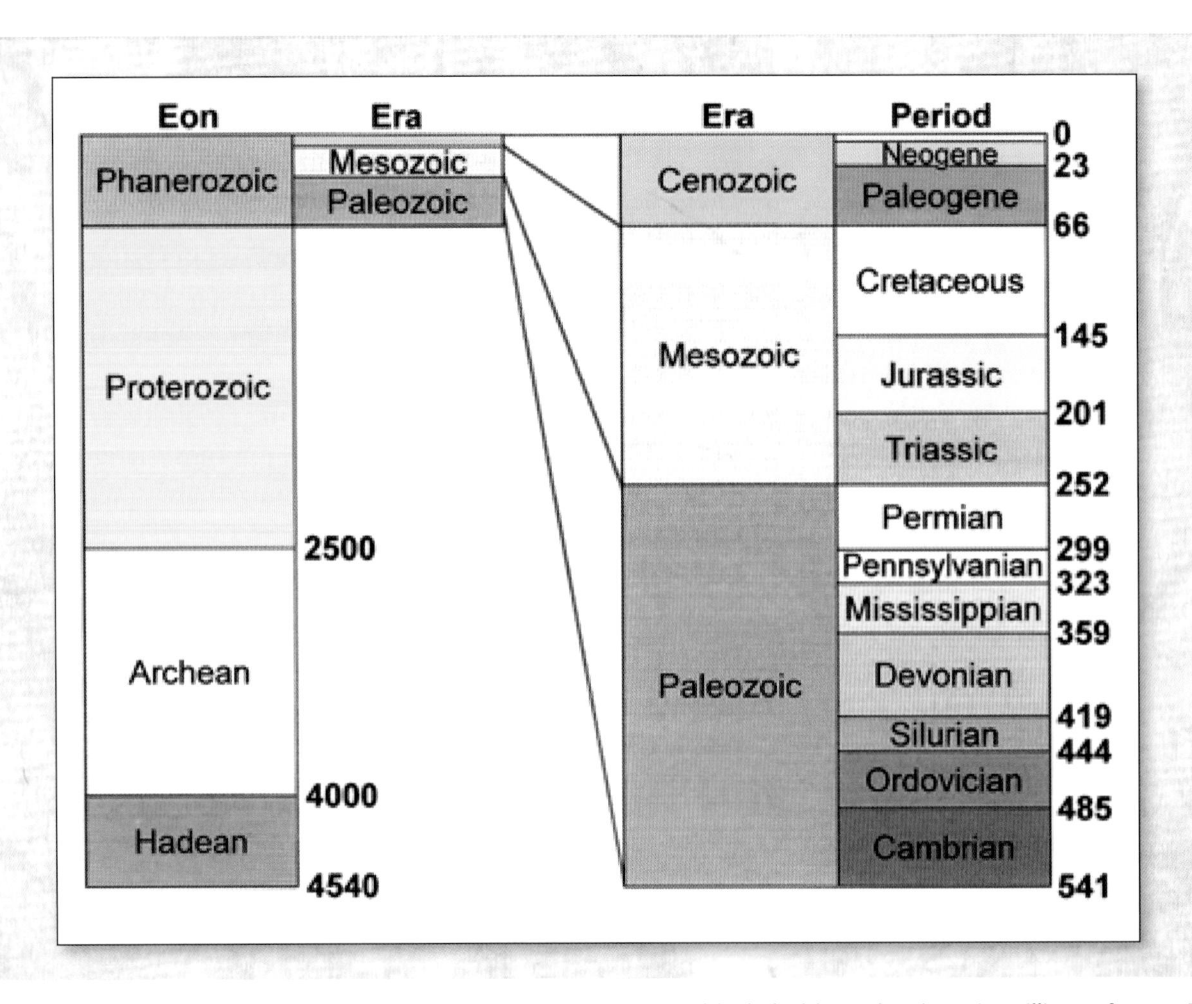

Figure 7-1. Geologic time and the geologic column. The absolute ages (black, bold numbers) are in millions of years. The ascending order of the rock periods through time is based on the relative dating method applied over the last 200 years; the absolute ages are based on modern radiometric-dating techniques. Note that 4,540 million years (4.54 billion years, the beginning of the Hadean Eon) for Earth's rocks is not quite the same as the 4.5-billion-year age of planet Earth when our solar system first formed out of the solar nebula. That is because it took about 20 million years or so for the Earth to cool enough on its surface to form solid rocks.

dating (also referred to as absolute dating) of its rocks. Relative dating of rock layers is based on the *Principle of Superposition,* which states that in a stack of rock layers, such as seen in the Grand Canyon, the layer on the bottom is older than the one on top of it. This method gives the relative age of the layer but not its actual age. For example, in the stack of Earth's layers shown in Figure 7-1, Paleozoic Era rocks are older than Mesozoic Era rocks, and within the Paleozoic Era, Devonian Period rocks are older than Mississippian Period rocks. When the relative sequence of Earth's rock layers was worked out in the 1800s, the absolute ages of these rocks were unknown, but when radioactivity was discovered in the early 1900s and applied to the dating of rocks, radiometric (absolute) dates could then be assigned to all of Earth's rock layers (this entire sequence of layers is called the *geologic column*; Fig 7-1). The most important point of this column is that the absolute dates obtained much later in time *fit* with relative time determined previously; for example, rocks assigned to the Permian Period according to relative time do not date to, say, the Cambrian or Cretaceous Periods, they fit within the Permian range of ages – strong evidence that radiometric dating is reliable.

There are now so many pieces of evidence that support a 4.56-billion-year age for our Earth that it is tedious to mention them all; I have listed twelve

Box 7-2. Twelve Independent Lines of Evidence for an Old Earth (by=billion years)

- Relative dating fits with radiometric (absolute) dating
- Non-radiometric dating methods fit with radiometric dating methods
- Fits with the age of the universe as determined by astronomers (13.82 by)
- Fits with radiometric dating of oldest carbonaceous-chondrite meteorites (4.56 by)
- Fits with the age of oldest Moon rocks collected by the astronauts (4.47 by)
- Fits with the age of oldest known Earth rocks (4.1 by)
- Fits with plate tectonics/continental-drift rates (millions of years)
- Fits with erosion rates and compaction rates of sedimentary rock (millions of yrs)
- Fits with rates of oil and gas formation from organic matter (millions of yrs)
- Fits with annual layers in Earth's ice caps (tens to hundreds of thousands of years)
- Fits with annual layers in evaporite rocks (tens to hundreds of thousands of years)
- Fits with carbon-dating of annual tree rings in bristlecone pines (back to 12,000 yrs)

of the major reasons in Box 7-2. In fact, the evidence for an old Earth is now so *overwhelming* that believing in a young Earth is akin to believing (like the church maintained before and after Copernicus and Galileo) that the Earth, rather than the Sun, is at the center of our solar system, with the Sun revolving around the Earth rather than the other way around.

The Fossil Record

The fossil record is another important argument against the claim of flood geologists that all (or almost all) sedimentary rock formed in Noah's flood. This record does not contain a hodgepodge of fossil types all mixed up together, such as should be expected for a global flood where supposedly continents were rapidly splitting and moving apart and tsunamis were raging around the globe. Rather, successive rock layers contain different assemblages of fossils, where increasingly complex life forms are found in successively younger strata (Table 7-1).

A brief summary of the progression in life forms as displayed by the fossil record in Earth's sedimentary rock layers is shown in Table 7-1. The

Geologic Era/Eon	Time (Ma)	Types of Animals/Plants	Important Events
Cenozoic	65-0 Ma	Mammals	Birds, flowering plants and fruit trees, mammals, including humans
Mesozoic	~65 Ma		Extinction of dinosaurs at end of the Cretaceous
	235-65 Ma	Reptiles, dinosaurs, birds	First appearance of birds in the Jurassic; rise of angiosperms (flowering plants) in the Cretaceous
Paleozoic	~235 Ma		Mass extinction of 95% of invertebrates at end of the Permian
	540-235 Ma	Invertebrates such as trilobites & vertebrates such as fish	"Cambrian explosion" of life; all of the different phyla existing today start
Proterozoic	~570-540 Ma	Ediacaran life	First evidence of multicellular life in Earth's biota
	~2000-570 Ma	Stromatolites	Only single-cellular organisms; no multicellular organisms
Archean	~4000-2000 Ma	Algae, bacteria, single-cell organisms	Firm evidence of first life = ~3.7 billion years ago

Table 7-1. A condensed table of the different kinds of animals and plants in Earth's fossil record over time. The subject of human origins will be covered in Chapter 9. (Ma = millions of years ago.)

order of this table cannot be denied: *it is simply there* and thus must be accounted for by whatever position is favored – flood geology or modern geology. The first thing to note in Table 7-1 is how the complexity of life increases over time: from the simplest organisms like bacteria up to more complex forms like mammals. Major extinctions do occur, and these usually herald the appearance of new kinds of creatures (for example, the extinction of the dinosaurs and the "take over" by mammals). In this progression of life forms, later species are never found mixed with earlier species in rocks of earlier age. For example, in the Mesozoic Era, whales (mammals) are never found in the same rock with ichthyosaurs (sea dinosaurs). However, fish are found with dinosaurs and with mammals because fish (as an entire class), once they appeared early in the Paleozoic Era, never became extinct. The first flowering plants appeared late in the Cretaceous Period (Table. 7-1), and pollen grains from flowering plants and trees are never found in rocks older than this. Also, one never finds humans with dinosaurs, despite claims to the contrary (i.e., the so-called Paluxy "man tracks"). These are *not* human footprints as some flood geologists have claimed. They are dinosaur footprints in the Cretaceous Glen Rose Limestone – the tracks of the dinosaur genus *Irenesauripus*, distinguishable by its long and relatively slender toe marks and an acute "heel." Bones of this dinosaur have been recovered along the Paluxy River in the area of the so-called "man tracks."

How do flood geologists defend this undeniable order in the fossil record? In their classic book *The Genesis Flood*, authors Whitcomb and Morris speculated that a certain general semblance of order should be anticipated in a global flood, considering that submarine lavas, tsunamis, and cataclysmic upheavals of the Earth's crust supposedly accompanied the flood. According to this hypothesis, the creatures of the deep-sea bottoms would have been universally overwhelmed by the toxicity and violence of the volcanic emanations and bottom currents and, hence, would have been the first to be deposited from floodwaters. Fish and other organisms living nearer the surface would have subsequently been entrapped either by debris washing down from the land surface or by materials upwelling from the depths. Thus, they would have been deposited next, over the creatures of the deep sea bottom. Raging rivers would have carried the bones of animals and reptiles to the sea along with great rafts of vegetation, depositing them over the fish, while land animals and humans would have escaped temporarily to higher ground, thus being the last to survive and so be deposited higher than reptiles in the sequence of the world's rocks.

There are a number of objections to this flood geology explanation from the actual fossil record (Table 7-1). Only the most general objections will be mentioned:

(1) Creatures of the deep include fish as well as algae and bacteria, so why don't *any* fish occur in Proterozoic sedimentary rock? All that exists in Precambrian rock are simple, single-celled life forms, except for late Ediacaran multicellar life.

(2) Why wouldn't a raging flood have circulated and deposited tiny pollen grains uniformly throughout sedimentary sequences of rock, rather than only depositing this pollen in rock of Cretaceous age and younger?

(3) If raging rivers had carried the bones of *land* animals and reptiles to the ocean, then why do these bones occur in terrestrial (land) deposits and not with *marine*-type fish?

(4) Why couldn't animals and reptiles (such as dinosaurs) have "beaten" man and mammals (like cattle) to higher ground? Many types of dinosaurs were fast and agile creatures.

(5) Most important, the Bible itself speaks against Noah's flood accounting for the actual progression of life as shown in Table 7-1. Genesis 1 lists major taxonomic groups of plants and animals in the pre-flood creation that match those alive today: seed-bearing plants and trees that bear fruit with seed in it (v. 11); birds (v. 20); and livestock, cattle, and wild animals (v. 24). Why then aren't there *any* of these fossils in rock older than late Cretaceous in age over planet Earth if these were deposited in a worldwide flood? And, surely in a global flood situation *some* humans should have been deposited

Figure 7-2. The famous *Tyrannosaurus rex* (T. Rex) dinosaur of Late Cretaceous age. Does he look like he might be an herbivore? *3D rendered illustration ©Orlando Florin Rosu | Dreamstime.com.*

chaotically along with dinosaurs or marine fish if they all lived at the same time before the flood. Hydrologists (geologists that study water and floods) will tell you that one of the most diagnostic criteria for recognizing flood deposits is that different kinds of plants, pollen, animals, sediments, and rocks get all mixed up together. The most logical conclusion for Genesis 1 plants and animals not being encased in ancient sedimentary rock is that they did not exist until recent times.

Was *Tyrannosaurus Rex* a herbivore? Another flood geologist claim impossible to believe from the fossil record is that there was no death of animals before Adam sinned. The fossil record *is* a record of death, and animals in all states of death are preserved. Evidence of carnivore activity (animals eating other animals) is preserved in the fossil record from Cambrian time up to the present. If dinosaurs and other classes of animals were not carnivores, then what were their sharp teeth and claws used for (Fig 7-2)? The fossil record does not bear this claim out, and neither does the Bible. Where does the Bible say that no animals died before Adam sinned? It is a theological *supposition*. Not only is it *not* obvious from Genesis 3 that Adam's "fall" introduced death to all creatures, but the Apostle Paul offers contrary commentary on the matter in Romans 5:12-14:

> Therefore, just as sin entered the world through one man, and death through sin, and in this way death came to all men, because all sinned – for before the law was given, sin was in the world. But sin is not taken into account when there is no law. Nevertheless, death reigned from the time of Adam to the time of Moses, even over those who did not sin by breaking a command, as did Adam, who was a pattern of the one to come. (NIV)

In this passage, Paul is specific that death from sin applies to *humans,* and he does not consider the death of animals as consequential or relevant to his doctrinal point.

So what are the consequences of claiming that ferocious dinosaurs like *Tyrannosaurus rex* were herbivores? It automatically places Genesis in the mythological (fairytale) category, and for many people (especially young people) it makes *all* of the Bible suspect so that they give up their Christian faith entirely. Theology based on untruth can in, the long run, only be harmful to the spread of the gospel.

I will now present two case studies of how modern geology conflicts with flood geology.

Case Study #1: Mount Ararat

The geology of the Mount Ararat region will be the first of our two case studies because it directly relates to the Genesis story of the flood and because it will paint a simple picture for the untrained (in geology) as to why flood geology is implausible. Mount Ararat (Cudi Dag) is an almost 17,000-foot-high volcano that is still intermittently active (its last eruption was reportedly on July 2, 1840). The mountain rises above the high (~6,000 ft) plateau of eastern Turkey (Fig 6-4), which is crossed by a broad, east-west belt of folded mountains formed by the Armenian Taurus and Zagros systems that separate the plateau from the Mesopotamian alluvial valley or depression (Fig 3-3). As shown on the geologic map of Turkey, the volcanic Mount Ararat has intruded into Permo-Carboniferous, Cretaceous, Jurassic, and Miocene sedimentary rock (Fig 7-3). The volcanoes have erupted along a southwest-northeast trending linear structure, which formed at the beginning of

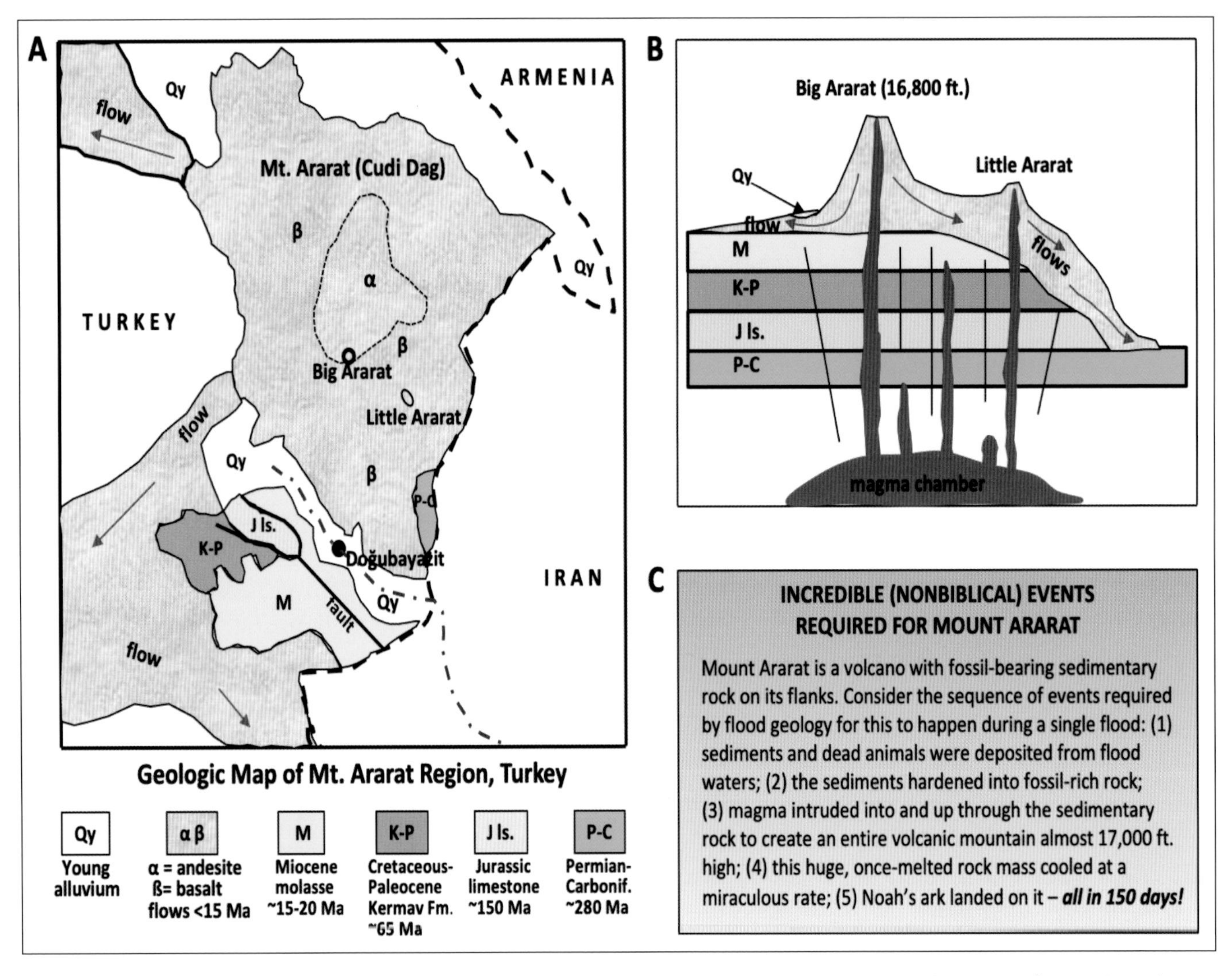

Figure 7-3. (A) Geology of the Mt. Ararat (Cudi Dag) region of Turkey. Geology of Armenia and Iran not shown since they are outside of the map of Turkey. (B) Schematic cross-section of Mt. Ararat showing the volcanic construct cross-cutting sedimentary rocks, thus implying the sequence of events in (C). Hot lava was ejected to the surface from a magma chamber below, thus creating the two volcanic cones, Big Ararat and Little Ararat. The black lines are fractures and faults in the sedimentary rock created by the injection of magma into the rock. Ma = millions of years ago. *Thanks to Davis Young and Larry Collins for helping with this figure.*

the Miocene Epoch (about 20 million years ago). Vast lava flows from then to the present cover many of the older sedimentary rocks of the region.

The claim of flood geologists is that all (or almost all) of the sedimentary rock on Earth formed at the time of Noah's flood, and this includes the sedimentary rock of the Ararat region. But Mount Ararat itself *cuts across* sedimentary rock, and so it must be *younger* than this sedimentary rock (Fig 7-3B). The flood geology scenario thus implied, according to the *actual* stratigraphic relationships present in the Mount Ararat region, is: (1) sediments (and dead animals) were deposited out of floodwater; (2) these sediments were compacted into fossil-rich sedimentary rock; (3) molten volcanic lava erupted, intruding across and flowing over this sedimentary rock to construct the volcanic cones, Big Ararat and Little Ararat; (4) the entire huge volcanic construct cooled; so that (5) the ark could land on Mount Ararat – *all* of which had to have happened in only *150 days* since, according to Genesis 8:4, the ark landed upon the mountains of Ararat on the 7th month and 17th day after the flood started (Table 5-1).

This flood-geology scenario proposes a series of geological impossibilities because it takes millions of years for sediments to deposit and solidify into rock. Then it takes many more millions of years for a volcanic construct the size of Mount Ararat to build up to a height of almost 17,000 ft and to cool to its present temperature. Furthermore, the *Bible claims none of this*! Genesis simply states that the ark *landed on the mountains of Ararat*; that is, on mountains that already existed in the region known to the ancient Mesopotamians as the land of Urartu, or what is now the area of southeastern Turkey.

Case Study #2: The Grand Canyon

The Grand Canyon is one of the most spectacular places on planet Earth! The canyon has incised a gorge over a mile deep through both sedimentary and crystalline basement rocks (Fig 7-4), which

Figure 7-4. Grand Canyon looking north along the Bright Angel Fault from Yavapai Point, South Rim. The dark-colored Precambrian rock of the lower, narrow Inner Gorge is overlain by flat-lying Paleozoic sedimentary rock. Phantom Ranch is located where the trees and path can be seen in the lower middle of the photo. *Photo by Mike Koopsen, Trails Traveled Photography.*

date from the Proterozoic (about 2 billion years) to the Mesozoic (about 200 million years) (Table 7-1). The flood geology view of the Grand Canyon has been featured in the Young-Earth Creationist book *Grand Canyon: A Different View*, which explains how the Grand Canyon formed during Noah's flood, both from the viewpoint of its rocks and canyon incision. The modern view of Grand Canyon geology, which refutes flood geology, is presented in the book *Grand Canyon: Monument to an Ancient Earth – Can Noah's Flood Explain the Grand Canyon?*

Many Christians cannot understand what is wrong with a flood geology view of the Grand Canyon – it sounds perfectly reasonable to them. This case study is intended to point out a number of reasons (among many) why modern geologists reject this view.

A Modern Geology Interpretation of Rock Layers in Grand Canyon

Modern geologists divide the more than 5,000 ft sequence of sedimentary and crystalline rocks in the Grand Canyon into rock layers spanning almost 2 billion years of time (Fig 7-5). Three different basic rock types exist on planet Earth

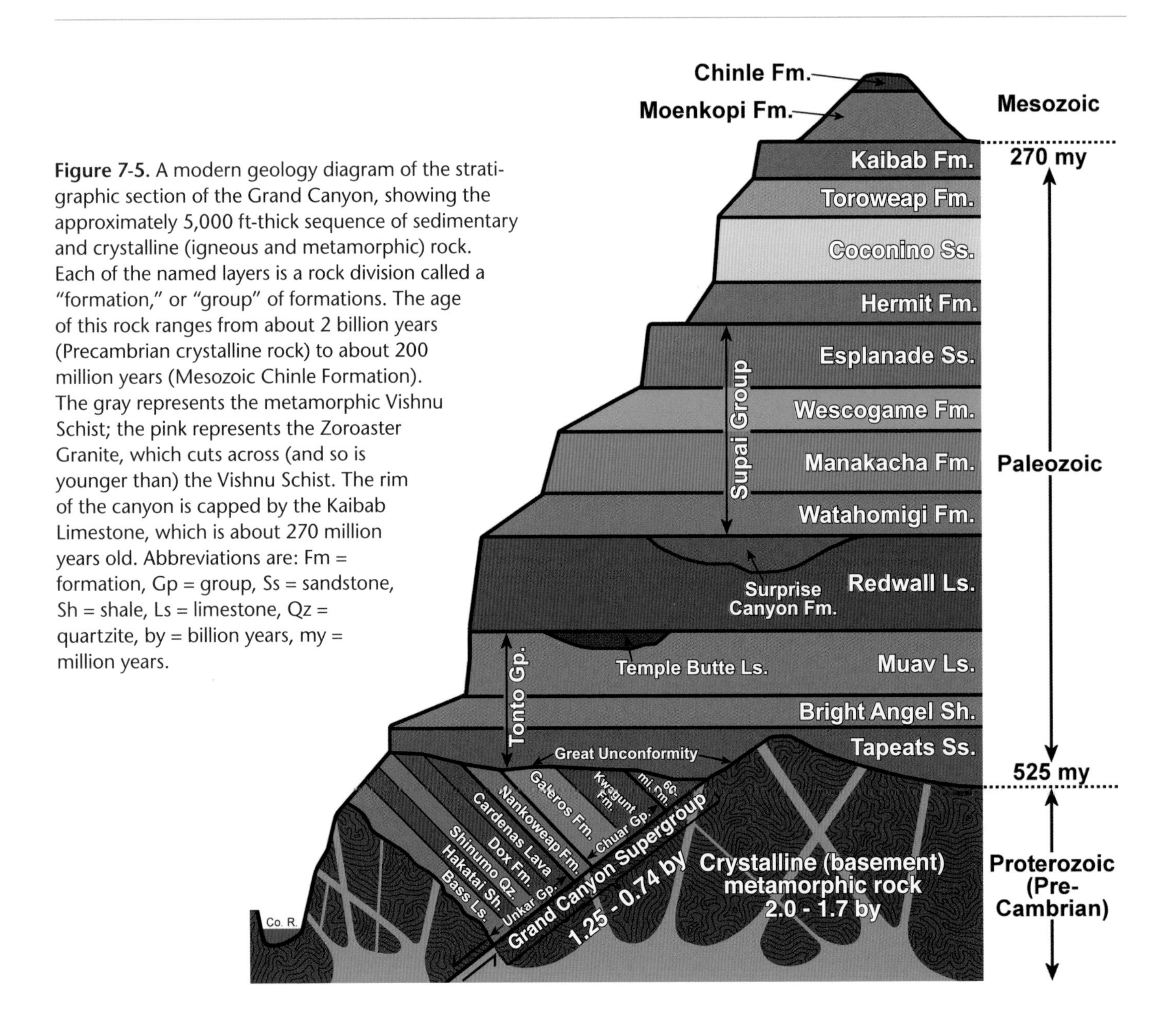

Figure 7-5. A modern geology diagram of the stratigraphic section of the Grand Canyon, showing the approximately 5,000 ft-thick sequence of sedimentary and crystalline (igneous and metamorphic) rock. Each of the named layers is a rock division called a "formation," or "group" of formations. The age of this rock ranges from about 2 billion years (Precambrian crystalline rock) to about 200 million years (Mesozoic Chinle Formation). The gray represents the metamorphic Vishnu Schist; the pink represents the Zoroaster Granite, which cuts across (and so is younger than) the Vishnu Schist. The rim of the canyon is capped by the Kaibab Limestone, which is about 270 million years old. Abbreviations are: Fm = formation, Gp = group, Ss = sandstone, Sh = shale, Ls = limestone, Qz = quartzite, by = billion years, my = million years.

Box 7-3. The Three Different Basic Rock Types

- **Igneous Rock** forms from melted material (magma). Igneous rock can form quickly when magma erupts onto the surface of the earth, either as volcanic lava flows or as explosive material. But other igneous rocks form very slowly when they cool beneath the Earth's surface.
- **Sedimentary Rock** forms from sediments deposited by wind or water. Sediments are eroded off the land, blown by the wind, carried to the oceans by rivers, deposited on the ocean floors, and then slowly turned into rock. Marine sediments can also include the shells and exoskeletons of marine invertebrate animals.
- **Metamorphic Rock** forms when igneous and sedimentary rocks are buried to great depths and are subjected to high temperatures and/or pressures over a long period of time. These processes cause these rocks to undergo a metamorphosis and become new rocks with different minerals, appearance, and structure that are compatible with their new pressure-temperature regime.

(Box 7-3), and all three are present in the Grand Canyon.

Igneous rock. Igneous rock forms from melted material (magma or lava). This melted material can cool slowly beneath the Earth's surface (called *plutonic rock*), or it can come to the surface and quickly cool there (called *volcanic rock*). The granite plutonic rock of the Inner Gorge has been dated at 1.7 billion years, and since this granite intrudes metamorphic rock, the metamorphic rock must be even older than this. The western part of the Grand Canyon contains many volcanic flows and cinder cones, such as Vulcan's Throne. These volcanic rocks have been dated from about 20 million years ago to less than a half a million years ago.

Sedimentary rock. Sedimentary rock forms by the compaction and hardening of sediments. The Grand Canyon contains an almost one-mile-thick sequence of sedimentary rocks (Fig 7-4). These rocks include limestone (e.g., the Redwall Limestone), shale (e.g., the Hermit Shale), sandstone (e.g., the Coconino Sandstone), and evaporites (e.g., gypsum beds in the Toroweap Formation). These sedimentary rocks do *not* consist of a jumble of rock types and fossils, such as might be expected for a tumultuous global flood, but represent many different sedimentary environments and ecosystems that existed over a long period of time.

The sedimentary rock sequence in the Grand Canyon also displays breaks between rock layers and types. These breaks are called *unconformities* and represent gaps in time where either erosion or non-deposition occurred. In the Upper Granite Gorge area of the Grand Canyon, where the Cambrian Tapeats Sandstone rests on Precambrian metamorphic Vishnu rock (Fig 7-5), the time gap between these two rock types is about one billion years. This erosion surface is called the "Great Unconformity" because of the immense elapse in time that occurred between the erosion of the lower rock formation (Vishnu) and the deposition of the upper formation (Tapeats). A number of shorter-time unconformities also exist in the Grand Canyon rock sequence. In the overall sequence of Paleozoic sedimentary rocks in the Grand Canyon, roughly 75 percent of the rock record is missing, primarily due to having been eroded away over time during periods when the land rose above sea level.

Metamorphic rock. Metamorphic rock forms when igneous and sedimentary rocks are buried to great depths and are subjected to high temperatures and/or pressure over a long period of time. The metamorphic rocks of the Grand

Canyon lie at the very base of the canyon and represent the core of a very ancient mountain range. Crystalline basement rocks are present in the Inner Gorge of the canyon as the Vishnu Schist (metamorphic rock derived from precursor sedimentary rock) and the Zoroaster Granite (igneous rock which was intruded into the metamorphic rock and which cooled deep beneath the ground surface).

A Flood Geology View of Rock Layers in Grand Canyon

Since flood geologists, from their so-called "literal" interpretation of Genesis, assume a one-year-long time frame for Noah's flood (Table 5-1), they must "squeeze" all of the sedimentary rock in the Grand Canyon (and planet Earth) into this very compressed time frame. Figure 7-6 shows a flood geology model of how the rocks in the canyon correspond to their time divisions established for Noah's flood.

A flood geology interpretation of the crystalline basement rocks in the Grand Canyon would be that these are rocks formed about 6,000 years ago (~4000 B.C.) during the creation week (seven literal days). "Pre-flood" rocks are sedimentary rocks formed between the creation week and Noah's flood – or during a time span of 1,656 years, if one adds the ages of the patriarchs when their first son was born until the flood started (Table 4-2). In other words, over 15,000 feet of Precambrian sedimentary rock exposed in the lower levels of the Grand Canyon would have had to have been deposited in 1,656 years from the creation to the flood.

Above the pre-flood and creation week rocks, according to flood geologists, the "early flood" sequence of sedimentary rock – from the Tapeats Sandstone up to the Kaibab Limestone – had to have been deposited in 150 days before the fountains of the deep and rain stopped (Table 5-1). The "late flood" episode, or the last 215 days of the flood (since the entire flood took 365 days, Table 5-1), is supposedly when all the Mesozoic rocks, and most Cenozoic rocks, on planet Earth formed (Table 7-1). These Mesozoic sedimentary rocks are no longer present over the Kaibab Limestone in the Grand Canyon area (except as remnants; Fig 7-5), but they are present in the region north of the Grand Canyon in what is called the Grand Staircase and in other places on planet Earth and, therefore, they must be accounted for in a global-flood model.

Two Major Problems with a Flood Geology Interpretation of the Grand Canyon

The problems for a flood geology model of the Grand Canyon rocks are similar to those discussed in our previous Mount Ararat case study – only on a much grander scale!

One-year time frame. The most acute problem of flood geology relates to the time involved for geologic processes to occur. Consider the time is would have taken for almost 5,000 feet of sediment to have settled out of flood water, the time for flood sediments to have turned into sedimentary rock, the time for sedimentary rock to have turned into metamorphic rock, the time for erosion to have created unconformities between rock layers, and the time involved for the canyon to have been carved to its present one-mile depth. How much time? Millions upon millions of years. Yet flood geologists insist that all of these processes happened in the last 6,000 years before, during, or since Noah's flood, with most sedimentary rock forming in the *one year* of Noah's flood.

Catastrophic plate tectonics. Since flood geologists must compress the deposition of sediment in the Grand Canyon (and planet Earth) into a one-year time frame (Fig 7-6), they also must invoke processes that could have moved those sediments from one place to another (such as to the Grand Canyon) very quickly over the globe during Noah's flood. So, they have proposed a model of "catastrophic plate tectonics" which invokes whole mountain ranges rising in a few months and Earth's crustal plates moving rapidly around the globe in

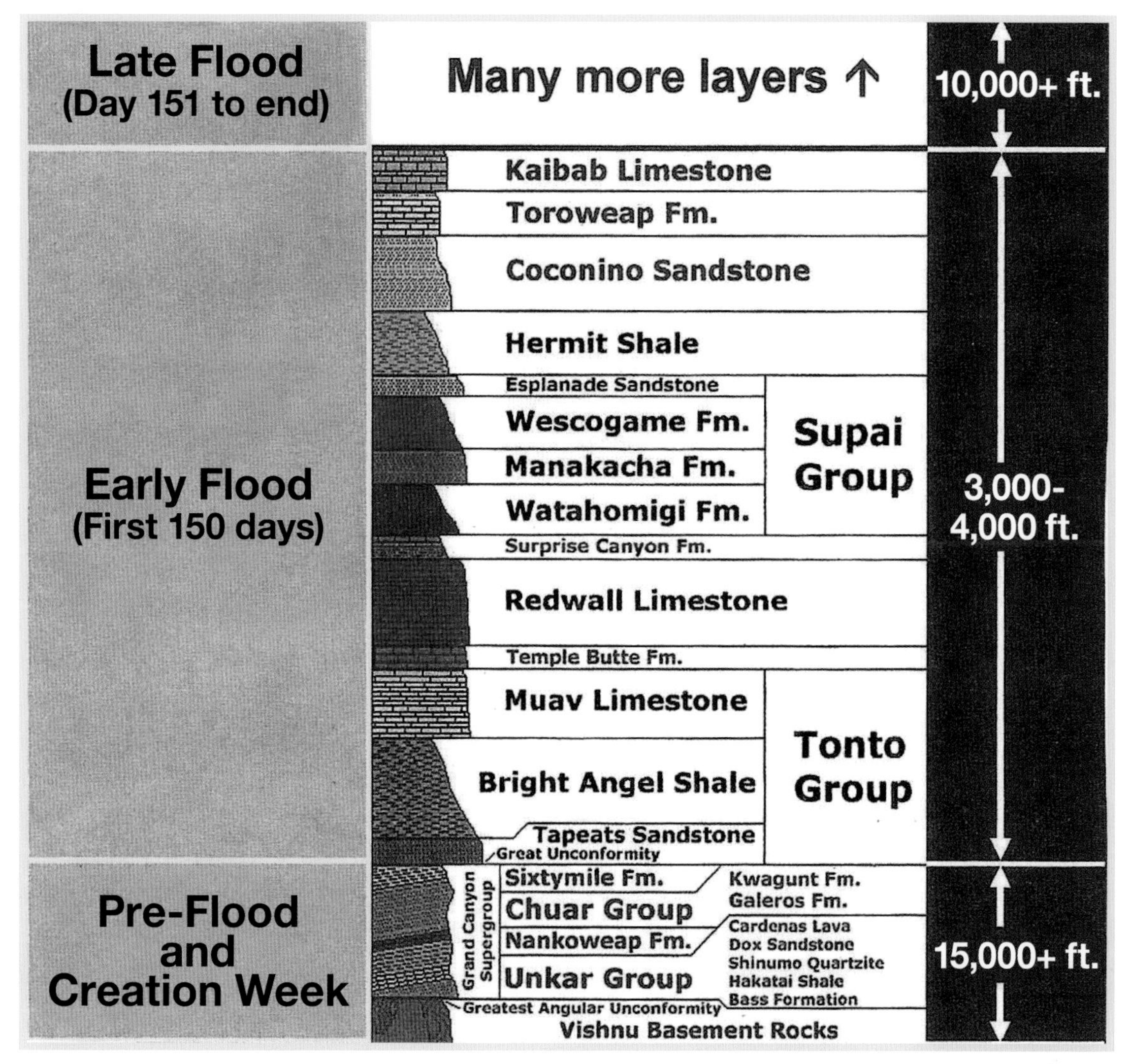

Figure 7-6. A flood geology interpretation of the stratigraphic divisions in the Grand Canyon with respect to the times specified by Genesis 7-8 for Noah's flood. Compare these times with the ages in Figure 7-5. *Modified from a drawing by Tim Helble.*

less than one year, while generating humongous earthquakes and tsunamis – a process that would have generated so much heat that it would have been impossible with the laws of physics as we know them. In addition, flood geologists propose that the entire carving of the Grand Canyon into the rock layers that we see today occurred catastrophically after Noah's flood due to the flood waters "returning off the earth continually" (Gen. 8:3; Table 5-1) – or over the last 5,000 years or so if the flood happened at about 2900 B.C. (Chapter 5). It is beyond the scope of this book to refute all of the many problems with the flood geology model in this brief discussion, and the reader is again referred to the book *Grand Canyon: Monument to an Ancient Earth, Can Noah's Flood Explain the Grand Canyon?* for an extensive coverage of why Noah's flood *cannot* explain the Grand Canyon.

Sedimentary Structures in Grand Canyon Rocks

In addition to major time and tectonic problems, a number of small features in the sedimentary rocks of Grand Canyon attest to their non-flood origin. The reason geologists know the origin of these structures is because we can *witness* how these features form today. Geologists call this the *Principle of Uniformitarianism*, or the "present is the key to understanding the past." The next series of figures show only a few (of many) sedimentary structures that illustrate this principle.

Figure 7-7A. Ripple marks forming in a shallow beach setting off of Bimini Island, a marine protected area. Ripple marks in ancient sandstones formed in the same type of setting. *Photo by Grant Johnson. Facebook.com/Bimini; public domain.*

Figure 7-7B. Ripple marks preserved in the rock of the 550 million-year old Cambrian Tapeats Sandstone, Grand Canyon. *Photo by Alan Hill.*

Ripple marks. Ripple marks are typically (but not always) generated by the to-and-fro motion of waves in shallow water at depths of a few tens of feet at the most. Figure 7-7A shows some ripple marks forming in a modern beach environment, compared to ripple marks that formed in the Cambrian Tapeats Sandstone 550 million years ago (Fig 7-7B).

Mud cracks. Mud cracks are sedimentary structures that form by the shrinkage of wet mud when it dries out. Usually, mud cracks are preserved by being filled with sediment that covers the mud-cracked layer or by calcite that later fills the cracks. Invariably mud cracks imply baking under the Sun. Figure 7-8A shows mud cracks forming today in the Grand Canyon along

Figure 7-8A. *Left.* Mud cracks forming in wet mud along the Little Colorado River near its confluence with the Colorado River in Grand Canyon. *Photo by Bob Buecher.*
Figure 7-8B. *Below.* Mud cracks in the 550 million-year old Tapeats Sandstone, which over time have filled with white calcite. *Photo by Doug Powell.*

Figure 7-9A. *Left.* Fossil tracks in the 275 million-year old Coconino sandstone cliffs of Marble Canyon. The reptiles that made these tracks ranged from iguana size to about half the size of a Komodo dragon. *Photo by Bill Hatcher, Bill Hatcher Photography.*

Figure 7-9B. *Below.* Close-up of a track made by a small reptile as it made its way up a rain-moistened dune surface in the Coconino Sandstone, Grand Canyon. Note the 5 delicate claw marks. *Photo by Cyndi Mosch.*

Figure 7-10. *Above.* Fossil raindrop prints in the Coconino Sandstone, caused by a pounding rain on the surface of sand dunes in Coconino time (about 270 million years ago). *Photo by David Elliott.*

the Colorado River near its confluence with the Little Colorado River, compared with ancient mud cracks formed in the 550 million-year old Cambrian Tapeats Formation (Fig 7-8B).

Tracks. Tracks are impressions left in soft mud or wet sand by the feet of birds, reptiles, or other animals. Reptile footprint tracks are common in the Coconino Sandstone, and that is one of the reasons why geologists are certain that the cross-bedding in this rock formation is of non-marine, sand-dune origin rather than of marine origin. The reptile tracks in the Coconino were made by small to large reptiles that crossed the sand dunes about 265 million years ago (Fig 7-9A). Delicate features, such as claw marks, are even preserved in tracks, and attest to very gentle and optimal preservation conditions (Fig 7-9B).

Raindrop prints. Raindrop prints are made when droplets of pounding rain impact wet mud, silt, or sand, thus creating small depression imprints of those drops in the sediment (Fig 7-10). This can only happen when moist sediment is exposed to the air, because if the sediment is underwater, it cannot be impacted by raindrops. In other words, this feature *could not* have formed in a rapidly rising floodwater environment or even in a body of water deeper than a few inches.

Importance of Sedimentary Structures. Why are sedimentary structures pertinent to a critique of flood geology? Because they occur throughout the entire sedimentary rock sequence of the Grand Canyon, from the earliest Precambrian sedimentary rocks up to the canyon rim. Ripple marks and mud cracks are preserved in the Precambrian Bass, Hakatai, Dox, and Nankoweap Formations. Further up in the Grand Canyon rock sequence, the Permian Supai Group and Hermit rocks contain mud cracks, ripple marks, and reptile footprints/tracks. The above-lying, cross-bedded Coconino Sandstone represents now-lithified sand dunes that display reptile tracks and raindrop prints. Well-developed mud cracks (polygons six inches or more in diameter) have also been observed in the overlying Toroweap Formation, and marine fossils typical of normal (not flood) conditions occur in the Muav, Redwall, and Kaibab Limestones.

On the basis of the above evidence, certain critical questions can be asked: If all of the sedimentary rock in Grand Canyon was deposited in a raging, miles-deep, global flood that lasted only one year, then why do the sedimentary structures in these rocks indicate a long series of environments that represent marine, to shallow-water, to drying-out, to sand-dune conditions? Why don't *all* of the fossils throughout the 5,000-foot sedimentary sequence in the canyon attest to chaotic flood conditions? How could the tiny claw marks and footprints of reptiles (Fig 7-9B) and delicate raindrop prints (Fig 7-10), be preserved when subjected to the turbulent, tsunami-like conditions claimed by flood geologists during a global flood? Answer: they couldn't!

Summary of Grand Canyon Case Study

Sedimentary structures and rock types are some of the many lines of evidence supporting a long history of sedimentation, compaction of sediment into rock, and then incision of the Grand Canyon into this rock. The evidence shows that the rocks displayed in the canyon do *not* represent a one-year-long, miles-deep, global flood. Rather, they record the comings (transgressions) and goings (regressions) of an ancient sea. When the sea transgressed over the Grand Canyon area, it left behind limestone; when it regressed, it left behind shale and sandstone. And when it left the area completely, sand eroded from sandstone piled up into dunes. Then the sea would transgress again – back and forth, back and forth, over time, leaving behind the rocks, fossils, and sedimentary structures in the canyon's rocks that attest to its long geologic history.

While the Grand Canyon is the "geologic showcase of the world," similar long and complex histories are also written into the rocks of the rest of the world. This consistent and planet-wide evidence is what has convinced geologists over the course of 400 years that Earth's sedimentary rocks are *not* the product of a year-long biblical flood. Young-Earth Creationists tell their adherents that their position of flood geology is based on Scripture, so they need to believe it. But is their position based on a poor hermeneutical rendering of Scripture that does not take into account the cultural and worldview backdrop of the Genesis stories?

Worldview and the Young-Earth/Old-Earth Debate

Young-Earth/flood geology adherents often claim that we are all looking at the same data but that our different worldviews cause us to "see" the data as evidence for vastly different and conflicting processes. The underlying assertion is that we are all practicing good science but are arriving at different interpretations of the data because of the biblical or humanistic "glasses" each person wears. Adherence to the Bible is said to draw attention to the natural evidence that supports a young age for the Earth, and rejection of the Bible causes support for an old age. Hence, based on the same data, flood geologists come to a young-Earth conclusion while modern geologists come to an old-Earth conclusion. As discussed in this chapter, this flood geology assertion is fallacious because it ignores or rejects the *overwhelming* amount of evidence from geology and *all* other branches of science.

However, from the perspective of this book, *both* camps – those that favor an old Earth and those that favor a young Earth – usually ignore the differences in worldview between us and the *biblical authors* with regard to science. Both camps assume, and argue from, an interpretation of Genesis that derives from our *modern* scientific concept of planet Earth and its geology, rather than from interpreting Genesis from the limited worldview of "earth" held by the biblical author(s). To ignore this point is to miss the key factor (worldview) in resolving the debate between science and Scripture on the age of the Earth.

WHERE DID ALL OF THIS COME FROM?

All explanations by flood geologists are said to be based on the Bible. So where in Scripture do we find references to Noah's flood linked with earthquakes, shifting continents, rising mountains, tsunamis, and rapid rates of sediment deposition? The answer is nowhere; so, how can this position be based on a "literal" interpretation of Scripture? What did the biblical authors and people living at that time know about these geological processes? Absolutely nothing. Where then does the flood geology view come from? It comes from "leaps of logic" applied to biblical verses taken out of context in order to force-fit them with modern science. It comes from the theological position stated in Chapter 1 that God's revelation in Scripture was given in terms of his omniscient knowledge of history and science, and the pre-scientific understanding of the biblical authors is not a prime factor in judging science-Scripture issues. In other words, it comes from not considering worldview to be essential in the interpretation of Scripture.

Creation of plant and animal species by fiat,
according to the traditional understanding of "after its kind" in Genesis 1 (KJV).
© Pemaphoto Naništa | Dreamstime.com.

CHAPTER 8

EVOLUTION AND THE "NEW" GENETICS

Genesis 1:11. And God said, Let the earth bring forth grass, the herb yielding seed, and the fruit tree yielding fruit after his kind, whose seed is in itself, upon the earth: and it was so. (KJV)

Ever since Charles Darwin wrote *On the Origin of the Species* in 1859, the scientific world and the Christian church have been in conflict over evolution. Evolutionary theory is widely accepted among secular scholars, and most academic theologians and scientists in the church accept the scientific consensus. Nevertheless, a highly vocal minority of pastors and lay Christians (especially in the U.S. and Britain) accept Young-Earth Creationism, are anti-evolutionists, and hold to flood geology. According to these fundamentalist Christians, the ideology of evolution – or evolutionism – has been responsible for naturalism, modernism, communism, secular humanism, and the breakdown of the family, among other societal woes. Even the word *evolution* can make some Christians fighting mad.

Evolution: What Does This Word Mean?

Because evolution can be used in different ways, we must start our discussion by defining what the term means to biologists compared to what it means to many Christians.

The word *evolution* can be used in a general sense for any change or development over time. You could describe how stars evolve – that is, they change their size, explode, become novas or supernovas over time, and so on. Or you could trace the cultural evolution (or change) of a people group. You could even say that the Bible has "evolved" (developed or unfolded) as God has revealed himself to humankind over time. (This topic of Progressive Revelation will be covered in Chapter 10.) Most Christians – but not all – accept this general sense of the word "evolution."

Special Theory of Evolution (Microevolution)

The word *microevolution* is not recognized by biologists as being separate from macroevolution, but the term is included here because many Christians consider it to be a part of their understanding of evolution. Microevolution became "standard Christian fare" through Pattle P. T. Pun's 1982 book *Evolution: Nature and Scripture in Conflict?*, and, since then, it has been used in publications by a number of Christian biologists and authors. This so-called "special theory" describes how changes in biological populations occur over time, especially in response to changes in environmental conditions; for example, viruses that become resistant to drugs. Microevolution is usually viewed as the change among offspring that comes from the genetic variability that occurs *within* species (e.g., the diversity of breeds observed among dogs). These changes in organisms due to the process of natural (or human-caused) selection and other evolutionary mechanisms can be readily observed or verified in the natural world, and thus, are not controversial for most Christians.

General Theory of Evolution (Macroevolution)

The general theory of evolution, or macroevolution, is the position that all living species evolved from simple microorganisms. From this common ancestor, all other living things such as plants, animals, and even humans evolved over vast amounts of time. This principle of *common descent* is a central tenet of evolutionary theory and was one of the conclusions that Charles Darwin came to in his book *On the Origin of Species*. Many Christians have been taught that the evolutionist's concept of macroevolution embodies the philosophy that everything (living and non-living) evolved solely by *chance*, even though the primary mechanism of species formation via natural selection is a distinctly non-random process. As a biological theory, evolution says nothing about whether a creator God is part of the evolutionary process: it simply describes the *process* of evolution. However, the Bible does explicitly claim that all we see in the natural world (which includes evolution) was created and is sustained by God, and, therefore, proponents of evolutionary theory who claim that the process requires the non-involvement of God as being part of the evolutionary process is what makes it so controversial and seemingly diametrically opposed to the Bible.

The general theory of evolution includes several important concepts:

Species. The most common definition of "species" among lay persons is: "animals or plants that look alike and which can interbreed and produce fertile offspring." However, in reality the term "species" is difficult to define because every imaginable gradation of change exists between species, and for this reason there is no universal agreement among biologists as to what constitutes a species. In the microbial world the concept of species is semi-meaningless because microbes such as bacteria are asexual and are capable of swapping DNA between "species."

Mutations. For populations to change over time, they must obtain new genetic variation, and mutations are the means by which new genetic variation is added to a population. It is this genetic component that is passed on to succeeding generations, forming the impetus for evolution. Most mutations are neutral, but sometimes they can be harmful or advantageous to an organism. Genetic mutations have traditionally been considered by most evolutionists to be without direction; that is, mutations occur randomly and natural selection determines the value of the mutation for the overall good of the organism.

Natural selection. Natural selection occurs when the environment acts as the selective agent by which combinations of characteristics best fit that particular environment; that is, it determines the "survival of the fittest." A classic example of natural selection can be observed with moth populations, which darkened in color during England's Industrial Revolution corresponding to an increase in industrial dirt and grime. In this case, the species didn't change into another species, just some characteristics of the moth population changed by adaptation to the changing environment. Another example of natural selection can be seen in the beaks of finch populations on the Galápagos Islands changing in response to wet and dry years and to the expanding or dwindling of food supplies.

After Its Kind: What Does This Phrase Mean?

People who oppose evolutionary theory on biblical grounds usually quote the phrase "after its kind" used in Genesis 1:11-12, 21, 24-25 to support their position that God created each species (page 108) separately, and that each species has been fixed since its creation – a position referred to as the "fixity" or "immutability" of species. However, what did the phrase mean from the worldview of the biblical author(s)?

"After Its Kind" in the Old Testament

The phrase "after its kind" is not only used in Genesis 1 (Gen. 1:11-12, 21, 24-25), but also in Genesis 6 (Gen. 6:20) and Genesis 7 (Gen. 7:14), and in Leviticus (Lev. 11:14-6, 19, 22) and Deuteronomy (Deut. 14:13-15, 18). In addition, the phrase "after their kinds" and "according to their kinds" (plural) is used in Genesis 8:19 and Ezekiel 47:10, respectively. It is only in Genesis 1 that the term "after its kind" is used when referring directly to God's creation of the living world. In Genesis 7 and in Leviticus and Deuteronomy, it is used in reference to dietary restrictions, and in Genesis 8:19 and Ezekiel 47:10 the term applies to general categories of animals where the propagation of species is not implied. So, this is the first thing to note about the use of the phrase "after its kind": it conforms to the repetitive and prosaic literary style of ancient Near East texts. This type of speech is similar to what was discussed in Chapter 6 with regard to such phrases as "all the earth" or "every nation under heaven."

The second thing to note is how the phrase "after its kind" fits within the overall numerological harmony of the Genesis text. The phrase is used in seven different verses of the Masoretic Hebrew text (Gen. 1:11, 12, 21, 24, 25; Gen. 6:20; Gen. 7:14) – seven being the sacred number of the Mesopotamians (see Chapter 4). It is used only once in Genesis 1:11; it is repeated twice in each of the verses of Genesis 1:12; 1:21; and 1:24; and three times in Genesis 1:25, adding up to ten, the decimal base of the Mesopotamians' combined sexagesimal-decimal numbering system. Furthermore, in Genesis 6, "after its kind" is repeated three times and in Genesis 7 it is repeated four times, adding up to seven. Thus, this phrase was intentionally woven into the fabric of the text by the biblical author(s) in keeping with the numerological style of sacred narratives.

The word *kind* (*mîyn* or *mîn*) in "after its kind" comes from an unused root meaning to portion out or separate as in a "splitting off." From an etymological study of the word *mîn*, J. B. Payne concluded that the original word referred to subdivisions within the types of life described and not to the general quality of these types themselves. However, if *mîn* does refer to subdivisions, what subdivisions was the biblical author alluding to? Scholars holding various points of view (Young-Earth Creationists as well as theistic evolutionists) have equated "kind" with species, genus, family, and even higher orders in the Linnaean classification scheme (Box 8-1). But are these subdivisions what the original biblical author(s) had in mind? Can a "splitting off" be construed to support the splitting of species in divergent evolution, as has been done by some theistic evolutionists? Does *mîn* refer only to fixed species, or can it also refer to new species that have evolved over time? Technically speaking, in evolution each

Box 8-1. The Linnaean Classification System

Domain = Eukaryota
Kingdom = animals, plants, archaea
Phylum = 21 groups; e.g., chordates (backbones)
Class = e.g., reptiles, mammals
Order = e.g., primates
Family = e.g., hominids (fossil "man")
Genus = e.g., Homo (man-like)
Species = e.g., *Homo sapiens* (modern humans)

generation exhibiting a slight genetic change *does* reproduce after its kind (the mutated gene is passed down to its offspring).

"After Its Kind" and Folk Taxonomy

The plant and animal classification scheme of Linnaeus is part of our modern scientific tradition, one that characterizes a complex society. However, pre- or proto-scientific peoples in ancient times had botanical and zoological classification schemes of a more shallow hierarchical structure, similar to those of primitive societies today. That is, these "folk" classification schemes are based on the most distinctive species of a local habitat and on the characteristics of plants and animals that are readily observable.

Folk zoological classification schemes are exhibited by cultures all around the world. These "native" taxonomies characteristically break down the classification of animal life forms into one or more of the following categories: (1) *Fish* (aquatic animals, mostly fish but even whales); (2) *Birds* (animals that can fly, usually birds but occasionally bats); (3) *Snakes* (creeping life forms such as snakes and lizards); (4) *Wugs* (worms plus bugs, or, in general, the insect world); and (5) *Mammals* (in general, large animals, mostly mammals, but even large forms of reptiles). How many of these categories a society embraces (or embraced) depend on that society's stage of complexity. Primitive societies may have only one to three of these categories. More highly organized, proto-literate societies (like the ancient biblical ones) may have had three or four of these categories, whereas modern societies have all five (or more). It is apparent that native categories do not correspond to our modern scientific classification schemes based on comparative anatomy and DNA, but rather they are based on the most obvious features or functions (actions) of these animals.

The four life-form categories mentioned in Chapter 1 of Genesis (and elsewhere in the Old Testament) are typical for a proto-literate society, both in number and in their non-scientific nature: fish (water-dwelling creatures including fish, but also whales; Gen. 1:21); birds (flying creatures, Gen. 1:21; but also bats, Lev. 11:19); snakes (every creeping thing, *remeś*; Gen. 1:24, 25); and mammals (*bĕhēmāh*, meaning "dumb beasts" like domesticated cattle, but also *chay*, meaning "wild beasts"; Gen. 1:25). If compared to our modern Linnaean system (Box 8-1), the four native life forms of Genesis 1 mostly equate with *class*, the subdivisions of which are order, family, genus, and species. In reality, the meaning of "kind" (*mîn*) in the Old Testament can refer to any of these divisions, and in some cases even to phyla. Which taxonomic subdivision is used depends on a number of factors, especially on the type and size of the plant or animal and its importance (for example, edibility). Usually for mammals and birds – but also for the larger reptiles and amphibians – the word *mîn* ("kind") in the Old Testament corresponds to genus or species. For example, in Leviticus 11:15 and Deuteronomy 14:14, the "raven after its kind" presumably includes all six species of the genus *Corvus*. However, with owls, each of Palestine's eight owl species is mentioned, from the huge eagle owl down to the tiny scops owl (mistranslated in early versions of the Bible as ibises, water hens, swans, seagulls, and cuckoos for one or another of these owl species).

The most important point of this folk classification discussion is that the ancient Mesopotamians were still at the stage where they were using a *native* classification system. Therefore, the subdivisions of *mîn* are for this native system and cannot be made equivalent to our Linnaean categories of order, genus, species, and so on. The text of Genesis 1 fits *historically* within a proto-literate society having no knowledge of our modern-day classification system, which shows that God did *not* impart advanced scientific knowledge to the ancient Mesopotamians but gave his revelation to people with a pre-scientific view that was commensurate with their understanding of the world (see Chapter 1).

In order to see how the biblical authors understood the classification of living things, we must look at the worldview of the people of that day and their close relationship to the land and pastoral way of life. The ancient Mesopotamians were savvy when it came to plant and animal reproduction because their livelihoods depended on raising crops and herding livestock. Even as early as Uruk time (ca. 3800-3100 B.C.; Table 3-1) archaic cuneiform texts from Uruk document the categorizing of sheep into breeding bulls and rams, an indication that these people had some knowledge of selective breeding. Genesis 1:11 breaks the general class of "vegetation" (*deshe'*) into separate categories of plant-bearing seed and fruit-bearing seed, both kinds of seeds representing important food sources for a farm-based economy.

As with their concept of the cosmos (Fig 1-1), the ancient Mesopotamians knew only what they could *observe* of the natural world. They knew that when they planted a barley seed (the main food crop of ancient Mesopotamia), a flax plant didn't grow from that seed – barley grew from it. This is the connotation of "after its kind" of the plant yielding seed in Genesis 1:12. They knew that a fig tree couldn't bear olive berries and that a grapevine couldn't bear figs (James 3:12). This is the connotation of "after its kind" of the tree yielding fruit, whose seed is in itself in Genesis 1:12. Also, people in the ancient Near East observed that animals produce after their own kind (Gen. 1:21): a whale doesn't give birth to a fish, a crow doesn't produce a dove, and a donkey doesn't produce a lamb. So in this respect, the biblical author(s) *did* mean procreating species, and their definition of "species" would probably have been about the same as the popular definition of today: animals or plants that look alike and which can interbreed and produce fertile offspring. This is the *practical* way of defining species, one in accordance with the everyday experience and knowledge base of the ancient people groups living in the Near East in proto-literate time.

"After Its Kind" and the Theory of Evolution

What then does "after its kind" have to do with today's controversy of evolution versus the fixity of species? Nothing, because the biblical author(s) of 2500-1500 B.C. had no concept of the evolution of one species into another. Whether evolution does or does not occur slowly over time is not something they could have observed. Therefore, from a Worldview Approach, the biblical phrase "after its kind" cannot imply either the viability or non-viability of evolution. The biblical author(s) of Genesis 1 were completely pre-scientific in their methods and motives, and Christians are reading their modern scientific worldview into the text when they take the phrase "after its kind" and apply it to evolution.

"AFTER ITS KIND" AND THE FIXITY OF SPECIES

What then does "after its kind" have to do with today's controversy of evolution versus the fixity of species? Nothing, because the biblical author(s) of 2500-1500 B.C. had no concept of the evolution of one species into another. Whether evolution does or does not occur slowly over time is not something they could have observed. Therefore, from the perspective of a Worldview Approach, the biblical phrase "after its kind" cannot imply either the viability or non-viability of evolution. The biblical author(s) of Genesis 1 were completely pre-scientific in their methods and motives, and Christians are reading their modern scientific worldview into the text when they take the phrase "after its kind" and apply it to evolution. These attempts to force the implications of modern scientific knowledge onto the ancient biblical text results in a distorted interpretation.

The distortion of the meaning of *mîn* arises not from the subdivision of the word into "species," "genus," or whatever taxonomic category is championed, but from forcing the implications of modern scientific knowledge onto the ancient biblical text.

Developments in Evolutionary Biology

Much of the evidence which eventually convinced scientists that organisms change over time was collected in the 1700s, *before* Darwin lived and thus before he proposed his theory of evolution. It started with observations from the fossil record and comparative anatomy and has progressed over the last 300 years to include modern genetics. Let us now take a brief survey of some of the evidence that has been critical in formulating many aspects of evolutionary theory over the years.

Orderly Arrangement of the Fossil Record

Before Darwin (1809-1882), paleontologists (scientists who study fossils) realized that the fossil record is not a random hodge-podge of fossil plants and animals such as might be expected if they had been deposited in a catastrophic flood. Rather, there is a particular order to this record, where successive rock layers contain different assemblages of fossils and where increasingly complex life forms are found in successively younger strata (this order is called *the Principle of Faunal Succession*). Simple, single-celled organisms (bacteria, algae) are found in the lowest and oldest sedimentary rock layers, while more complex organisms (mammals) are found in the topmost and youngest layers (Table 7-1).

Comparative Anatomy

In 1735 the Swedish botanist Carolus Linnaeus (Carl von Linné) proposed the first taxonomic classification system for all living things based on comparative anatomy (sometimes called *homology*, or the identification of similar anatomy in fish, amphibians, reptiles, and mammals). For example, you have five digits on each hand and so do your cat and dog, all species of the cat family look and act the same, and so on. You probably learned the Linnaean classification system in high school biology. As you move from the most general (domains) to the most specific (species), the more closely related living things become more alike anatomically (Box 8-1); for example, hominids are more closely related to other primates than to bacteria. Homologous bone structure was used as evidence supporting evolution during Darwin's time, and this anatomical sequence strongly suggests a common descent connection between all of life.

Vestigial Structures

Vestigial organs and structures are body parts no longer used by an animal – or at least not used for the same function – that attests to its evolutionary relationship with common ancestors that did use them. For example, whales are descended from a four-legged mammalian ancestor that walked on land about fifty million years ago, and the ancient legged whales *Ambulocetus* and *Rodhocetus* are transitionary forms between modern whales and this land ancestor. Such an evolutionary connection is not obvious to most of us who are not paleontologists. However, on rare occasions a "throwback" is encountered that we can see; for example, a modern whale is hauled in with a partially-formed hind leg, complete with thigh and knee muscles, sticking out of its side (Fig 8-1). The reappearance of vestigial organs is controlled by genes that are generally switched off yet are still present in the genome of individuals and, thus, have the ability to be switched on. Thus, every once in a great while an entire sequence of a suppressed genetic code of an animal can reactivate – such as in the case of a modern whale reverting back to producing legs from the pelvic area where legs used to grow millions of years ago in its whale ancestors.

Embryonic Organs

Embryonic organs are those that emerge during the early stages of embryonic development, but

Vestigial Bones: Evidence for Descent

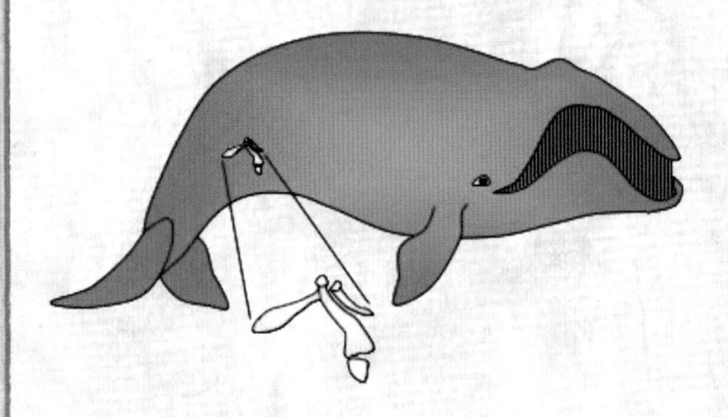

Figure 8-1. Some whales contain a small pelvic bone, which is but an evolutionary remnant of bones its land ancestors used when they had legs. In very rare cases, as shown in the photo on the right, whales still have the genetic "throwback" ability to grow rudimentary legs. *Google Image.*

The appendage in the photo is 4.2 ft long. From R.C. Andrews, 1921, "A Remarkable Case of External Hind Limbs in a Humpback Whale." *American Museum Novitates,* No. 9, June 3. *Photo and labels were supplied by Gregg Davidson.*

which disappear before or after the organism's birth. Human embryos, for example, form a yolk sac during the early stages of their development. In the case of birds and reptiles, a similar sac surrounds a nutrient-rich yolk, but in the human embryo, the yolk sac is empty. Such organs that are present in embryos, but not in adult organisms, have been viewed by many biologists – even before Darwin – as evidence of an evolutionary ancestry. Like vestigial organs, embryonic organs support the idea of common descent.

Comparative Genetics (DNA)

How far animals are genetically apart from each other generally correlates well with their comparative anatomical structure; that is, usually the more similar animals are to other animals morphologically on the Linnaean system, the more similar their DNA sequences are. Genetic comparisons between animal species – and thus their probable ancestry – are now being made for all kinds of plants and animals, and DNA studies are refining the Linnaean classification system. If one considers the overall genetic sequence identity – the total similarity of the coding regions of their genomes – between humans and other animals, the following approximate differences are observed (Box 8-2).

It is this relative amount of sequence identity, and not the number of genes, that is crucial to the morphological and behavioral differences between organisms. For example, sponges harbor between 18,000 and 30,000 genes, or roughly the same *number* as humans and mice (about 20,500 genes), yet sponges display a completely different level of biological complexity than higher organisms. And, while chimps share approximately 99 percent of our DNA, yet that tiny 1 percent portion of unshared DNA makes a world of difference in the behavior and intelligence of chimps and humans. Besides the strong sequence identity between chimps and humans, if our entire sequence of genomic homologies – including "derelict" genes (*pseudogenes*) – is compared to that

Box 8-2. Genetic Coding Similarities	
E. Coli (bacteria)	~45% genetically identical to that of humans
Chicken (bird)	~90% genetically identical to that of humans
Rabbit (mammal)	~95% genetically identical to that of humans
Chimpanzee (primate)	~99% genetically identical to that of humans

of chimpanzees, the similarities are found to be remarkably identical. It is extraordinary to view large segments of chimp and human DNA, aligned side-by-side, and see the same sequence of genes and pseudogenes – yet another correspondence that strongly implies that both species (and other higher primates) are products of a common ancestral lineage. Or, as Francis Collins, former head of the Genome Project, has remarked: "Certainly this kind of evidence is strongly in support of evolutionary theory." The genomic evidence for common descent is vast, complex, and beyond the scope of this book. The interested reader is referred to Collins's book *The Language of God* and other publications on this topic.

The "New" Genetics

Since the human genome project was completed in April of 2003, a remarkable explosion of genetic knowledge has occurred. Words like *epigenetics* have recently been introduced to the public, and since many of these terms are new (at least to Christians), a few key terms will be briefly defined for the reader. Many of these mechanisms have been addressed by James Shapiro in his book *Evolution: A View from the 21st Century* and by Sy Garte in his condensed article "New Ideas in Evolutionary Biology." The following descriptions of genetic terms come mainly from these two sources.

Genome. The complete DNA molecule within a cell or other biological entity, such as a virus. Various degrees of gene duplication and reorganization events can comprise whole genomes.

Transposons. Mobile segments of DNA that can make copies of themselves and "jump" into new positions in the genome, thus altering the activity of full-length genes. Transposons are sometimes referred to as *jumping genes* or *transposable elements*. The basic characteristic of transposons seems to be their ability to help organisms adapt quickly to rapidly changing environments by taking on new characteristics.

Phenotypic plasticity. The ability of an organism to change its behaviors or features in response to the environmental factors it experiences during its lifetime. The speed at which evolution can occur has suggested to some biologists that changing environments could re-orchestrate the genes an organism already has and cause new forms to emerge *without* mutations.

Interspecific hybridization. The mating between parents of closely related but distinct species to produce a hybrid offspring. Such mating encounters cause chromosomal breakage and rearrangements that allow new gene variants to flow across species boundaries. The probability of genetic change in hybrids is high and can exceed 100 percent (in other words more than one change can be found in every progeny of interspecies mating). Not only does a young species retain DNA from its ancestors (as expressed in vestigial structures), but its old genes may still be similar enough to function successfully between closely related species to form hybrids (different species).

Horizontal DNA transfer. Horizontal DNA transfer may occur when one organism transfers a large chunk of genetic material to another organism. This transfer has been found to apply to bacteria-bacteria, bacteria-plant, bacteria-yeast, and bacteria-insect interactions.

Epigenetics. The old idea of evolution was that who you are is a matter of which genes you have. Epigenetics (*epi-* meaning "above" or "over" genetics) does not refer to inherited changes in the genetic molecule DNA, but to chemical alterations of DNA and its associated proteins that influence how genes *behave* (their so-called "expression levels") due to the effects of different environments. This idea that heritable changes can result from environmental factors is a radical one because it favors a Lamarckian-type mode of evolution rather than a gene-based Darwinian mode. What is being referred to here is not the Lamarckian view of use and disuse (like giraffes' necks get longer over time because they reach high into the trees), but rather the idea that exposure to environmental signals can alter the gene arrays that are turned on and off.

A key concept of genetics that has emerged in recent years from all of this data is that *living organisms actively change themselves*. James Shapiro's position is that the insistence of *randomness* and *accident* with respect to evolution is still the prevailing and widely accepted worldview of biologists, and the perceived need to reject supernatural intervention has led to an *a priori* philosophical distinction between the "blind" processes of hereditary variation and all other adaptive functions. Rather, *natural genetic engineering* – which represents the ability of living cells to manipulate and restructure the DNA molecules that make up their genome – is a better way of looking at evolution. Living cells do not operate blindly (i.e., randomly or by chance). They have been "programmed" to continually acquire information about the external environment and to monitor their internal operations accordingly.

A similar concept that has recently emerged is *evolutionary convergence*. Instead of the huge role that accident and random chance supposedly plays in controlling the evolutionary process, convergence suggests that there may be specific directions in evolution. Certain common biological features that arise in unrelated lines (such as mammal and marsupial "lions") suggest that such features are controlled at a molecular level and are inevitable. Or, as stated by Sy Garte: "Evolution does not allow for anything and everything: if it isn't in the tool kit it doesn't happen....These constraints have profound implications for the idea of evolutionary direction, and even of teleology."

Supposed Problems for Evolution in the Fossil Record

We are now going to cover three aspects of the fossil record that are considered to be problematic by many Christians: (1) the appearance of life on planet Earth in the Archean Era, (2) the so-called "Cambrian explosion" at the end of the Proterozoic Era, and (3) the claim that the fossil record is devoid of intermediate transitional types, such as should be expected from evolutionary theory.

Appearance of First Life on Earth

The oldest known fossils have been found in Archean rock dating to about 3.7 billion years ago (Table 7-1). Algal and bacterial fossils have been recovered from the ancient sedimentary rocks of South Africa and western Australia, and there is also evidence from rocks in western Greenland that life may have existed on planet Earth as long ago as 3.8 to 4.0 billion years. These findings challenge the supposed long time spans needed for evolution. If the Earth formed about 4.56 billion years ago from a coalescing solar system, then how could life have evolved from non-life in perhaps less than a billion years? This problem has caused some scientists to conceptually transfer the origin of life to outer space (called *panspermia*), with life inhabiting the Earth from space as soon as its surface and atmosphere became fit to support life.

The reason that having only a billion years for life to originate is a problem is because the probabilities against it happening (by chance) are astronomical. Astrophysicist Fred Hoyle and his colleague Chandra Wickramasinghe calculated that the probability of life evolving by pure chance was $10^{-40,000}$ – a statistical improbability that was expressed by Hoyle in his now famous metaphor: that a living organism emerging by chance from a pre-biotic soup is about as likely as "a tornado sweeping through a junk yard might assemble a Boeing 747 from the material therein." And transferring the origin of life to outer space doesn't help because if these statistics are correct, life would not have had enough time to evolve in the universe in 13.82 billion years any more than it could have evolved on Earth in less than 1 billion years.

Besides the time problem, there is also the serious biochemical problem of how to bridge the gap between complex organic chemistry and a replicating *living* system? As elaborated on by Simon Conway Morris in his book *Life's Solutions: Inevitable Humans in a Lonely Universe*, conceiving of how such a transformation could occur is most formidable! Moreover, even if humans are ever able to make living organisms in the laboratory (by whatever means), such organisms would have been constructed by an *intelligent* and *directed* intervention.

The Cambrian Explosion

The term *Cambrian explosion* refers to the so-called "explosion" of animal life during the Cambrian Period (500-540 million years ago). During the Ediacaran period of the late Precambrian (~540-550 million years ago), fossils of the first soft-bodied, multicellular, macroscopic animals appeared (Table 7-1), while by the Cambrian most Ediacaran fauna had become extinct. Within about fifty million years – a blink of an eye by geologic timekeeping – nearly all of the major phyla known today appear in the fossil record. Early Cambrian life is best represented in Canada's Burgess Shale (Fig 8-2) and China's Qingjiang biota, and the body plans that first appeared in the Cambrian have, by in large, served as the original blueprints for all animals that have lived since then.

The evolutionary questions that the Cambrian explosion raises are: Why did these body plans seemingly appear so suddenly and simultaneously, and then almost no new body plans develop for over a half a billion years? A number of explanations have been offered: perhaps there was an increase in seawater oxygen levels at this time or so many ecological niches were then available that a number of phyla innovations filled them. Or perhaps animals developed hard shells or the first image-forming eyes and, thus, had an immediate advantage in natural selection. The Cambrian explosion is one of the most remarkable and puzzling events in the history of life on Earth,

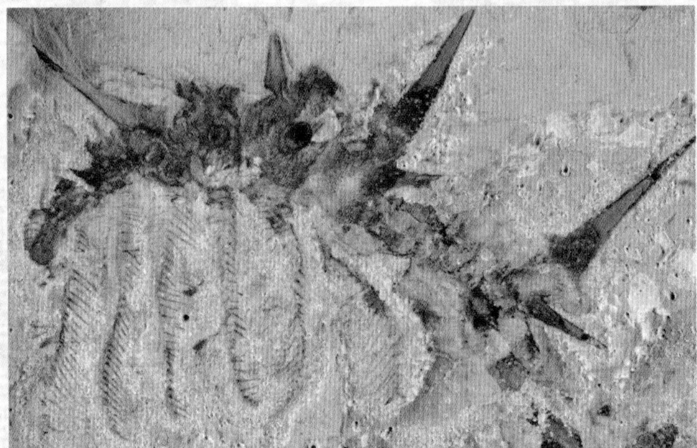

Figure 8-2. *Hallucigenia,* the fossil oddball of the Burgess Shale, now known to be one of a very diverse and abundant group of organisms called lobopods. *Left:* Drawing of *Hallucigenia* from its fossil *(above)*, by Dirk Wachsmuth, Collinsium-Ciliosum Art Work. *Fossil photo: Yang et al. (2015).*

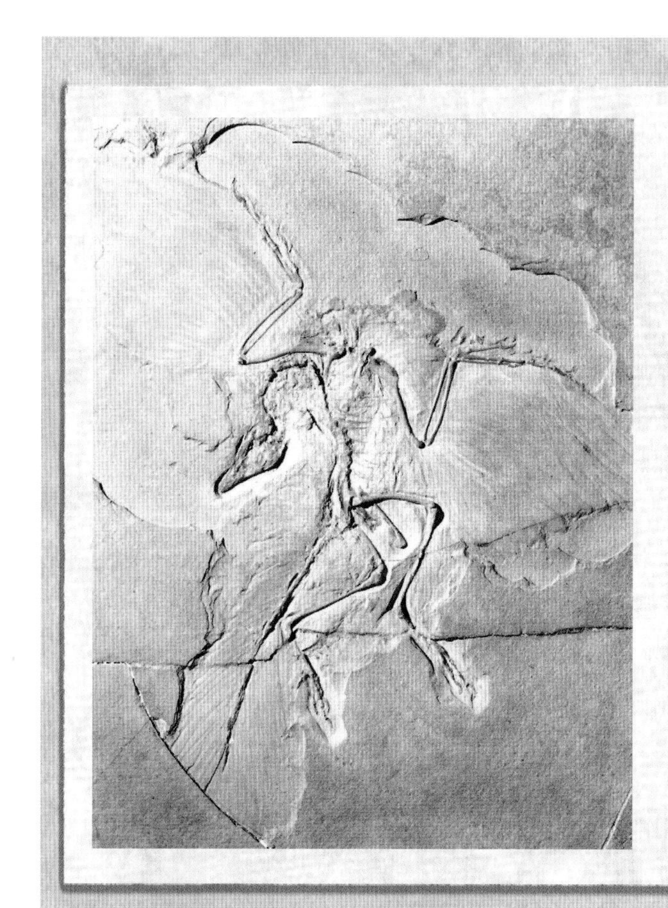

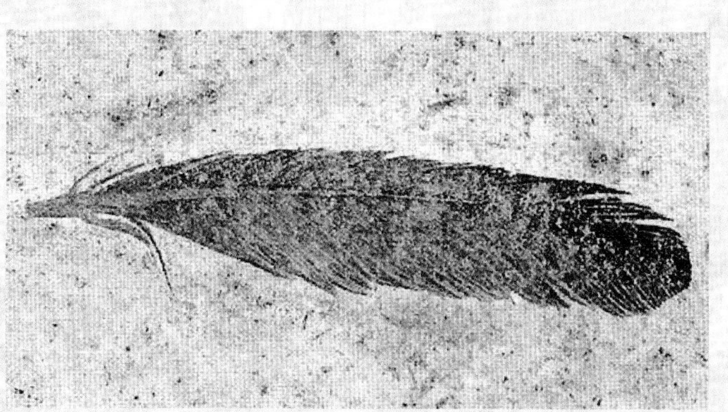

Figure 8-3. *Archaeopteryx (left)*, a prehistoric bird that lived in the Jurassic about 150 million years ago. The skeleton is reptilian, yet it displays feathers. *Archaeopteryx* feathers *(above)* are virtually identical to the feathers of modern birds. From its wing structure, paleontologists have concluded that the wings of *Archaeopteryx* were made for flapping, similar to pheasants and quails. *Google Images.*

and it is critical to fully understanding the process of evolution. Ongoing discoveries by paleontologists of Ediacaran-Cambrian life in rocks all over the world are helping to substantiate this early record of animal life and why it happened.

Transitional Fossils

The traditional premise of macroevolution is that species gradually change into other species over time. If this is so, the question then arises: Why don't we see innumerable transitional forms everywhere in the fossil record? Darwin himself asked this question 150 years ago; he assumed that transitional forms were present in the fossil record, but reasoned that these fossils had not yet been found because so little of the Earth's sedimentary rock had been examined. Today, practically all of the sedimentary rock on Earth has been field checked, yet transitional forms are still relatively scarce in the fossil record.

Many people within the evangelical Christian community have heard presentations stating as fact that "there are no transitional forms" in the sedimentary record. This is not true! There are numerous examples of fossils with transitional morphologies at all taxonomic levels, from species to phyla, and more "gaps" in the fossil record are being filled every year by paleontologists. The famous *Archaeopteryx* fossils, first discovered in Germany in 1861, display both a reptile skeleton with a full set of teeth and a long bony tail, and also wings and avian feathers (Fig 8-3), and today many more examples exist of the remarkable fossil record of dinosaurs that led to birds. Also, in the case of invertebrates, there are numerous fossils that record evolutionary trends. In fact, the pace of recent discoveries of new fossil forms that "gap-fill" the paleontologic record has sped up considerably.

Here is the norm for the fossil record: the sudden appearance of a species and then "stasis," where the fossil form can remain unchanged in layer upon layer of rock for millions of years. Notable examples of static species are the one-billion-year-old fossils of blue-green bacteria, which look exactly like their modern "pond scum" counterparts, or ancient stromatolites that closely resemble stromatolites living today. An invertebrate example of a static species is the horseshoe crab, which still appears upon New England shores every year to spawn, and which has remained relatively unchanged (a "living fossil") for over 450 million years (since the late Ordovician; Fig 8-4). A notorious vertebrate example is the coelacanth, a "fossil fish" dredged up from the deep waters of the Indian Ocean and off of South Africa, which has remained relatively unchanged from its Late Cretaceous fossilized predecessors.

Such "stasis" in the fossil record is what prompted Niles Eldredge and Stephen Jay Gould in 1977 to propose their "punctuated equilibrium" model for evolution, which characterizes evolution as consisting of long periods of little change punctuated by relatively brief periods of rapid change. Punctuated equilibrium is not a new theory of evolution, it is just a new label for a mode of evolution that recognizes stasis in the fossil record, in contrast to "gradualism," or slow, uninterrupted evolution, which has been prevalent since Darwin's time. The importance of punctuated equilibrium is that it tries to take the actual fossil record into account.

At the same time, it is important to realize what is *not* implied by the geologic record for the sudden appearance of a species followed by stasis. The theory of punctuated equilibrium does not imply that species supposedly appearing instantaneously in the fossil record are "fully formed" in an "after its kind" fiat biblical sense (page 108). To a geologist, "sudden, brief periods of rapid change" may amount to only tens of thousands or a few million years, rather than hundreds of millions of years. This record also does not necessarily imply that species do not evolve into other species. Pond-scum bacteria could have remained in their particular habitat for a billion years unchanged, whereas other of their bacterial descendants could have evolved into other species occupying different ecological niches.

Figure 8-4. A horseshoe crab, one of many that still emerge along New England shores to breed. Fossil horseshoe crabs have been found in rock as far back as 450 million years ago (Table 7-1). Thus, these crabs are referred to as "living fossils" or "stasis species." *Google Image.*

Furthermore, a number of reasons can be invoked as to why transitional forms have not been preserved in the fossil record. For example, with regard to the marine record, it could be that transitional populations remained small as they adapted to ecological niches. Then, once they occupied an available niche, they quickly multiplied to fill that niche, and then remained static after that. In this way, their appearance would seem to be instantaneous, punctuating an otherwise static record. With regard to the land record, it is not surprising that transitional forms have not been discovered since this environment is not conducive to fossilization. Land reptiles, birds, and mammals only become fossilized under special and optimal conditions.

After almost 150 years since Darwinian evolution was proposed, the specific mechanisms for forming transitional populations are still not fully understood. However, the "new" genetics presented earlier in this chapter may help unlock this mystery.

Contemplating the Evidence: My Story

The Case of Cave Fish

I am a geologist by training – specifically, a geologist who has specialized in caves for over 40 years. Although I am not a biologist, I am familiar with cave life and the remarkable adaptations that different forms of cave life have made in a complete-

Figure 8-5A. *Above: Orconectes pellucidus,* troglobitic eyeless crayfish, Mammoth Cave, Kentucky. **Figure 8-5B.** *Top right: Amblyopsis spelaeus,* troglobitic fish, Mammoth Cave, Kentucky. **Figure 8-5C.** *Right: Typhlomolge rathbuni,* Texas blind cave salamander.

All three photos by Smithsonian Institution photographer, Chip Clark, deceased; permission granted by his wife Deborah Clark.

ly dark environment. This familiarity with the cave environment has influenced my thinking on the subject of evolution.

Many different kinds of animals are known to be "cave adapted." That is, they have lost their eyes and have become blind, their skin pigment has turned white, and their sensory appendages have grown longer and more tactile (Fig 8-5 A, B, C). This phenomenon is called "regressive evolution" because living in a dark cave environment has caused these animals to lose some of their bodily features and gain others. Animals that have completely lost their pigment and eyes are called *troglobites*, or cave dwellers. In contrast, animals that are infrequent visitors to caves are called *trogloxenes*, while animals that frequently inhabit the twilight zone and which have become partly cave-adapted (e.g., developed small eyes) are called *troglophiles*. Typical cave-adapted troglobites are worms, millipedes, spiders, insects, crickets, shrimp, crayfish (Fig 8-5A), fish (Fig 8-5B), and even amphibians (Fig 8-5C). No troglobitic reptiles or mammals are known from caves; however, they are known from other completely dark environments such as near the bottom of oceans (Fig 8-6) or under the surface of the ground (e.g., blind-mole rats). The mechanism by which regressive evolution occurs has been a subject of controversy from the time blind cave animals were discovered up until the present. Regressive evolution is by no means unique to caves (e.g., loss of flight in birds in the outside world), but it is especially pronounced in this environment.

Caves turn out to be good field locations for testing some of the fundamental assumptions of evolution. This is because caves are simple, isolated

Figure 8-6. A completely dark-adapted, transparent, deep-sea turtle from the Kermadec Trench off of New Zealand. Kermadec Sea Creatures image. *Public Domain.*

environments where selective pressures remain relatively constant (no light, nutrient-poor, constant temperature, and high moisture and relative humidity). Here are some of the notable features of cave-adapted life:

(1) In all known cases, cave-adapted species are clearly related to surface species.

(2) The time that it takes for a complete adaptation to a cave-environment and for new species to emerge can be remarkably short.

(3) Lineages of the same species (for example, *Astyanax mexicanus*) long separated in different caves unconnected to each other may all end up blind, but with different genes having mutated to converge on the same result. This type of *convergent evolution* – where selective environments can cause completely different lineages to converge towards the same kind of adaptations – is ubiquitous in the natural world.

(4) The mode of regressive evolution for life in totally dark caves is almost exactly the same as it is for deep-sea creatures where there is no light (e.g., white, eyeless sharks), and for creatures that burrow underground in the dark (e.g., blind, white mole rats). That is, adaptation to a no-light environment, wherever it may occur, results in the loss of skin pigmentation and eyes, and in the acquisition of increased sensory organs that make up for blindness. Hence, the evolution of adaption to caves is not solely due to the loss of characteristics but also requires the acquisition of novel features that allow for better survival in this challenging environment.

Although a number of different kinds of cave-adapted creatures exist, this discussion will focus on cave fish. Let's trace what happens to a surface-stream fish species as it becomes cave-adapted. When a fish such as *Amblyopsis* regresses, it does so in stages (Table 8-1). The family *Amblyopsidae* is comprised of seven species in five genera, and the whole range of cave adaptation is represented by this one single small family. *Chologaster cornutus* is the outside-world amblyopsid species. It is well-pigmented and there is a longitudinal stripe along each side of the body. Its eyes are normal. As *Amblyopsis* colonizes the twilight zone of a cave, the optic lobe progressively decreases in size and the eyes become reduced, such as is the case for the troglophile *Chologaster agassizi*, who spends time both underground and aboveground. Then, in the three cave-adapted troglobitic species, *Typhlichthys subterraneus*, where the eyes are mere vestiges, and in *Amblyopsis spelaeus* and *Amblyopsis rosae*, where the eyes are completely gone, the ocular system becomes degenerate while the number of sensory papillae increases, and there is a loss of skin pigmentation. The sequence of *Amblyopsis* shown in Table 8-1 represents the amount of regressive evolution with respect to the length of time that generations of *Amblyopsis* have inhabited the cave environment. *Typhlichthys*, the youngest troglobite inhabitant of the three, is normally without pigment but is able to gain pigment back if kept in the light for a few months. *Amblyopsis rosae* is completely cave-adapted, and when taken out of a cave, it will not regain any of its original characteristics. It is a new species, unable to breed with its relatives.

The adaptation sequence illustrated in Table 8-1 documents the *progressive* evolution of one species into another species. Or, in other words, this example shows that "microevolution" *can* lead to macroevolution! If "after its kind" in Genesis 1 refers to separate acts of creation, such acts must be applicable on a genus, family, or higher level than species because, otherwise, cave-adapted species would not always correlate with their surface counterpart species. To me, this series of cave-adapted species is *compelling* evidence that species *do* evolve into other species under the right selection pressures (in this case complete darkness). Why else would there be transitional species from the surface (trogloxenes), into the twilight zone (troglophiles), and then into the completely dark cave zone (troglobites), each transition showing a greater adaptation to darkness?

Another important characteristic of regressive evolution – and evolution in other extreme environments – is that it can happen very rapidly in small populations. For example, populations of the freshwater Mexican cave fish *Astyanax mex-*

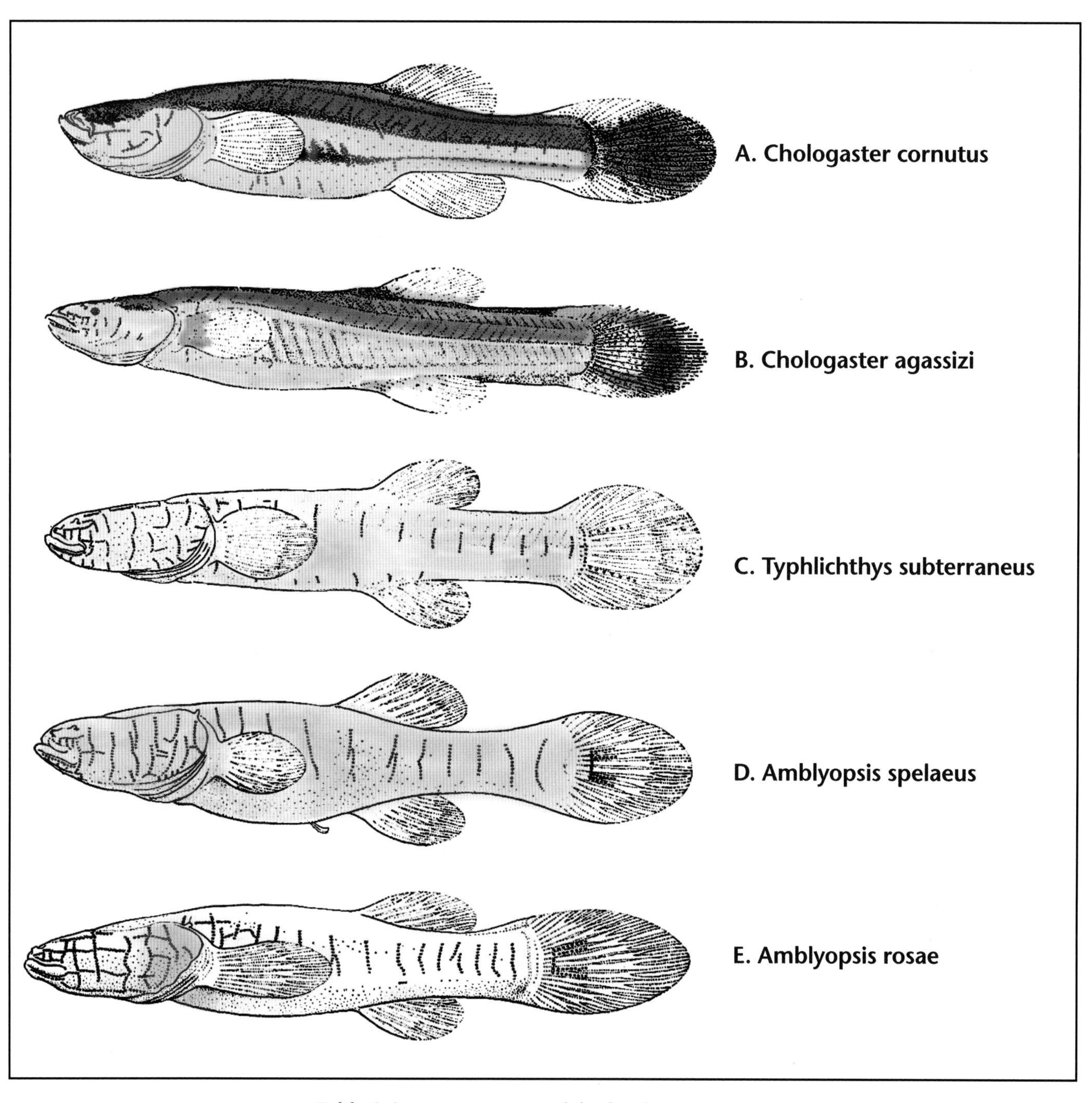

Table 8-1. Representatives of the family *Amblyopsidae*.
Copyright permission from the American Midland Naturalist, University of Notre Dame.

icanus may have lost their pigment and eyesight in 10,000 years or less. Furthermore, it has been found that this process of extensive evolutionary simplification in *mexicanus* has occurred not just once, but independently several times when surface stream populations have become isolated in different caves. Even though it is more likely for a small isolated population to change its genetic makeup faster than a large stable population, the question can still be asked: How can the evolution of species happen so fast if speciation is dependent on purely random mutations?

An even more astounding tale of rapid speciation is the blind, flightless, pigmentless Hawaiian planthopper *Oliarus polyphemus* of the arthropod cixiid genus, which are obligate troglobites (meaning they are confined to life underground) (Fig 8-7). As is common for all cave organisms, surface relatives of cave-adapted *Oliarus* species – which are known to colonize lava flows soon after they cool – are characterized by large compound eyes, long wings, and dark pigmentation. All planthoppers – including underground species – find mates by song, with both sexes singing until they find each other. The difference in songs between the troglobite species *Oliarus polyphemus* in different lava tubes suggests that they are slightly different species which are not related to each other by dispersal between lava tubes through cracks or other means, but species that have regressed quickly and independently within separate lava-tube caves.

The Big Island of Hawaii itself is known to be about one million years old, and, since that time, lava has piled up from northwest to southeast to form Mauna Kea, Mauna Loa, and the latest volcano, Kilauea, where an estimated 90 percent of its surface area has been replaced within the past 1,500 years. The Kazumura lava tube formed on the western slope of Kilauea starting about 300 to 500 years ago, and, if this age is correct, it would mean that *Oliarus polyphemus* has completely regressed from the surface planthopper species in less than about 500 years – an astonishingly rapid rate of evolution that exceeds the highest recorded speciation rate of any taxon known.

Figure 8-7. Planthopper *Oliarus polyphemus* in the Kazumura lava tube, Big Island of Hawaii. Its "tail" is composed of a waxy substance which these troglobites spread over rootlets to form "nests" (the white filaments) on which their eggs are laid. *Photo: Peter and Ann Bosted.*

The most important question to ask regarding regressive evolution and its significance for evolution in general is: Can genetic changes in organisms be *caused by the environment itself*? *Amblyopsis* and *Astyanax* cave fish, and other organisms that live in complete darkness, appear to have lost their pigment and eyes as a *direct response* to the lack of light. However, this whole line of inquiry smacks of Lamarckism – that is, the inheritance of acquired characteristics, where an animal's use or disuse of an organ affects that organ's development in the animal's offspring. Lamarckism has been out of favor for most of the twentieth century, but, in fact, in the latter part of the nineteenth and early twentieth centuries regressive evolution as seen in cave biota played an important role in the rise of neo-Lamarckian ideas. The problem of regressive evolution also puzzled Darwin, who ascribed the loss of eyes in cave creatures as "wholly to disuse."

The intriguing subject of regressive evolution brings up many questions about the process of evolution in general. If eye loss in cave fish is tied to multiple mutations, then do these have to occur simultaneously for the macroevolution of species to occur? How exactly does the *need* to see light facilitate a whole eye to evolve (or in complete darkness to regress to no eye)? Could transposons (jumping genes), phenotypic plasticity, interspecific hybridization, horizontal genome transfer, or epigenetic factors be responsible for both regressive and progressive evolution? If the process of natural genetic engineering were to be reviewed, scrutinized, and controlled at various stages by an "engineer," then could such intelligent feedback produce the complexity seen in life forms?

Christian Views on Evolution

As stated in the first paragraph of this chapter, evolution has been a controversial topic for those who believe in God as well as for those who don't. Not only is it divisional between the secular and Christian worlds, today it is also one of the principal battlegrounds within the church itself. This battle shows no signs of relenting or coming to a consensus. Rather, the division among Christians over the topic of evolution has segregated different church denominations into separate factions.

Before we discuss the various positions on evolution, we need to clarify the meaning of the terms *creationist* and *evolutionist*. Are these terms mutually exclusive? In the broadest sense, a "creationist" is simply someone who believes that the universe has been created; in a teleological sense, it is a person who believes that the world and everything in it was *designed* and exists for a *purpose*. An "evolutionist" can be an atheistic evolutionist who believes that the whole universe formed by *chance* (i.e., *naturalistic evolution)*, or he/she can be a theistic evolutionist who believes that evolution was one way that God used to bring about the universe and life to the point where it is today (*theistic evolution*). The answer to the above question thus depends on if the two terms are broadly or narrowly defined. The terms *are* mutually exclusive to many in the evangelical Christian community. If a person even so much as leans towards being an "evolutionist," that person is automatically considered to be a non-Christian. The same narrow, judgmental view is also true of many scientists. If a person says that he or she is a "creationist," then that person is automatically labeled as a Young-Earth Creationist. In reality, Christians – including evangelical Christians – hold a wide range of positions on evolution, but most of them don't state their position for fear of being ostracized by other church members or of creating division within the church.

We are now going to discuss the four major positions that Christians take on evolution (Box 8-3). Note that three of these positions (Young Earth, Progressive, and Evolutionary Creationism) are positions discussed previously in this book, while the fourth (Intelligent Design) is a teleological position that relates to evolution and, only marginally, to the other subjects we have talked about.

Young-Earth Creationism

The traditional view of Young-Earth Creationists is that "after its kind" in Genesis 1 means that God created each species separately and that no evolutionary connection exists between different species. Rather, there was a "fiat" creation of all species in their present form over the literal six-day period specified in Genesis 1 (page 108). All species in the fossil record were formed during this original six-day creation period, with many becoming extinct during Noah's flood. For some Young-Earth Creationists, this fiat position also prohibits the use of the word "evolution," even in its general sense. Nothing evolves – not the universe or stars or civilizations or life or anything in the natural world. All was created around 6,000 years ago with the *appearance of age* (that is, the creation is not old – it only *looks* old), and any further development or change after that time has been by fiat and not by process.

While Young-Earth Creationists emphatically reject evolution, there is a strong sub-movement within this camp to broaden the scope of the word *kind*. Should it be fixed at the species level or at a higher level? Many Young-Earth Creationist leaders and adherents now opt for a hyper-fast "adaptation" of animals since the time of Noah's flood because the basic animal types carried on Noah's ark had to

Box 8-3. Four Christian Views on Evolution

Young-Earth Creationism	Progressive Creationism	Intelligent Design	Evolutionary Creationism
"After its kind" means each species was instantaneously and separately created by God during the creation week of Genesis 1.	God has intervened in life processes by creating new species over time; then these species remain static until his next intervention.	God used common DNA designs over and over for all of his specially created species.	Evolution by common descent was the mechanism that God used to bring life forms to the point where they are today.

somehow diverge into all of the species we have today in the last 5,000 years or so. Some Young-Earth Creationists even argue that God created "basic kinds" (e.g., "cat kind"), and then allowed for limited evolutionary adaptation within these kinds to produce the many types of cats that we have today (e.g., lions, tigers, cheetahs, domestic cats, etc.). Thus, belief in evolution is disguised by calling it *adaptation*, but it is still new species deriving from old species – although at rates far faster than suggested by any evolutionist.

The biggest objection to the Young-Earth Creationist view on evolution is that it ignores or denies the overwhelming scientific evidence that plants and animals have existed on Earth for millions of years, and that the rock and fossil record does not support a single, worldwide, catastrophic flood (Chapters 6 and 7). Also, this position dismisses the very strong DNA genomics evidence that supports the position that all life on Earth is interconnected anatomically, paleontologically, and genetically, as was discussed earlier in this chapter.

Progressive Creationism

Progressive Creationism is the view that God has intervened in the development of life at various stages, and that, intermittently, there has been the miraculous creation of new species, such as during the Cambrian explosion and after major extinction events. According to this view, God created the first organisms supernaturally, and he has also supernaturally introduced major new types of organisms over geologic time. A somewhat different version of Progressive Creationism is that God has divinely introduced or manipulated genetic material intermittently through time to bring about new life forms.

The Progressive Creationist view of evolution is often held by the Day-Age view advocates discussed in Chapter 2. Its strongest advantage is that it allows for long periods of geologic time in which animals lived and became extinct according to the fossil record. It also attempts to explain stasis and the sudden appearance of animals in the fossil record by attributing their miraculous creation to God; then these types became fixed over subsequent long periods of time. The Progressive Creationist view does not endorse evolution; it favors the miraculous periodic creation of new animal species over geologic time.

Like the Day-Age view itself, this position lacks empirical support and raises lots of questions. Does "new creatures" refer to species, genera, families, orders, or classes? Did these stock species evolutionarily branch into other species? If so, how far do evolutionary processes proceed before new creatures have to again be miraculously created? When and how did God interfere in life's history? By creating the first DNA or the first cell? This view sounds like a variation on a "god-of-the-gaps" theology, where when something can't be explained, God is envisioned as "jumping in" to do a required miracle. But is this unexplainable gap a *gap in our knowledge* or a true *gap in nature* in which God intervenes? The great danger in this theology is that it is always liable to become a casualty of further scientific advances. Over and over in the history of theology, what was once attributed to God has been explained by natural processes.

Intelligent Design

Akin to the Progressive Creationist position of special creation events is another popular view called "common design," which arose out of the Intelligent Design movement. Intelligent Designers reject biological evolution and, like Progressive Creationists, argue that miraculous interventions placed design in living organisms. C. John Collins, in his book *Science and Faith: Friends or Foes?* is one of many authors who have presented a teleological (design) argument for the origin of life, one which challenges the Darwinian view that valid science must appeal *only* to natural causes and assume an unbroken chain of natural causes from beginning to end. In his book, Collins quotes atheist Richard Dawkins, author of *The Blind Watchmaker*, as saying "Biology is the study of complicated things that give the appearance of having been designed for a purpose," which immediately raises the question: If biology gives the *appearance* of being designed for a purpose, should science ignore this evidence?

In further support of Intelligent Design, many of the constants of physics related to cosmology are so finely tuned as to make the universe *appear* to be designed to support life (these appearances, all taken together, are referred to as the *anthropic principle*). For example, if the constants of gravitational force, electromagnetic force, rate of expanding universe, or speed of light were even a tiny bit larger or smaller, it would preclude the universe and life from even being here. Or, as succinctly stated by physicist-engineer Walter Bradley:

> The universe is such a remarkable place of habitation for complex, conscious life that it is extremely difficult to believe that it is the result of a long series of cosmic accidents. The elegant mathematical forms that are encoded in nature, the twenty-two universal constants with values within very narrow ranges of exactly what they need to be, and the multitude of initial conditions that must be within a very narrow bandwidth, which they are, would seem to suggest a universe that has been carefully crafted for our benefit.

Intelligent Design advocates have developed a "common design" argument that goes something like this: Why couldn't a designer (not explicitly stated, but presumably God) have used similar DNA genomics and body structure for different organisms without the evolutionary "common descent" of all species being mandatory? That is, why couldn't God have followed the same "blueprint" and created all life to only *appear* as if it is connected. For example, in the case of vestigial structures that we discussed earlier in this chapter, an intelligent design argument could be: if whales were created through successive generations from the same stock, they could retain some of the same anatomical features adapted for use in different environments.

Yet, what geneticists observe time and again, is that genetic sequences in organisms thought to be close evolutionary relatives match at all genomic levels (which is precisely what one would expect from a common ancestry, since the hypothesis is that similar organisms were once the same species with identical genomes). This also brings up the question: Why is there the overwhelming genomic appearance of a shared ancestry if all species are separately created organisms? Is this logic similar to the "appearance of age" argument used by Young-Earth Creationists for a 6,000 year universe and Earth? In both cases, why would God deceive us by making the data appear one way, but with reality being unrelated to the evidence?

So is Intelligent Design an alternative to evolution? The Bible affirms the reality of Intelligent Design, as in Psalm 19:1 (NIV): "The heavens declare the glory of God, and the skies proclaim the work of his hands." Design is also observed in nature, both cosmologically and biologically. However, Intelligent Designers also claim that design in nature is *scientifically* detectable, and that is where the design argument breaks down. As Denis Lamoureux has

INTELLIGENT FEEDBACK WITHIN EVOLUTION?

The intriguing subject of regressive evolution brings up many key questions about the process of evolution in general. The most important question is: Can genetic changes in organisms be *caused by the environment itself?* If eye loss in cave fish is tied to multiple mutations, then do these have to occur simultaneously for the macroevolution of species to occur? How exactly does the need to see light facilitate a whole eye to evolve (or in complete darkness to regress to no eye)? If the process of natural genetic engineering were to be reviewed, scrutinized, and controlled at various stages by an "engineer," then could such intelligent feedback produce the complexity seen in life forms? Or did God just cause the whole DNA evolutionary program to progress by itself when he spoke all life into being?

pointed out in his book *Evolution: Scripture and Nature Say YES!*, intelligent design is a *belief* that the world's beauty, complexity, and functionality point toward an Intelligent Designer. In other words, science deals only with the physical, not the spiritual, and a religious belief cannot be verified by scientific proof. So, should Intelligent Design be taught in secular science classrooms or only in theology classes as evidence of design by a creator God? Or should students be encouraged to express their opinions and concerns about both evolution and various types of creationism in any classroom setting? In some studies, acknowledging the students religiosity actually increases their acceptance of evolution.

Evolutionary Creationism

The position of Evolutionary Creationism on evolution is aligned very closely with what is known by most Christians as "theistic evolution." This view maintains that not only was God the originator of creation, but he also sustains and upholds it with the word of his power (Heb. 1:3). Rather than being a random, directionless process, evolution has a goal-directed purpose and design behind it that stems from God. Or, as Francis Collins remarked in his book *The Language of God*, the book of the human genome "was written in the DNA language by which God spoke life into being." Evolutionary Creationism is the view favored by most scientists who are Christian, especially Christian biologists and geologists. It is a position rarely held by Protestant evangelicals, but is commonly the view of Roman Catholics and more liberal Protestants.

The main advantage to this theological position is that it satisfies the scientific requirement that all creatures are connected by common descent and the biblical requirement that this descent was by divine plan. In this view, natural processes and divine action are not in competition with each other, but are complimentary. What we call a natural process is not outside of his realm, but rather the universe and all living things within it have slowly evolved under the sovereign control of God's Holy Spirit, in the sense of the first creation account in Genesis where we read, "And the Spirit of God moved upon the face of the waters" (Gen. 1:2, KJV).

God's biological direction includes the evolution of animal life to *Homo sapiens* until such time as humans were capable of a relationship with God. The end result of this evolutionary process was a creature "made in God's image," which includes the idea that humans have the spiritual capacity for communion and fellowship with their Creator (see Chapter 9). The key idea here is that, even though humans evolved over a long period of time, God created a longing for himself as part of their evolving human nature.

The Evolutionary Creationism position sounds reasonable for natural processes, but it also raises important theological questions. The hardest to answer is: What about miracles? Christianity seems to demand miracles and the supernatural – the virgin birth, the incarnate Christ, Christ's resurrection, the miraculous healings of Jesus, and so on; that is, it also involves the *spiritual dimension*, which falls outside the laws of nature.

There are also theological problems with the time dimension of evolution. Did God just start life and let it evolve on its own? And if God let evolution run by itself, doesn't this limit the creative activity of God? Is this view supported by the Bible, which indicates that God has intervened in his creation over time? Do humans have a distinct origin in theistic evolution, or are they but just another "let run" phenomenon? This position also has the weakness of straining Genesis 2 and 3 in terms of Adam and Eve being historical figures. Because populations – not individuals – evolve, one alternative theistic evolution subview of Adam and Eve is that they are only symbols or archetypes of the human race instead of being real historical persons.

Another huge problem with theistic evolution is: How does it actually work? Exactly what is the interaction between God and the macroevolution of species? If God does work through evolution – such as maintained by Michael Behe in his book *The Edge of Evolution* – has God manipulated genetic material in some way impossible for science to detect? Or is this process of God working in evolution a variation of a "nature miracle" such as we discussed at the end of Chapter 6? In nature miracles, one does not have to invoke the notion of the suspension or violation of natural laws; divine action can simply be understood as higher-order laws (God's ultimate purpose) working seamlessly with lower-order laws (God's physical laws). In the examples given in Chapter 6 of the Jordan River parting, Jesus rebuking the winds and sea, and Joshua's long day, it was the *timely* intervention of God into natural processes. But in the case of evolution, could this intervention be a continuous, ever-abiding process? In other words, could evolution be a self-organizing process that God directs from within – feedback that includes small "tweaks" in genes that affect protein changes? Or, as Simon Conway Morris asked: Is the evolutionary process "an emergent property that is wired into the biosphere," with the final result being a sentient species able to comprehend a creator God?

A Personal Perspective

There are good reasons why Christians have traditionally rejected "naturalistic" (chance) evolution. Most important, it is the basis of Christian faith that God created humans – they did not evolve by chance. But there are other reasons why it has also been rejected. One reason is the ubiquitous design feature of the biological world, where plants and animals have seemingly miraculously adapted to their specific environments. And, since examples of these (often amazing) adaptations are based on *scientific observations*, their design features should not be discounted in understanding their underlying origin. (In other words, perhaps it is like the proverbial duck: If the natural world *appears* like it was designed, and if it *functions* like it was designed, maybe it *was* designed.) Then there is the time-factor reason, which was discussed earlier in this chapter; that is, random mutations seem inadequate to explain the very rapid regressive evolution of species such as cave-adapted fish and insects. Most important, the genetic *mechanisms* behind evolution have not yet been fully demonstrated and the ultimate question has not yet been answered: How can the *need* for adaptations to different environments *cause* these to happen so as to *appear* as if they were designed?

In concluding this chapter, I would like to state my personal opinion on evolution. There is very strong scientific evidence in favor of evolution, even though the mechanisms behind it are not yet fully known. This is rapidly changing, however, as modern genetics continues to unravel the dynamics of the genome. In the final analysis, whether one is an Evolutionary Creationist or a Naturalistic Evolutionist is based on *faith*. This is true for people of *both* persuasions because neither position can be proved with scientific certainty. For me personally, it takes more faith to believe that all matter in the universe, including life, came about by pure chance than to believe that God created it and has sustained it throughout all time.

WHY CHRISTIANS REJECT EVOLUTION

There are good reasons why Christians have traditionally rejected "naturalistic" (chance) evolution. Most important, it is the basis of Christian faith that God created humans – they did not evolve by chance. But there are other reasons why it has also been rejected. One reason is the ubiquitous design feature of the biological world, where plants and animals have seemingly miraculously adapted to their specific environments. And since examples of these (often amazing) adaptations are based on *scientific observations,* their design features should not be discounted in understanding their underlying origin. Most important, the genetic *mechanisms* behind evolution have not yet been fully demonstrated and the ultimate question has not yet been answered: How can the need for adaptions to different environments cause these to happen so as to appear as if they were designed?

The Tower of Babel by Brueghel, Jan the Elder ("Velvet"), around 1620. This painting shows the tower ascending into heaven and is typical of the Renaissance style of European artists. The real tower of Babel was the most famous ziggurat of ancient time; its mound still exists in Babylon (Fig 3-3). *Pinacoteca Nazionale, Siena; Art Resources #69565.*

CHAPTER 9

ADAM AND EVE AND ORIGINS

Genesis 3:20. And Adam called his wife's name Eve; because she was the mother of all living. (KJV)

This chapter is about Adam and Eve and Origins. *Origins* means the origin of humankind and embraces both the science of evolution (which we discussed in the last chapter), the science of anthropology, the science of archeology, and the scriptural origin of humans in Genesis. Since pre-Adamites and Adam and Eve fit early in the timeline of Figure 1-4, it would seem reasonable to have covered this chapter where Genesis chronologically puts it: after the six days of creation (Chapter 2) and directly before or after the garden of Eden (Chapter 3). However, since the subject of Adam and Eve and human origins is the most difficult and complicated of the many science-Scripture issues, and since its understanding relies on many of the topics we have covered in preceding chapters, I decided to discuss it last among the eight science-Scripture controversies.

That Adam and Eve are the father and mother of the whole human race has been the traditional view of the church for centuries, and many church "professions of faith" have this belief-requirement in them. Therefore, most evangelical Christians are not about to change their views on this topic. Furthermore, the church has not, in general, had this traditional view challenged by science because it has only been in the last fifty years or so, and especially in the last twenty, that the preponderance of scientific evidence has mounted against it.

It has been my experience that Christians exhibit one of the following attitudes on the subject of human origins, and I, personally, know Christians that fall into each of these categories:

Ignorance: ***People who don't know.*** Many Christians, and perhaps the majority, fall into this category. The scientific evidence is relatively recent, in contrast to the subject of evolution, which has been debated for almost 150 years. Therefore, many Christians are unfamiliar with this subject as an essential science-Scripture issue.

Apathy: ***People who don't care.*** Other Christians are aware that there is scientific evidence for the position that Adam and Eve are not the parents of us all, but they prefer to remain apathetic about it, their mindset being: "If we ignore the evidence, maybe it will go away." So these people choose not to think about the subject. They leave it where it is, in relative obscurity, because it is simpler to do so than to invite controversy.

Denial: ***People who deny the evidence.*** Still other Christians deny the evidence by minimizing it or by discrediting it as being unreliable. I have heard Christians dismiss the anthropological evidence as being "just a few old bones here and there" or "that Piltdown man thing was shown to be a hoax, and if you ask me, all of this early-man stuff is bogus."

Hostility: ***People who are hostile to the evidence.*** Some Christians are hostile to the evidence because it upsets their already-set belief system. They have investigated the evidence (or at least part of it), and this evidence is upsetting to them because they realize that it challenges their theological worldview. They feel backed into a theological corner, and thus they exhibit a threatened response.

Harmonization: ***People who attempt to harmonize the evidence.*** There are relatively few Christians who fall into this category because it is a difficult task – both from a theological perspective and from the psychological risk of being ostracized by the church. Or, as stated by Mark Noll in his book *The Scandal of the Evangelical Mind*: "The issue of the origin of humankind is especially sensitive. It seems that the church is afraid to look into paleoanthropology." Despite these difficulties, a harmonization position is advocated in this chapter. It is sincerely believed that, since this is a major apologetics issue crucial to many people's faith, it is time to squarely face the issue instead of "sweeping it under the rug."

What Is the Scientific Evidence?

Anthropological and Archeological Evidence

Since most Christians have little knowledge of the substantial anthropological and archeological evidence regarding human origins, I will first present an overview of this evidence, as depicted in Table 9-1; only later in this chapter will I cover the scriptural aspects of this subject. Some general comments on this table are:

(1) As you go forward in time, more fossil and artifact evidence becomes available.

(2) Not all of the fossil record is shown on this chart – just the most well-known hominid or *Homo* divisions.

(3) The span of ages in the date column represents anthropological or archeological time periods, and these are not necessarily all-encompassing for the hominid or *Homo* species listed in the far-right column. For example, *Homo erectus* is thought to have lived from about 1,500,000-1,000,000 years ago over a large part of Africa, Europe, and most of Asia, but not to have become extinct in Israel until about 600,000 years ago. *Homo erectus* "Peking man" from the Zhoukoudian, China area is thought to have lived in that vicinity from at least 780,000 to about 400,000 years ago. Another member of the *Homo* genus, the now-extinct *Homo floresiensis* (sometimes called *hobbits* for their small size), may have lived as late as ca. 50,000-60,000 years ago on the Island of Flores, Indonesia. This *Homo* group may be a dwarfed descendant of *Homo erectus* instead of *Homo sapiens* – although this interpretation still remains highly controversial.

(4) The abbreviation YBP = **Y**ears **B**efore **P**resent is used for ages greater than 10,000 years ago, whereas the abbreviation B.C. = years **B**efore **C**hrist is used for ages less than 10,000 years ago. To convert the two ages, simply add 2,000 years to the B.C. age to get the YBP age (for example, 5000 B.C. + 2000 years since Christ = 7000 YBP). The approximate (~) sign means the dates may change as more data becomes known.

Pre-Paleolithic. The fossil record for the Pre-Paleolithic Period is sparse, and you would be correct in saying that "there are just a few old bones here and there." Because the scientific evidence is based mostly on a small number of partial skeletons, interpretations and classifications of the data are, at best, sketchy. It's like trying to visualize the picture on a 1,000 piece jigsaw puzzle when you have only 20 or 30 pieces. You may know where some pieces go relative to other pieces (a blue sky piece goes

Period	Date	*Homo* Species/Artifacts
Pre-Paleolithic		
Hominids	~6,500,000-2,500,000 YBP	*Australopithecus* ("Lucy").
Homo	~2,500,000-1,500,000 YBP	*Homo habilis;* earliest tool maker; flaked tools.
	~1,500,000-400,000 YBP	*Homo erectus;* flaked and chopping tools, fire control. Found in Europe, Israel, Africa, Asia ("Java man," "Peking man").
Paleolithic (paleo = old, lithic = stone)		
Lower Paleolithic	~1,000,000-45,000 YBP	*Homo neanderthalensis* (~650,000-45,000 YBP), *Homo denisovan* (~300,000-50,000 YBP); *Homo naledi* (~300,000 YBP); ritual burials, flint tools, fire, spears, pendants, carvings. *Homo sapiens* located in Africa (~200,000 YBP); stone hand-axes, huts, bone markings, use of ocher, "Mitochondrial Eve," "Y-Chromosome Adam."
Middle Paleolithic	~120,000-45,000 YBP	*Homo sapiens* migrate out of Africa in two waves: a minor one at ~100,000 YBP (Nubian), and a major one at ~60,000 YBP (Fig 9-6); *Homo floresiensis* (~100,000-50,000 YBP).
Upper Paleolithic	~45,000-20,000 YBP	*Homo sapiens* appear abruptly in Europe at ~45,000 YBP *(Cro-Magnon).* Neanderthals coexist and interbreed with *Homo sapiens* in Europe from ~45,000-40,000 YBP. Cave art, sculptures, beadwork, weaving, spears, ritual burials, use of primitive boats; animism and shamanism (?).
Mesolithic (meso = middle, lithic = stone)	~20,000-10,000 YBP	*Homo sapiens:* Natufian, Kebanan cultures in Europe; bow-arrow, cave art, "Venus" figurines. Use/trade of obsidian and bitumen in Middle East. Animism and shamanism.
Neolithic (neo = new, lithic = stone)		
Pre-Pottery	~10,000 YBP-5000 B.C.	*Homo sapiens:* beginnings of agriculture and domestication of animals. Animism, beginnings of polytheism. "Cheddar Man" in Great Britain at ~8000 YBP.
Pottery	~5500-5000 B.C. to present Adam, Cain	Mesopotamian culture; irrigation, first cities, temple building, polytheism; early pottery.
Chalcolithic (chalco = copper; use of copper)	~5000 B.C.-3200 B.C. Tubal-Cain	Metallurgy (copper), city-states, warfare between cities; "Ötzi the Ice Man" in Europe.
Bronze Age (use of bronze)	~3200 B.C.-1200 B.C. Noah (~2900 B.C.) Abraham (~2000 B.C.)	Metallurgy (bronze = copper + tin); boat making; import and export of goods; city-states consolidated into countries.
Iron Age (use of iron)	~1200 B.C.-600 B.C. Solomon-David	Manufacture of iron, larger-scale warfare. Biblical history well founded.

Table 9-1. Chart of anthropological and archeological periods, including where Adam and his descendants fit in time according to Genesis. **YBP** = **Y**ears **B**efore **P**resent, **B.C.** = Years **B**efore **C**hrist.

toward the top of the puzzle, a green grass piece goes toward the bottom), but you don't know how they fit together as a composite whole. Therefore, relationships of hominids to *Homo* are not clear, and "family trees" always seem to be changing.

Paleolithic. More evidence exists starting in the Paleolithic (Old Stone Age), and fossil evidence of the lower Paleolithic species *Homo neanderthalensis* (often spelled and pronounced Neandertal by scientists) has been found all over Europe: they are the brutish-looking "cave men" you see unrealistically portrayed in movies, drawings, and cartoons. Neanderthals inhabited Europe and Asia from around 650,000 to 40,000 years ago. Remains of Neanderthal are almost always found in caves because the cave environment is conducive to preserving human remains and tracks (Fig 9-1). Over 100 Neanderthal sites have been found in Europe, the Near East, and the Middle East, from Portugal on the west to Uzbekistan on the east, and over 500 individuals, ranging from a few isolated teeth to complete skeletons, have been excavated – clearly more than just "a few old bones." Included in the ritual burials of Neanderthal are flint tools and evidence of fire, spears, pendants, and carvings.

Figure 9-1. Neanderthal footprint in Vârtop Cave, Romania. The Vârtop individual lived in Romania sometime before 62,000 years ago, long before the appearance of *Homo sapiens* in Central and Eastern Europe. The earliest records of *Homo sapiens* in Europe date from around 45,000 to 35,000 years ago. *Photo by Bogdan Onac.*

Later in the Paleolithic, many fossils of our species *Homo sapiens* (modern humans) have been found in Africa and elsewhere. In Europe, *Homo sapiens* overlaps in time with Neanderthal, and recent DNA studies – where a significant part of the genome has been analyzed – have shown that Neanderthal and *Homo sapiens* (us) did interbreed and that some non-African human populations still have a small percent of Neanderthal genes in their gene pool. In addition, *Homo sapiens* DNA has also been retrieved from Neanderthal bones. In the Middle Paleolithic, it is important to note that hunter-gatherer tribes inhabited the Persian Gulf area at least by 100,000 yrs ago (Table 9-1), and they continued to intermittently occupy this area as nomads until around 7000 YBP (5000 B.C.), or about the time of Adam and Eve (Fig 3-1), when agriculture begins to replace hunting and gathering.

Homo sapiens in Europe, sometimes referred to as *Cro-Magnon* man, are the humans thought to be responsible for the fantastic cave paintings in Spain and France (Fig 9-2). Over 120 such sites have been discovered, and most of these paintings date from about 35,000 to 12,000 YBP. These ancient Europeans not only drew fabulous cave-art scenes, but they also created such items as figurine sculptures, beadwork, textiles (woven cloth), spears, and bone flutes. Because their cave-wall scenes resemble those of some ancient and modern tribes of Africa who are known to practice animism in the form of shamanism, it is suspected that Cro-Magnon man in Europe may have been practicing some sort of animistic religion by about 30,000 years ago (Table 9-1).

Mesolithic to Early Neolithic. Within the time frame of the Mesolithic (Middle Stone Age) to Early Neolithic (~20,000-10,000) there are hundreds upon hundreds of human-occupied sites, including numerous cultures in Europe, the Near East, and Middle East, but also in North America, South America, Australia, China, and other parts of Asia. Figure 9-3 is a map of sites only in the eastern Mediterranean region.

Chalcolithic (~5000-3200 B.C.). In the Chalcolithic (chalco = copper), the mining, transportation, and metallurgical working of copper ore began in the Middle East and Europe at about 5000 B.C. or a little before (Table 9-1). This was also the time when larger city-states arose in the region, and when foreign trade relations began to range far and wide. Raw materials were acquired from all over the Near East and Middle East, and objects, techniques, and artistic styles of various origins began flowing into Chalcolithic settlements. The growing and processing of olives and olive oil also began in the Chalcolithic. This was the time period when the famous "Ötzi the Iceman" lived in the mountainous border between Italy and Austria 5,250 years ago (Fig 9-4); Ötzi carried a copper axe, which probably places him near the beginning of the Chalcolithic.

It is into a Neolithic-Chalcolithic time frame that Genesis places Adam and Eve and Cain and Abel. How do we know that Genesis places them there? Because Genesis 4:2 says so: "Now Abel kept flocks, and Cain worked the soil" (NIV), which implies that agriculture and the domestication of animals were already the main means of livelihood in southern Mesopotamia (Table 9-1). It also says in Genesis 4:17 that "Cain was then building a city, and he named it after his son Enoch" (NIV), implying that the building of the first cities in Mesopotamia had also begun around this time. Genesis also places Tubal-Cain near the end of the Chalcolithic – or perhaps closer to the beginning of the Bronze Age, because in Genesis 4:22 (KJV) it says that "Tubal-cain was an instructor of every artificer [craftsman] in bronze and iron." Sometimes "bronze" is translated as "copper," so the time of Tubal-cain

Figure 9-2. Great Hall of the Bulls, Grotte de Lascaux, Vézère, France. In animistic/shamanistic cultures, natural objects and phenomenon are thought to possess souls. Many early cultures, such as those in Africa and Australia, depicted animals on rock walls as part of their religious activities. Since these paintings in France date up to about 35,000 years ago, it is thought that shamanism may have existed in parts of Europe by that time. *Wikimedia Commons.*

Figure 9-3. Seventy-eight Early Neolithic Natufian sites in the eastern Mediterranean area, ~12,000 to 10,000 years ago, by which time the Natufians were already cultivating wild wheat and barley. Within the red circled areas, many of the less prominent sites are not named. This map shows that people groups inhabited the Near East long before 6,000 years ago, the age of the Earth claimed by Young-Earth Creationists. *Modified from James Mellaart,* The Neolithic of the Near East, *Thames and Hudson, London (1975).*

Figure 9-4. Reconstruction of Ötzi the Iceman, showing his clothes, copper axe, and bow and arrows; also note his clothes of animal hides and plant fibers. He also has the oldest known tattoos on his body (not shown). Ötzi was roughly contemporaneous with Adam (around 5000 B.C.; Tables 3-1 and 9-1), but his world in Europe was primarily one of hunting and gathering, not one of agriculture and husbandry. Ötzi's body and his belongings are on display in the South Tyrol Museum of Archeology in Bolzano, northern Italy. *Google Image.*

WHEN DID ADAM AND EVE LIVE?

It is very important that the Genesis chronologies place Adam and Eve at approximately the same time as does the anthropological and archeological evidence (Table 9-1). As discussed in Chapter 4, gaps may exist in the Genesis chronologies so that these chronologies cannot be used as an absolute time scale. However, these gaps probably do not represent more than a few hundred years of time (at the most), and so these chronologies roughly place Adam in a time frame just before the Chalcolithic or just after it began (~5500-4000 B.C.; Tables 3-1 and 9-1). While the ages of the patriarchs back to Adam may represent exaggerated numerological numbers, clan ages, and/or abridgement ages, the biblical authors/scribes still seem to have had an approximate idea of how long before them their forebearers Adam and Eve and their descendants lived. This is further evidence for the historical framework of the Genesis account.

cannot be absolutely fixed, but most probably he lived around 3200 B.C. when iron working was just beginning. All of the occupations mentioned in Genesis 4:20-22 are associated with a civilized mode of life, and not with a hunter-gatherer society where stone tools were used. Furthermore, pottery has been found at the earliest southern Mesopotamian archeological sites, Tell el-Oueili (Fig 9-5) and Eridu, near to where the garden of Eden was supposedly located (Figs 3-3, 3-4), so this places these sites in the Late Neolithic (Table 9-1). Therefore, these descendants of Adam could *not* have lived before the Neolithic in the Paleolithic/Mesolithic, as has been proposed by some Progressive Creationists.

It is very important that the Genesis chronologies place Adam and Eve at approximately the same time as does the anthropological and archeological evidence (Table 9-1). As discussed in Chapter 4, gaps may exist in the Genesis genealogies so that these chronologies cannot be use as an absolute time scale. However, these gaps probably do not represent more than a few hundred years of time (at the most), and so these chronologies roughly place Adam in a time frame just before the Chalcolithic or just after it began (5500-4000 B.C.; Tables 9-1 and 9-2). While the ages of the patriarchs back to Adam may represent exaggerated ages, clan ages, and/or abridgement ages, the biblical authors/scribes still seem to have had an approximate idea of how long before them their forbearers Adam and Eve and the descendants of those two lived.

Bronze Age (~3200-1200 B.C.). Bronze is a metal composed of copper mixed with small amounts of tin – the tin making the copper harder and more durable. It was at this time in the Middle East that boat making, warfare between city-states, and the import and export of goods began on a large scale. As discussed in Chapter 5, Genesis places Noah and the flood in this time frame, around 2900 B.C., and it also places Noah's descendents, including Abraham, Isaac, Jacob, and Joseph, in the Bronze Age. Moses and the Exodus also fit into the Bronze Age, with the Exodus probably occurring around 1260 B.C. during the reign of Ramesses the Great (Fig 1-3; also see Chapter 10).

Iron Age (~1200-600 B.C.). The Bible documents the Iron Age with great precision, as this was the time of Solomon and David. Biblical history is well founded during and after this time.

Archeological Period	Archeological Assigned Age	C-14 dates (calibrated)	Biblical Person/Event
Ubaid	~5500-3800 B.C.	ca. 6000-4000 B.C.	Eridu, Adam and Eve?
Uruk	~3800-3100 B.C.	ca. 4000-3350 B.C.	Tubal-Cain, Jabal, Jubal?
Jemdet Nasr	~3100-2900 B.C.	3350-2960 B.C.	Shuruppak, Noah and flood?
Early Dynastic I	~2900-2750 B.C.	2960-2760 B.C.	Nimrod?
Early Dynastic II	~2750-2600 B.C.	2760-2655 B.C.	Tower of Babel?
Early Dynastic III	~2600-2350 B.C.	2655-2260 B.C.	
Dynasty of Akkad	~2350-2150 B.C.		
Third Dynasty of Ur	~2150-2000 B.C.		
Old Babylonian	~2000-1600 B.C.		Abraham = ~2000 B.C. Joseph = 1800 B.C.

Table 9-2. Archeological periods of Mesopotamia and their possible correlation with people, places, and events in Genesis. If Adam lived in southern Mesopotamia, where Genesis says the garden of Eden was located, it would have been at the beginning of the Ubaid Period, since that is the earliest archeological period identified for that area. The radiocarbon (calibrated C-14) dates are from a variety of sources. *(Table 3-1 is repeated here as Table 9-2 for the reader's convenience.)*

Archeological Evidence from Southern Mesopotamia

Genesis places Adam and Eve within the geographical and archeological framework of Mesopotamia – specifically with respect to southern Mesopotamia, where the garden of Eden was located (Chapter 3). Scholars have broken up the ancient history of this region into nine periods, and Table 9-2 shows how some of the main people and events in Genesis might fit within these archeological periods. (Note: Table 3-1 is repeated here as Table 9-2 for the reader's convenience).

The Ubaid Period records the first agricultural (non-nomad) occupation of southern Mesopotamia. Tell e-'Oueili, Eridu, and al'Ubaid (now all mounds; Fig 3-3) are the oldest cities of this time period (~6000-5000 B.C.), and these cities exhibit the following characteristics:

(1) A fully developed agricultural way of life existed without local antecedents, and this agriculture was combined with fishing and the herding of domesticated animals. Stone hoes and clay sickles (Fig 9-5) are common artifacts found at these early sites.

(2) No Pre-Pottery Neolithic settlements have been discovered in southern Mesopotamia (Fig 9-1), and a fully developed style of pottery has been found even at the earliest Ubaid levels; for example, at Tell e-'Oueili (Fig 9-5) and in the lowest levels excavated at Eridu and Ur (called "Eridu-ware" pottery). Even later, the red and gray "Uruk ware" signified the arrival of the Sumerians. Eridu-ware pottery has not been traced to any other known ancient pottery type in the region, although some similarities have been noted between the coarse Ubaid ceramics at Eridu and the earlier Hassuna and Samarra cultures to the north.

Figure 9-5. Painted pottery shards found at Tell el-'Oueili, considered to be one of the oldest settlements in southern Mesopotamia (ca. 6000-5500 B.C.). Middle curved artifacts are sickles made of clay. *Google image.*

(3) Irrigation agriculture was based on canal construction. It was these canals, along which water was diverted from the Euphrates and Tigris Rivers, that allowed the early civilization of southern Mesopotamia to flourish.

(4) The first ziggurat temples were constructed in the Ubaid Period, with bitumen (pitch) used in the construction of these temples and other buildings, and in the caulking of boats.

(5) A fully developed language (Sumerian) was spoken, a "language isolate" that is not known to be related to any other language. The Sumerians (southern Mesopotamians) were even linguistically distinct from the Semitic Akkadians who lived in northern Mesopotamia.

All of these characteristics suggest a culture that arose "full-blown." That is, the culture did not stem directly from a nomad, hunting, or subsistence agricultural society, but from a more highly sophisticated culture surrounding the Mesopotamian area. However, what culture that might have been still remains a mystery. As mentioned in Chapter 3, Eridu is the most likely candidate for being located near the garden of Eden, both from the locality specified by Genesis 2:10-14 and from cuneiform texts, which state that the descent of kingship from heaven was supposedly to the first city founded in Mesopotamia; that is, Eridu. It could be speculated, from a biblical point of view, that Adam and Eve could have founded Eridu after being dispelled from the nearby garden of Eden.

DNA Evidence of Human (Homo sapiens) *Populations*

In the past twenty years or so, the science of DNA molecular biology has advanced to the point where scientists are now able to genetically determine which human populations are most closely related to other humans, and where those populations most likely came from and when. Since this science is probably unfamiliar to most readers, some terms will be briefly defined here.

DNA. Deoxyribonucleic acid is an extremely long nucleic acid molecule that is the main constituent of chromosomes and that carries genes as segments along its strands. Each of the forty-six chromosomes contains a single double-stranded DNA molecule composed of different sequences of the letters A, C, G, and T.

MtDNA (Mitochondrial DNA). Unlike the DNA in the chromosomes of the cell's nucleus that is inherited from both parents, mitochondria are inherited from only one parent – the mother. MtDNA genetic information is passed down through the female line, but not through the male line. After tracing back genetic information several thousand generations, geneticists have reached a woman referred to as "Mitochondrial Eve," who represents the maternal root of the family tree of all modern humans.

Y-DNA. Y-DNA is genetic information that is passed down only through the male line. The male-transmitted DNA carries many more nucleotides than mitochondrial DNA, enhancing the ability to distinguish one human population from another. After tracing back genetic information for hundreds of generations, one comes to "Y Chromosome Adam," representing the male root of human populations.

Mutations. Mutations are "glitches" in the DNA record – inherited markers that allow the ancestry of organisms to be traced back in time. Most mutations in mitochondria occur in a short stretch of DNA called "the control region," thus making it relatively "easy" to trace genetic changes over time among closely related organisms (for example, within species).

DNA Clock. The "DNA clock" is based on the hypothesis that random, neutral genetic mutations accumulate at a steady rate. However, neutral mutation rates have been found to vary among different lines of descent and between different times. These rates can only be defined in certain cases, contrary to radioactive decay rates, which are known to be invariant. In addition, as covered in the last chapter on evolution (Chapter 8), random neutral mutation studies do not take into account the new genetics where mutations may be caused by epigenetic factors.

Are Mitochondrial Eve and Y Chromosome Adam the Biblical Adam and Eve?

Some DNA studies show that Mitochondrial Eve and Y Chromosome Adam were roughly contemporaneous (~150,000 YBP), but that does not mean that these persons necessarily lived at the same time or same place. Also, early humans lived in groups where populations never dropped below a few thousand; therefore, they are *not* equivalent to the biblical Adam and Eve and should not be considered as such! For non-geneticists, tracing our human origins back to a single mother *and* to a larger contemporaneous population at the same time may seem contradictory, but actually it is quite plausible. The explanation is that Mitochondrial Eve's offspring mated with members of the larger population present at the time, but within subsequent generations, only Mitochondrial Eve produced an unbroken line of *daughters*. Lineages from other females living at the same time as Mitochondrial Eve eventually passed through a generation of all male offspring (or no offspring). The same logic applies to Y Chromosome Adam: only one male's lineage produced an unbroken line of *sons* leading up to the present.

When Did Ancient Homo sapiens *Live?*

How do scientists know when ancient *Homo sapiens* and other *Homo* species lived? They do this in two ways: (1) by the radiometric dating of human bones and artifacts found in rocks within or surrounding the area of human occupation (e.g., volcanic ash), or those found within caves, sinkholes, and cave formations (speleothems) in which the bones are encased; and (2) by using mtDNA and Y-DNA clocks.

It is significant that these two completely different types of dating methods roughly agree. From the fossil record, the approximate 150,000-year-date for Mitochondrial Eve and Y Chromosome Adam seems reasonable, as *Homo sapiens* (modern human) fossils from Ethiopia have been found to date back to about 195,000 YBP, or even before.

The science of DNA and the DNA clock method just described seems suspect to many Christians, and it is easy to pass this off as just another crazy theory perpetuated by secular scientists to disprove the Bible. However, before this genetics approach is discarded, it should be mentioned that this DNA technique has also been applied to biblical persons. For example, DNA testing of Arabs and Jews has linked both groups to a common male ancestor several thousand years ago. While the exact DNA date of this common ancestor has not been determined, it could be that future DNA studies will confirm the biblical record of a common ancestor for the Arabs and Jews around 4,000 years ago (Genesis says that it was Abraham). Furthermore, other Y chromosome studies have confirmed a distinct paternal genealogy for Jewish priests, as claimed by the Bible in Exodus 40:12-16: "Bring Aaron and his sons to the entrance to the Tent of Meeting and wash them with water... their anointing will be to a priesthood that will continue for all generations to come" (NIV). Such DNA studies support the biblical record and, thus, this method should not be labeled as unreliable when applied to other human populations.

Migration of Humans around the World

From all of this anthropological, archeological, and DNA evidence, it now appears that at around 150,000 years ago, there was a Mitochondrial Eve who was the "mother" of all humans living today. This woman lived in northeastern Africa along with a small tribe of ancient humans. The earliest, but relatively minor, migration wave of humans out of Africa and into the Middle East is believed to have occurred about 100,000 years ago. Then, a second, major wave of a more extended human migration from both within and out of Africa occurred about 70,000 to 50,000 years ago, which gave rise to all non-African *Homo sapiens* populations alive today (Fig 9-6). A date of approximately 50,000 YBP is also significant in that there seems to have been a burst of creativity in human-occupied sites at this time. Instead of finding only primitive stone tools, archeologists begin to find cave art, beadwork, sculptured figurines, and the beginning signs of animistic beliefs.

We will now trace the migration of humans around the world (Fig 9-6). This exercise may seem a bit tedious to the reader, but it is extremely important for the apologetics of Origins because this overwhelming amount of evidence confirms that different human groups inhabited almost the entire planet Earth long before 6000 B.C., long before Adam and Eve lived in the Mesopotamian region, and long before the supposed dispersion of the human race after Noah's flood. But before we start, three points need to be made:

(1) From the DNA tracing of thousands of different people groups around the word, it has been conclusively shown that all humankind belongs to the same species, *Homo sapiens*, with but a very minor genetic input from Neanderthals.

(2) The DNA genetic data supports the conclusion that modern humans originated in Africa, and if one is looking for "Eden" based on a Paleolithic Adam and Eve, it will have to be in Africa. Time- and place-wise, Mitochondrial Eve and Y Chromosome Adam cannot be identified with the biblical Adam and Eve, nor do they prove their existence. According to Genesis and archeological radiocarbon dates (Table 3-1), Adam and Eve lived in southern Mesopotamia around 8,000 years ago (~6000 B.C.), not 200,000 to 50,000 years ago in Africa (Table 9-1).

Figure 9-6. *Facing page:* The *main* (~60,000 YBP) migration of humans throughout the world, starting in eastern Africa with the !Kung people (black star). The gray denotes the extent of the Table of Nations; i.e., the area where the descendants of Noah migrated after the flood (Genesis 10). All the dates are **Y**ears **B**efore **P**resent **(YBP)**. ***B*** =**B**asque.

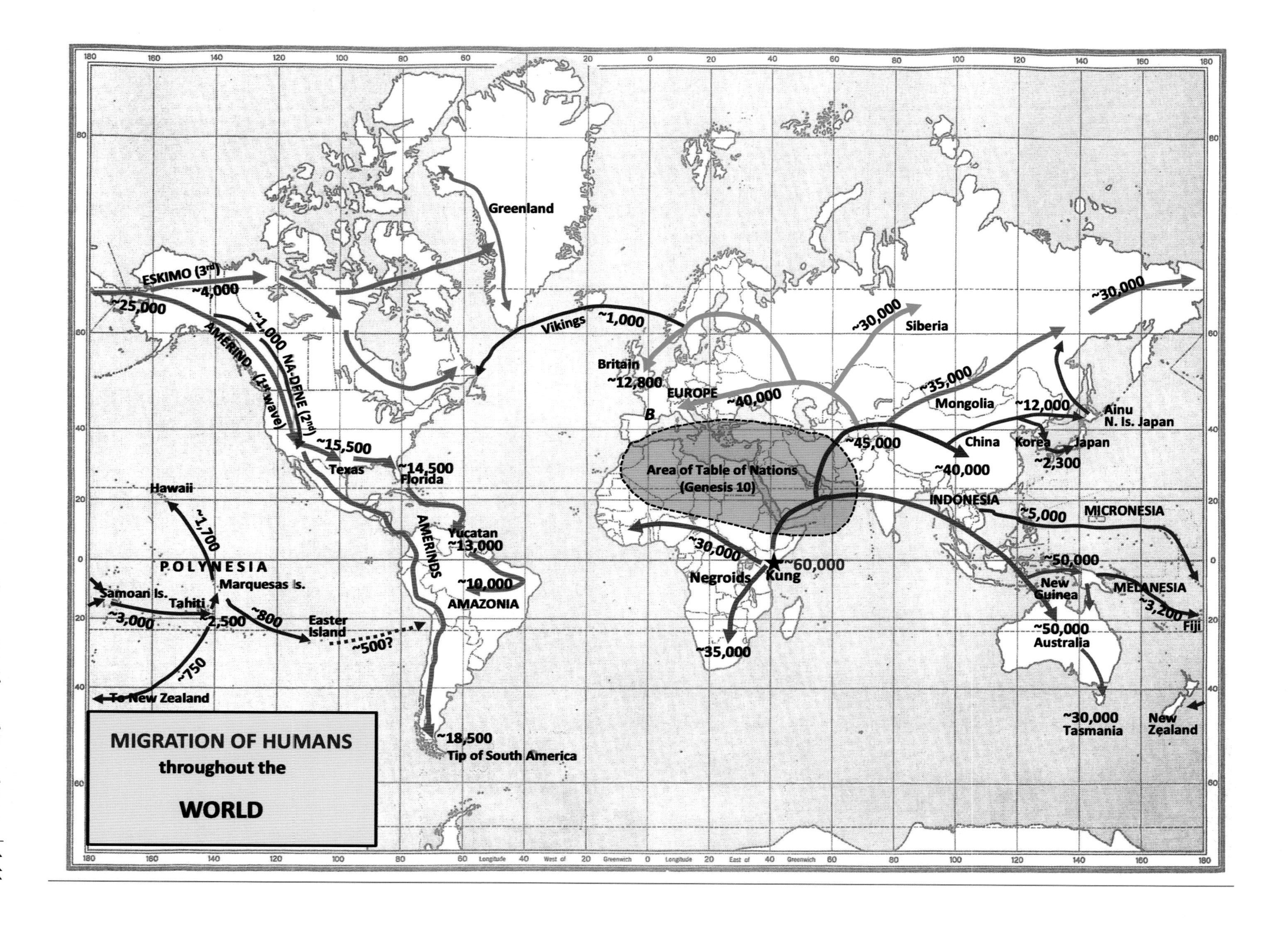
MIGRATION OF HUMANS
throughout the
WORLD
Greenland
ESKIMO (3rd)
~4,000
~25,000
AMERIND (1st wave)
~1,000 NA-DENE (2nd)
~15,500
Texas
~14,500
Florida
Vikings
~1,000
Britain
~12,800
EUROPE
B
~40,000
~30,000
Siberia
~35,000
Mongolia
~12,000
Ainu
N. Is. Japan
~45,000
China
Korea
Japan
~2,300
~40,000
Area of Table of Nations
(Genesis 10)
INDONESIA
~5,000
MICRONESIA
~30,000
~60,000
Kung
Negroids
~35,000
~50,000
New
Guinea
MELANESIA
~3,200
Fiji
~50,000
Australia
~30,000
Tasmania
New
Zealand
Hawaii
~1,700
POLYNESIA
Marquesas Is.
Samoan Is.
Tahiti
~3,000
2,500
~800
Easter
Island
~500?
~750
To New Zealand
AMERINDS
Yucatan
~13,000
~10,000
AMAZONIA
~18,500
Tip of South America

(3) DNA genomic data is not the only evidence on which this migration chart is based; it is also supported by the anthropological, archeological, and linguistic evidence. For example, over 50 years ago, linguists hypothesized that there were three separate waves of migration into the Americas based on the different languages of these people groups (Fig 9-6). However, archeological and linguistic evidence, by its very nature, can only hint at the past, whereas DNA genetics is the most important and confirming line of evidence.

(4) Similar to Table 9-1, the (~) signs in Figure 9-6 denote that these dates are approximate. The acquisition of DNA evidence for human populations is relatively recent, and each new finding can alter these dates. It is becoming apparent, however, for both Table 9-1 and Figure 9-6, these dates will probably not decrease – they will only increase.

Let us now follow the human migration map of Figure 9-6. According to Bryan Sykes, author of *The Seven Daughters of Eve*, the DNA trail for humans has been traced back to the !Kung (Khoisan, or San Bushmen) people of southern Africa. The !Kung people no longer live in the eastern region of Africa because they were pushed into the more arid regions of southern Africa during the expansion of the Bantu farmer people around 1000 B.C.-A.D. 1000 However, DNA tracing suggests that the !Kung are genetically the ancestors of all other human populations. The !Kung people use click sounds in their language (! = click), and some linguists have speculated that this click-type speech may be the remnant of a very ancient language used by early humans.

The first, relatively minor, wave of humans that left Africa arrived in the Middle East at about 110,000 YBP at a time when North Africa's vast Sahara Desert contained rivers and lakes; then *Homo sapiens* may have spread east as far as China by approximately 80,000 years ago. The second, *main* migration wave from Africa settled in the Middle East around 60,000 to 50,000 YBP (Table 9-1); interbreeding with Neanderthals probably occurred only during this second wave of *Homo sapiens* from Africa (~55,000 to 40,000 YBP). From there, one segment of the *Homo sapiens* population migrated west and northwest into Europe about 40,000 to 35,000 YBP, carrying their Neanderthal genes and tool-making techniques with them. It is thought that the Basque people in Spain and France may be part of this first ancient people group that reached Europe because they are genetically and linguistically distinct from surrounding European populations (Fig 9-6, ***B***); that is, the Basque language is unrelated to other languages in Europe or anywhere else. Humans reached the British Isles at least by 12,800 to 12,600 years ago.

Other *Homo sapiens* groups who left northeastern Africa about 60,000 YBP (Fig 9-6, black star) also crossed the Red Sea and skirted along the coasts of Arabia, India, and Indonesia during the last Ice Age when the sea level is known to have been much lower than it is today. Generations of these migratory peoples then made their way to New Guinea and Australia about 50,000 years ago, and into Tasmania about 30,000 years ago. This correlates with the fact that the Australian aborigines, New Guineans, and some Indian people groups look more alike and are more closely related DNA-wise to Africans than to other populations. But even at this early stage, the people who migrated to Australia and New Guinea may have "island hopped" there by boat (rafts, canoes?) because it is known that the continental shelf between Indonesia and Australia was then partly exposed due to an episode of low sea level.

From the Middle East, a later segment of humans migrated into Siberia, China, and Mongolia, reaching these locations at about 40,000 to 30,000 YBP. At a later date (~12,000 YBP), part of this population migrated to Japan (the Ainu), and then, much later, a larger population moved from Korea into Japan, replacing the ancient Ainu people except in the northernmost island.

From northeast Siberia, nomad Mongolian tribes, following the migration of reindeer and other animals, moved across the Bering Straight into North America beginning about 25,000 YBP during the Last Glacial Maximum when sea level was again low. This first major wave of migration into the Americas is called the *Amerind* migration, after the linguistic group of that name, and most of the Native American tribal groups of North America – and all of those in South America – fall into this group (Fig 9-6). Recent evidence suggests that these earliest Native Americans traveled south along the Pacific Coast, probably by boat, reaching the southern tip of South America by about 18,500 YBP. Then, these New World peoples spread eastward across North America, reaching the Texas area by about 15,500 YBP, Florida by about 14,500 YBP, Central America by about 14,000 YBP, Yucatan by about 13,000 YBP, and Amazonia by about 10,000 YBP or perhaps even earlier.

The second major wave of migration of Native Americans was from Mongolia into Alaska and Canada at about 12,000 YBP, and from there into the western United States (Arizona, New Mexico) at about 1,000 YBP. These are the *Na-Dene* peoples (the Navajos and Apaches). Na-Dene people do not speak the same language as the earlier Amerind tribes, except for some common root words traceable to Mongolia. For example, the Hopi, descended from the ancient Anasasi (Amerind) people, and Navajo (Na-Dene) people do not speak the same language even though they live in close proximity to each other in the southwestern United States.

The Aleut-Eskimos are the third and most recent immigrants into the Americas and Greenland (gray arrows, Fig 9-6). They first settled in North America about 5,000 years ago, with a second wave appearing about 4,000 years ago. However, all of these three major migrations of native peoples into the Americas have basically the same origin: they derive from Mongolian tribes in northeast Asia. That's what the DNA genetic tracing of these populations shows, and that is who these people resemble.

The last major wave of human migration was into the islands of the Pacific. Melanesians (darker-skinned people who had reached New Guinea approximately 50,000 YBP) migrated to Fiji by about 3,200 years ago (Fig 9-6). The Micronesians reached Samoa from Indonesia at about 3000 years ago, and then spread to Tahiti ~2,500 YBP, and then throughout Polynesia in all directions by canoe, reaching Hawaii by about 1,700 YBP (Native Hawaiians), Rapa Nui (Easter Island) by about 800 YBP ago (Rapa Nuis), and lastly New Zealand by about 750 YBP (Maoris). Amazingly, some recent studies suggest that the Easter Island Polynesians may have sailed back and forth to South America sometime around 500-800 YBP (~A.D. 1200-1500) – but this hypothesis is still controversial.

What about the Table of Nations and Tower of Babel?

If the above migration scenario is correct, then what about the Table of Nations described in Genesis 10 and the Tower of Babel described in Genesis 11 (page 130)? Doesn't the Table of Nations imply that all of the world's peoples are descended from Noah? And doesn't the Tower of Babel story say that Noah's descendants built a tower that reached to the heavens, so that then the Lord "scattered them abroad upon the face of all the earth" (KJV)? How does the DNA human migration map of Figure 9-6 fit with these two Genesis chapters?

This line of research is called *ethnology*, which is the branch of anthropology that deals with racial origins and distributions. The Table of Nations traces the lineages of Noah's sons Shem, Ham, and Japheth. Biblical scholars such as Speiser, Wenham, Hamilton, and Kitchen (and many others) have studied all three of these people groups mentioned in the Table of Nations. These regions correspond to the sons of Noah as follows (refer to the gray area of Fig 9-6 that roughly surrounds the Mediterranean-Near East area):

Japheth. The ancient nations descended from Japheth are Anatolia (western Turkey), Ionians (Greece, Crete, Italy?, Spain?), and the Aegean (Cyprus, Rhodes, Sardinia); in other words, the northern Mediterranean region.

Ham. The ancient nations descended from Ham are Cush (Ethiopia), Mizraim (Egypt), Put (Libya), Canaan (Phoenicia, Palestine, Lebanon); in other words, northern Africa and the Palestinian region.

Shem. The ancient nations descended from Shem are Joktan (various Arabian tribes such as Ophir and Havilah), Ashur (Assyria), Aram (Syria),

Elam (Iran), Lud (eastern Turkey), and Uz (Arabian Peninsula); in other words, the Arabic nations.

The last two regions speak what linguists refer to as "Semitic-Hamitic languages," where "Semitic" refers to the languages of peoples descended from Shem, and "Hamitic" refers to the languages of peoples descended from Ham. These languages include Egyptian, Hebrew, Arabic, and Aramaic, among others, plus a number of extinct languages such as Babylonian.

Linguistic and ethnological studies by biblical and other scholars have shown that Genesis is correct in its tracing of the sons of Noah – but *not* over the entire planet Earth! Geographically, the Table of Nations list goes as far east as Persia (Iran), as far south as Ethiopia and the Arabian Peninsula, as far north as Anatolia (Turkey), and as far west as Crete, Spain?, and Libya. By the time of the Tower of Babel (~2700-2600 B.C.; Table 9-2) the major human racial groups that we know of today had *already* spread around the world (Fig 9-6).

The Genesis 11:1 passage "And the whole earth was of one language, and one speech" (KJV), in connection with the Table of Nations and Tower of Babel, may refer to the ancient Sumerian language, with "the whole earth" referring to the Mesopotamian alluvial plain, as it did for Noah's flood (see Chapter 6). By about 2700-2600 B.C., the Sumerian language was gradually being replaced by Early Semitic and Old Akkadian dialects, and by about 2400 B.C., it been entirely replaced as a "living language," but still remained as a written language. A date of around 2700-2600 B.C. for the dispersion of languages story in Genesis 11 also correlates in time with other ancient Sumerian stories on the same topic. The Tower of Babel story may, thus, possibly be a different version of a tradition shared by Genesis 11 and other early Mesopotamian cultures that were also experiencing the same language change (thus the use of the phrase *the whole earth*, meaning the whole *land* of Mesopotamia).

All of the evidence points to the same conclusion concerning the Table of Nations: the Table of Nations is a record of the dispersion of the sons of Noah after the flood that pertains *only* to the covenant line of Adam and *not* to all of the nations of the world. The intent of the Table of Nations in Genesis 10 was to show the relative kinship of all the *known*

TABLE OF "ALL" NATIONS ON EARTH?

All of the evidence points to the same conclusion concerning the Table of Nations of Genesis 10: the Table of Nations is a record of the dispersion of the sons of Noah after the flood that pertains to the relative kinship of the *then-known* nations of the world to the people descended from Noah's sons, who were to *become* the ancestors of what would later become Israel. This is a very important point, not to be missed: the Old Testament is a story of covenants with Adam's line only. This story does *not* include (or only marginally includes) the rest of humankind. Also, the DNA, archeological, and paleontological evidence does not support the position of Young-Earth Creationists/flood geologists that the "all flesh" (all animals and humans) of Gen. 7:21-22 (KJV) died in a global flood except for Noah's family and the animals on the ark, and that only these eight people and animal species migrated from Mount Ararat over the entire planet Earth during the last 5,000 years or so since Noah's flood. Rather, the migration of humans happened long before this time over the face of planet Earth (Fig 9-1), journeys in which they encountered many indigenous species of animals.

nations of the world to the people groups who were descended from Noah's sons, who *were to become* the ancestors of what would later become Israel. This is a very important point, not to be missed: the Old Testament is a story of covenants *with Adam's line only*. This story does not include (or marginally includes) the rest of humankind. (We will elaborate on this theme later in this chapter and in Chapter 10.) The DNA and archeological evidence also does *not* support the position of Young-Earth Creationists/flood geologists that the migration of Noah's descendents (supposedly the only humans to survive a global flood) from Mount Ararat over the entire planet Earth happened during the last 5,000 years or so.

This conclusion of a limited extent of the Table of Nations is also further confirmation of the discussion we had in Chapter 6 on Noah's flood being a local flood covering only the region of Mesopotamia. "All," as used in Genesis 6-8, refers to the *known* world of Noah's time – not to the whole planet Earth. From the time of Noah's flood (~2900 B.C.) to the Table of Nations (~2700 B.C.), the known world of the biblical author(s) had expanded somewhat, but probably still revolved mainly around Mesopotamia, even though there was more contact with other surrounding nations by that time. The Genesis 10 text was probably written down in the patriarchal period (early second millennium) when narrative writing began, the text coming as a tradition with Abraham journeying northwestward from Mesopotamia. By this time (~2000 B.C.), the known world was nearly the whole Mediterranean and Near Eastern region, and, therefore, these are the regions mentioned in Genesis 11 as being occupied by the descendants of Noah's three sons.

The point of this discussion is that the Genesis 6-8 and Genesis 11 stories record different geographical extents for "the whole earth" because what was *known* by the biblical authors/scribes about the extent of planet Earth had expanded over this time period. Similarly, the "uttermost part of the earth" of Acts 1:8 (KJV) refers to the known Earth of around A.D. 60. Eventually over time, such phrases came to represent *all* of planet Earth (e.g., the gospel *shall* be preached to *all* nations; Matt. 24:14; Lk. 24:47) – that is, this directive was to be accomplished in the *future*, and so would someday include all nations. From a Worldview Approach, this past and future distinction is important. The Young-Earth Creationist and Progressive Creationist positions both interpret Genesis phrases such as "all the earth" from a twenty-first century perspective of planet Earth, whereas a Worldview Approach interprets such phrases from the ever-increasing (over time) geographical and scientific knowledge base of the biblical authors/scribes, and then of future generations of people (like us) reading Scripture.

Cheddar Man or Adam?

All of the anthropological and archeological evidence confirms that Adam and Eve were *not* the first people to live on planet Earth, and only some people like the Arabs and Jews can trace their ancestry directly back to Adam. To illustrate the problem that many of us have with connecting our ancestry to a historical Adam who lived about 5000 B.C., I will trace my lineage based on DNA analysis and genealogy records.

From analysis of my mitochondrial DNA along my mother's line, it was found that I belong to a specific lineage of the human family called *haplogroup U*. A haplotype is an entire shared region of DNA – share a haplotype, and you share an ancestor. From the study of haplotypes, it is possible to trace the origins of founder mutations and to track human populations. Specific lineages are traced back in time through genetic mutations called *markers*. After a mutation occurs in the mitochondrial DNA of a woman, she then passes this marker on to her daughters, and her daughter's daughters, and so on. My lineage starts with "Mitochondrial Eve," who likely originated in East Africa about 150,000 years ago. About 60,000-50,000 years ago, the female descendents of this woman left Africa and went to the Near East, and then about 40,000 years ago they moved from the Near East into northern Europe (Fig 9-6). Today, members of haplogroup U make up about 11% of the population of Europe, with U5

"THE WHOLE EARTH" OVER TIME

The Genesis 6-8 and Genesis 11 stories record different geographic extents for "the whole earth" because what was known by the biblical authors/scribes about the extent of planet Earth had expanded over this ~200 to 300-year time period (Table 9-2; repeat of Table 3-1). Similarly, the "uttermost part of the earth" of Acts 1:8 (KJV) refers to the known Earth of around A.D. 60. Eventually over time such phrases came to represent *all* of planet Earth (e.g., the gospel shall be preached to *all* nations; Matt. 24:14; Lk. 24:47) – that is, this directive was to be accomplished in the future, and so would someday include all nations. From a Worldview Approach, this past and future distinction is important. The Young-Earth Creationist position interprets Genesis phrases such as "all the earth" from their twenty-first century perspective of planet Earth, whereas a Worldview Approach interprets such phrases from the ever-increasing (over time) geographical and scientific knowledge base of the biblical authors/scribes (from the Old Testament to the New Testament), and then of later generations (like us) that lived after them.

inhabiting England (where many of my ancestors come from).

My mother's father's side of the family is named Read. As far back as we can trace, they lived in Axbridge, a small town in the picturesque hills of Somerset, located in southwestern England. About three miles from Axbridge is the town of Cheddar, famous for its cheese and also for the caves in Cheddar Gorge – especially Gough's Cave, where the Early Neolithic corpse of "Cheddar Man" was discovered in 1903. Cheddar Man is the most complete ancient skeleton ever found in Britain and has been dated to about 9,000 years ago (Table 9-1). This man was part of a population of hunter-gatherers who first inhabited the British Isles about 12,000 years ago (Fig 9-6). Cheddar Man again made news in 1998 when DNA was extracted from this ancient skeleton and genetically related to a modern man through his mother's mitochondrial line – Adrian Targett who lives just a half a mile from Gough's Cave (Fig 9-7). Since some of my ancestors came from this area, I may also be related to Cheddar man.

My father's mother's name was Lillie Anna Sommers. Lillie's mother, Christiana Sinn, immigrated to the United States in 1855 from Unterheimbach, Wurttemberg, Germany, a small town just north of the Switzerland border. The Sinns had lived in this area for at least four generations (as far back as can be traced). Christiana Sinn married Phillip Adam Sommers, who had immigrated in 1876 from Prussia (then a part of Germany, now located in north-central Poland). Therefore, I may be related to Germanic tribes who inhabited Europe during and since the last Ice Age (~30,000 to 10,000 YBP). Perhaps I am even related to Ötzi the Iceman (Fig 9-4)?

What I do know is that my earliest ancestors were *not* genetically related to any of the people groups outlined in the Table of Nations – people who migrated in and around the Mediterranean and Middle East sometime after about 2800 B.C. (Fig 9-6). I am also probably not related to the line of Adam – unless I consider that Adam and Eve lived ~50,000 years ago (the Progressive Creationist view described in the next section), *or* unless my more

Figure 9-7. A modern man – Adrian Targett, who lives in the village of Cheddar, Somerset, England – has been found to be genetically related to the 9,000-year-old "Cheddar man," the skeleton of which is shown in this photo. Recently, the whole DNA genome of Cheddar Man was obtained, which shows him to have been blue-eyed and dark-skinned. *Photo by Robert Wallis.*

recent ancestors (after about 5000 B.C.) interbred with Table of Nations people who were descended from Adam and Eve. The timing is just not right! Adam and Eve lived in southern Mesopotamia at about 5000 B.C. (Table 9-2) *because the Bible puts them there in historic time*. The Earth was already occupied by humans *way before this* on all continents, including Africa, Europe, Asia, Australia, and the Americas.

Of course, the most important theological question raised by this conclusion is: if I am not genetically related to Adam and Eve, then what about the doctrine of sin imputed to the whole human race by Adam? An attempt to answer difficult theological questions like these will be covered in the last section of this chapter.

Four Christian Views

Now that we have reviewed the scientific evidence, we are in a position to evaluate the different Christian views on Adam and Eve. These Christian views fall into four main categories (Box 9-1): our three usual categories – Young-Earth Creationist, Progressive Creationist, and Evolutionary Creationist – and an additional fourth "Historical Adam" category. The Historical Adam category was added because it is a significant and separate deviation from Evolutionary Creationism – not in the matter of evolution but in the matter of Adam and Eve and Origins.

Box 9-1. Four Christian Views on Adam and Eve and Origins			
Young-Earth Creationist	**Progressive Creationist**	**Evolutionary Creationist (strict)**	**Historical Adam**
Adam and Eve were the first humans who lived; they lived about 6,000 years ago in southern Mesopotamia.	Adam and Eve were the first *Homo sapiens* who ever lived; they lived about 50,000 years ago in Africa.	Adam and Eve and the garden of Eden were not real people or events. Real people start with Abraham. (This is not the position of all Evolutionary Creationists.)	Adam and Eve were real people and the first "elected" to receive redemption; they lived about 5500-5000 B.C. in southern Mesopotamia.

Young-Earth Creationist View

The Young-Earth Creationist position is that Adam and Eve were the ancestors of all other humans, and sin was *biologically* transmitted by them to the entire human race. These biological parents of us all lived about 6,000 years ago in the region of Mesopotamia and were created on the sixth day of Genesis 1. Again, as seen in earlier chapters, this Young-Earth view is the simplest reading of the text, and it has also been the traditional position of the Western Christian church for centuries. For example, this Augustinian interpretation of Adam and Eve is the official position of the Catholic church: "Roman Catholics should not believe that true humans existed after Adam who were not generated from him, since this position cannot be reconciled with regard to original sin, which proceeds from a sin actually committed by an individual Adam and...was passed on to all."

In addition, there are Old and New Testament verses that seem to support this view. For example, in Genesis 3:20: "Adam named his wife Eve, because she would become the mother of all the living" (NIV), and Acts 17:26: "From one man [Adam] he made all the nations, that they should inhabit the whole earth" (NIV). This position also fits with the archeology of southern Mesopotamia, in that Genesis places Adam and Eve within this area about 7,000 years ago (or ~5500-5000 B.C.; Tables 9-1 and 9-2). From the archeological record, the first advanced human populations also appear "full blown" in southern Mesopotamia with a knowledge of agriculture and the use of pottery at about this time (~5500-5000 B.C.), and they also possessed the Sumerian language, an obtuse language that is not known to be related to any other.

The main problem with this "literalist" Young-Earth position is that it denies all of the anthropological, archeological, and genetic evidence discussed earlier in this chapter and in previous chapters. Young-Earth Creationists believe that Adam was a literal person, but they deny all of the scientific evidence for Origins – that is, the origin of the human race.

Progressive Creationist View

The Progressive Creationist view – also sometimes called the "Mitochondrial Eve" view – considers Adam and Eve to be the ancestors of the whole human race, but to have lived sometime between 200,000 to 50,000 years ago in Africa, not ~7,000 years ago in Mesopotamia. In this view, gaps in the genealogies of Genesis are used to justify this greatly extended ancestry, and a position that the biblical flood was universal to *all humankind*, but not necessarily global over planet Earth, is required. This placement of Adam and Eve at ≥50,000 YBP also assigns the flood to about the time, or just after, when the main migration of *Homo sapiens* left Africa. The idea is that shortly before Noah, all humankind moved to Mesopotamia, the flood came, and then after the flood Noah's descendants dispersed outward from Mesopotamia to inhabit planet Earth.

Biblical Eve is made equivalent to Mitochondrial Eve, and biblical Adam is made equivalent to Y Chromosome Adam. This is the position that Hugh Ross takes in his book *The Genesis Question* – and it is also the position of many Christians who are concerned about the theological implications of Adam and Eve not being the first *Homo sapiens* based on the accumulating anthropological, archeological, and genetic evidence.

A number of serious problems exist with the idea of Adam existing approximately 50,000 plus years ago and a Noachian flood of somewhat lesser age. If the genealogies of Adam are to be believed at all, Adam is not far removed in time from the flood or the Table of Nations, and certainly not by tens to hundreds of thousands of years. This whole scenario simply does *not* fit with the evidence *specified by Genesis*, which places Adam and Eve in the Neolithic after the advent of farming and husbandry and *not before* (Table 9-1)! Not only is the timing wrong, but the place is also wrong. Genesis specifically places the garden of Eden in southern Mesopotamia, not in Africa (see Chapter 3).

Another Progressive Creationist argument for this view is that there seems to have been a "big bang" of creativity and spirituality coinciding with the rise of *Homo sapiens* just about the time when humans began their migration out of Africa at approximately 50,000 YBP (Fig 9-6). This "big bang" of a culture explosion is interpreted by Progressive Creationists as an indication that modern man was different from other *Homo* species – even from the Neanderthal, who is known to have used fire and spears, but who supposedly did not display the creative traits of *Homo sapiens*. According to this view, humankind's burst of creativity correlates with being specially created as a separate *Homo* species at this time (refer to Chapter 8 and the Progressive Creationist view on evolution). However, this Progressive Creationist position is contrary to recent DNA findings that *Homo sapiens* and Neanderthals in fact *did* interbreed, and to the evidence that *Homo sapiens* had already developed brains capable of fully modern behavior way before 50,000 years ago. Rather, there is substantive evidence that runs counter to the Progressive Creationist's "big bang" model of human advancement and also to the mental capacities of the Neanderthal.

Other difficulties of the Progressive Creationist view abound. First, can the "gaps" in the genealogies of Genesis possibly be stretched back this far? As discussed in Chapter 4, gaps of a few hundred years (at the most) are justifiable from Scripture, but gaps stretching back 50,000 to 150,000 years? And what about even earlier *Homo* and hominid species (Fig 9-1)? Can the gaps be stretched back to millions of years, as has been proposed by some? Could the ark described in Genesis have been constructed by a Paleolithic or Mesolithic Noah using stone scraper and chopper tools (see Chapter 6)? Where Noah "fits," *according to Genesis*, is in the Neolithic, where the technology was already sophisticated (Figs 5-2, 5-3). Furthermore, since literary writing was not invented until about 2500 B.C., these early dates imply that the Genesis stories had to have been transmitted orally for tens of thousands to hundreds of thousands of years. All of these stretches of credibility are insisted upon by the Progressive Creationist view in order to maintain Adam and Eve as the biological ancestors of the entire human race.

An additional problem with the Progressive Creationist view is its anthropological universality position that the flood was local but killed all humans on planet Earth except for Noah and his family. According to this theology, the geographical extent of Noah's flood was determined by the geographical extent of the human community at that time, and as one can see from Figure 9-6, one would have to go back at least 50,000 years for the human race to have been located in one place – and this would have been in Africa and not Mesopotamia where Genesis places Adam and Eve. Davis Young has responded thusly to this position:

> The anthropological universality of the flood is assumed without any engagement whatsoever with the archeological and anthropological data relevant to the flood's impact on the human race. It's as if the

> hundreds, perhaps thousands, of ancient human sites around the world didn't exist.

Figure 9-3 shows some of these sites in the Near East: hundreds to thousands more sites exist in Europe and around the rest of the world.

Finally, and perhaps most damaging of all to the Mitochondrial Eve view, is the DNA-genomics evidence. According to the Progressive Creationist view, biblical Eve is made equivalent to Mitochondrial Eve, and biblical Adam is made equivalent to Y Chromosome Adam. Yet the genomics data appear to be at odds with the speculation that human mitochondrial DNA coalesces to a common ancestor approximately ~150,000 years or so ago. Our species did *not* originate through an ancestral pair (Adam and Eve), but through small interbreeding populations of *Homo sapiens*. Furthermore, there is no evidence for a change in population size stemming from a specially created Adam and Eve at the time of significant cultural and religious development of *Homo sapiens* at about 50,000 years ago.

In short, all of the genetic, anthropological, archeological, historical, and especially *scriptural* evidence does not support a 150,000-50,000-year-old Adam and Eve. Scripture demands a *Neolithic* Adam and Eve and, therefore, should not be ignored or discarded!

Evolutionary Creationist View

As explained in Chapter 1, the Evolutionary Creationist view, first proposed by Denis Lamoureux in his 2008 book *Evolutionary Creation: A Christian Approach to Evolution*, has changed so that now under its positional umbrella are adherents who conform with Lamoureux's view as originally written, but also others who disagree with him on a number of topics, especially on the historicity of Adam and Eve. Since this fundamental split in theology seems to me significant (but with the umbrella label remaining the same), I have divided this view into two separate sub-categories, the (strict) Evolutionary Creationist view (this section) and the Historical Adam view (next section).

It is the (strict) Evolutionary Creationist view that considers the people and events in the early chapters of Genesis (before Abraham) to be unhistorical, legendary, or fictional. Specifically for Adam, Lamoureux has this to say:

> First, Adam never actually existed.... Second, Adam never actually sinned. In fact, it is impossible for him to have sinned because he never existed. Consequently, sin did not enter the world on account of Adam.... Adam never existed and this fact has no impact whatsoever on the foundational beliefs of Christianity.

"Strict" Evolutionary Creationists like Lamoureux take this view because they consider that aspects of the early Genesis stories are fictional; for example, in the Adam and Eve story, talking snakes, creating Adam from the "dust" of the ground, creating Eve from Adam's rib, and so on. By proclaiming that these stories are unhistorical and "only the theology matters," it relieves the Genesis text from trying to explain it's more unscientific and unhistorical aspects.

However, relieving the text from these aspects creates a number of theological problems. First, if the people of Genesis are not real, then why does the Bible go to such great lengths to establish the genealogies of Genesis, Numbers, Chronicles, Ezra-Nehemiah, Matthew, and Luke? First Chronicles begins with *nine* chapters of "begots"; that is, keeping exact track of their ancestors was culturally very important to the ancient Hebrews (and also to other ancient people groups). If these genealogies are not real, then where do the unhistorical people end and the historical people start? Do real people start with Abraham, as maintained by some Evolutionary Creationists? Second, since the New Testament refers to the people and events in the Old Testament as being historical, is this a denial of the reliability of the entire Bible? If Lamoureux's above statement is true and Adam was not a historical person, then what about the foundational Christian doctrines of the "fall," original sin, and Paul's entire theology of Christ as the new Adam and his dying to save us from sin?

Historical Adam View

While the Historical Adam view is (at least presently) a subcategory of Evolutionary Creationism, it disagrees with the (strict) Evolutionary Creationist view on the important matter of the historicity of Adam and Eve and other figures and events in the early chapters of Genesis. It also disagrees with Young-Earth theology that Adam was the progenitor of the whole human race and with Progressive Creationist theology that Adam lived around 50,000 years ago. Rather, it favors the position that Adam was a historical person who was a *special creation* among a population of pre-Adamite humans in Mesopotamia around 5000 B.C. There are two possible variations on this view:

(1) Adam was a historical person literally created from the "dust of the ground" (Gen. 2:7). This subview would have Adam *physically* created *ex nihilo* from inorganic material, and would be somewhat analogous to Christ being miraculously made a human by means of the virgin birth.

(2) Adam was a historical person, but instead of being physically created about 7,000 years ago, Adam was the first human *elected* to receive redemption from God (Box 9-1). That is, there were other *Homo sapiens* before Adam, but Adam was the first "Homo divinus" in the spiritual sense. According to this subview, Adam was the first human to be made a "living soul" (Gen. 2:7, KJV), and the first human with whom God directly interacted. In support of a spiritual and figurative interpretation of the creation of Adam over a literal view is Genesis 2:24 where God institutes marriage: "Therefore shall a man leave his father and his mother, and shall cleave unto his wife; and they shall be one flesh" (KJV). What sense would this have made to Adam if he did not have a physical father or mother?

Both of these varieties of a Historical Adam view are compatible with the science presented in earlier chapters, and they are also compatible with certain aspects of the Genesis account. Adam seems to have had a fully developed language; in Genesis 2:19-20, he names all of the animals and in Genesis 2:23, he names Eve. The fact that Adam had language implies a preexisting culture where language was learned from that culture. However, the Sumerian language has no known counterparts, so it is possible that Adam was specially created with a unique language. In either variety, Adam and Eve appear to have been biologically compatible with other people around them, since their son Cain found himself a wife from the surrounding population (Gen. 4:17).

There are serious theological problems with the Historical Adam view, and it is *this* theology that will be discussed for the remainder of this chapter, using a Worldview Approach as the basis of interpretation.

A Worldview Approach to Origins

The basis of a Worldview Approach is that it tries to interpret the culture and mindset of the Genesis author(s) who wrote the biblical text. From what has been discussed in previous chapters, consider the proposition that the authors/scribes of the first four chapters of Genesis lived in southern Mesopotamia in the time frame of 2500-2000 B.C., at the dawn of narrative cuneiform writing. Let's say that these first authors/scribes had the task of faithfully transmitting the oral sacred traditions of their ancestors into written narrative form. The main purpose of these authors was to write the story of *God's interaction with, and revelation to, their ancestors* – not to write the story of other people groups outside of their ancestral line. These ancient biblical authors/scribes probably had no conception that God's ultimate purpose for this genealogical account was to trace the line of Adam to a much-later Christ. Rather, their job was to faithfully write/copy the covenantal history of God's interaction with Adam's line as it was known in their time; and they did so from their *covenantal worldview perspective*.

Genesis Uses Figurative Language

Figurative language is compatible with a Worldview Approach to Scripture because it exemplifies the literary mindset of the ancient Hebrews and their surrounding non-Hebrew, Near-Eastern contemporaries. Genesis 2:7 provides a good example of the use of figurative language: "And the Lord God formed man of the dust of the ground, and breathed into his nostrils the breath of life; and man became a living soul" (KJV). The expression "dust of the ground" is a poetic figure of speech, one used by various people groups of the ancient Near East and an expression that always signifies mortality in the Old Testament. In a Near Eastern cultural context, among the materials used by the gods for the creation of man was the "clay" of the earth. Therefore, in using this poetic expression, the biblical authors/scribes were conjuring up the creation process in the minds of the people living at that time. In addition, when ancient Near Eastern texts speak of people being created from "dust" or "clay," they are not talking about just one individual, but are addressing the nature (mortality) of all humanity. For example, in Genesis 2:7, the phrase "breathed into his nostrils the breathe of life" is to be taken archetypically for humanity, and "living soul" is to be translated "spirit" as implied in 1 Corinthians 15:45, 47: "The first man Adam was made a living soul [spirit]; the last Adam [Christ] was made a quickening [life-giving] spirit....The first man [Adam] is of the earth, earthy ["dust"]; the second man is the Lord from heaven" (KJV).

A literary approach also relates to the story of Adam and Eve. As discussed in Chapter 2, the use of a̲dam (*adham* for humankind) and A̲dam (*ha adham*, or "the adam" as a proper name for an individual) represents a play on words, as does the fashioning of Eve, "the mother of all living" (Gen. 3:20), from the rib of Adam (Gen. 2:21). The Sumerian word for "rib" was *ti*, which could alternately mean "life." Eve (*hawwah*) sounds similar to the verb "to live" (*hayâ*) and the adjective "living" (*hay*) since she was the mother of life. In Sumerian literature, the "lady of the rib" came to be identified with the "lady who makes live" through what may be another play on words. Such ancient literary puns represent "historical memories" in Genesis that attest to its early authorship because such word plays were typical of the ancient Mesopotamians' literary style. So while Adam and Eve could have been real people, the method of *describing* them was commensurate with the literary style of that time and place. Thus, there are elements of the Genesis text that point to a figurative interpretation of Adam and Eve, and other elements that point towards a historical or literal interpretation. It is like our discussion of the garden of Eden in Chapter 3: Adam was a historical person and the garden of Eden a historical place, but the story was told within the literary framework of the ancient authors/scribes.

Genesis Hints of Pre-Adamites

Genesis doesn't specifically state that other people lived in Mesopotamia alongside Adam, but it strongly hints at it. Again, the Genesis author(s) were not interested in tracing other human lineages besides Adam's, and so they only casually allude to other people existing at that time. The following "hints list" is taken primarily from Dick Fisher's *The Origins Solution*.

Genesis 4:17: *Cain's wife.* Where did Cain's wife come from if there were no other people around? The Young-Earth Creationist view says that he married his sister, and biblical exhortations such as Leviticus 18:9 forbidding incest did not apply to the start of the human race. But a more reasonable explanation is that the Bible is saying Cain married a woman outside of Adam's line, one living among a human population already established in southern Mesopotamia *before* Adam and Eve lived.

Genesis 4:17: *Cain builds a city.* Whereas villages contain tens to hundreds of people, cities need a population of thousands. Where did all of these people come from in order for Cain to have built a city? The Young-Earth Creationist view is that there was an exponential population explosion at this time because Genesis 5:4 says that Adam begot other sons and daughters over his lifespan of hundreds of years. The archeological record substantiates that a number of cities arose in southern Mesopotamia around 3400 B.C. in the Uruk Period, or approximately when Cain lived and supposedly established the first city (see Chapter 5). This implies that Cain did not establish the *only* city in this region; that is, there were other populations living in southern Mesopotamia who were also congregating into cities at this time.

Genesis 4:14: *Who will slay Cain?* If there was no one else around, why was Cain worried that someone might slay him? Was he worried that one of Abel's sons (if he had any sons before he was murdered) or one of his brothers would kill him? Or was he worried that someone outside of his family line would kill him? As mentioned earlier in this chapter, hunter-gatherer nomad tribes are known from archeological remains to have inhabited the Arabian Peninsula-Persian Gulf area *long* before the time that Adam and Eve lived (as early as around 100,000 years ago, when *Homo sapiens* first migrated out of Africa).

Genesis 3:24: *The cherubim guards.* Why were the cherubim commanded to guard the way to the tree of life (Fig 10-3, page 167)? Guard it from whom? From other people besides Adam and Eve who would try to get into the garden to eat of the tree of life in order to live forever (Gen. 3:22)?

Genesis 4:16: *The land of Nod.* Throughout the Bible phrases like the "land of Canaan" or "land of Egypt" refer to an area populated by these particular people. Similarly, could the "land of Nod" have been populated by the "Nodites" who lived east of Eden? In Hebrew, "nod" means "vagrancy" or "wandering," and so another suitable translation of Genesis 4:14 could be: "...and Cain dwelt in a land of nomads." Therefore, is it possible that the people Cain feared would kill him were wandering bands of desert nomads who lived in tents; that is, other nomadic people besides Jabel's line who dwelt in tents (Gen. 4:20)?

Genesis 6:1-2: *The sons of God and the daughters of men.* "And it came to pass, when men began to multiply on the face of the earth, and daughters were born unto them, that the sons of God saw the daughters of men that they were fair; and they took them wives of all which they chose" (KJV). This is one of the most highly debated passages of all Scripture. The two most popular interpretations of this passage are: (1) the "sons of God" refer to angels, or (2) the "sons of God" refer to the godly line of Shem, whereas the "daughters of men" were from the ungodly line of Cain. The reason for favoring an angel interpretation is Job 1:6; 2:1; and 38:7, where the "sons of God" are mentioned as angels. The main reason for not favoring this interpretation is Luke 20:34-35: "The people of this age marry and are given in marriage. But those who are considered worthy of taking part in the age to come and in the resurrection from the dead will neither marry nor be given in marriage, and they can no longer die; for they are like the angels" (NIV). In other words, the text says that angels don't marry and cannot propagate human children.

A word study of all the "sons of God" and "sons of men" mentioned in the Bible leads to the following considerations. Adam is called the "son of God" (Luke 3:38), while Christ is the "Son of God" (Luke 1:35). The term "son of man" refers to an actual individual man; e.g., Ezekiel, where God addresses Ezekiel nearly 90 times by this title. Daniel, in his prophetic vision of Christ (Dan. 7:13), uses the term "the Son of Man" to show that an actual man (the Messiah) will come in the clouds of heaven to receive a world-wide kingdom. Jesus referred to himself as "the Son of Man" nearly 80 times in the Gospels.

The term "sons of God" applies to believers. For example, John 1:12 says: "But as many as received him, to them gave he power to become the sons of God, even to them that believe on his name" (KJV), and Romans 8:14 says: "For as many as are led by the Spirit of God, they are the sons of God" (KJV).

The term "sons of men" applies to the human population in general. In Numbers, Psalms, Proverbs, Isaiah, Daniel, and Ecclesiastes the term "sons of men" refers to any people, be they Israelites or non-Israelites. It also sometimes refers to ungodly men who are against the Lord's people; for example, in Psalm 57:4, King David says: "My soul is among lions, and I lie even among them that are set on fire, even the sons of men, whose teeth are spears and arrows, and their tongue a sharp sword" (KJV). Other times, it refers to Israelites who are out of favor with the Lord, as in Numbers 23:19: "God is not a man, that he should lie; neither the son of man that he should repent" (KJV).

If this is the proper interpretation of the "sons of God" and "sons of men," then a third option regarding Genesis 6:1-2 is possible besides the two traditional ones. The phrase "the sons of God" refers to the chosen line of Adam, while "the daughters of men" could refer to the population of humans that lived alongside the line of Adam. That would explain why they took these women as wives, and also why the flood account directly follows in Genesis 6. The unbelieving wives from the non-Adamite population were turning their believing husbands away from God and toward wickedness and idol worship. Among the children of these unions were the giants (Nephilim) of Genesis 6:4, who seem to have survived Noah's flood, as they reappear in Numbers 13:33 – still another reason for not interpreting Noah's flood as globally killing off all the people on planet Earth besides Noah and his family.

The Male and Female of Genesis 1

We discussed how a Worldview Approach applied to Genesis 1 in Chapter 2 where we mentioned *tôleˇdôt* as it relates to the days of Genesis and the Literary view (Table 2-1). Now we will discuss Genesis 1:26-27 as it relates to the creation of humans. From the worldview perspective of the biblical author(s) of Genesis 1, to whom do the "male and female" of Genesis 1:27 refer? Is this phrase specifically referring to Adam and Eve, or might it refer to pre-Adamite humans? To most biblical scholars, it seems certain from Hebrew syntax that this "generations of" formula in Genesis 2:4 (which is repeated ten times in Genesis) is intended to be a superscription to what *follows* in Genesis 2, and, therefore, the majority of scholars over the years have favored a parallel structure between Genesis 1 and 2 – with Adam and Eve of Genesis 2 being the specific "male" and "female" of Genesis 1. However, according to John Walton in his book *The Lost World of Adam and Eve*, the text does not necessarily require this interpretation:

> The second creation account (Gen. 2:4-24) can be viewed as a sequel rather than as a recapitulation of day six in the first account (Gen. 1:1-2:3)...if Genesis 2 is a sequel, it would mean that there may be other people (in the image of God) in Genesis 2-4, not just Adam and Eve and their family.

A third "compromise" view held by theologian Roy Clouser – one that follows rabbinical tradition concerning how to understand Genesis 2:7 – is that while Genesis 2:4 is certainly a superscription to what follows in Genesis 2, it is *not* a second creation story or two accounts of the same event, but only an indication that we are being introduced to a new story that is a more detailed description of the general statement in Genesis 1:27 that God created humans. Or, in other words, these two accounts focus on different *aspects* of the story:

> In short, we do not have two creation accounts, one in Genesis 1 and another in Genesis 2. What we have is one creation account followed by another account of the beginning of redemption.

Essentially, this is similar to the "critical scholar" view (see Chapter 10) that Genesis 1 and 2 were written by different authors (that is, they had different sources), and that the "Adam and Eve" in Genesis 2 does not necessarily correlate to the "male and female" of Genesis 1, who might portray humans who lived before, or alongside, Adam and Eve. Actually, the idea of pre-Adamite humans is not a new proposition. Isaac de La Peyrère in 1655 suggested that pre-Adamite people had been created during the first phase of creation recounted in Genesis 1, while Adam and Eve did not appear until the next phase, recounted in Genesis 2.

Let us now examine these and other Genesis verses more closely within the context of Genesis

1 referring to humankind as a created species and with Genesis 2 referring to Adam and Eve as two specific human individuals at the beginning of God's plan of redemption.

Genesis 1:26. "And God said, Let us make man in our image, after our likeness: and let them have dominion over the fish of the sea and over the fowl of the air, and over the cattle, and over all the earth, and over every creeping thing that creepeth upon the earth" (KJV). This definitely sounds like the text is talking about humankind in general. Can this statement apply to pre-Adamites? Neolithic man certainly had dominion over the animals; they were hunters, fishers, and herdsman. Note that *if* this passage is talking about generic mankind (a̲dam), rather than a specific man (A̲dam), then generic man is created in the image of God as well as the specific man Adam and his descendants.

Genesis 1:27-28. "So God created man in his own image, in the image of God created he him; male and female created he them" (KJV). This gets a bit confusing. The pronoun tense goes from plural (them) in Genesis 1:26, then to the singular (him) and back again to plural (them) in Genesis 1:27, and finally stays with the plural in Genesis 1:28: "And God blessed *them....*" According to the Hebrew scholar Cassuto, the true sense of these verses is that he created *them* (in the plural). He blessed *them* and named *them* man (male and female) when they were created. Cassuto also goes on to say that two different words for God are used in Genesis 1 and 2. In Genesis 1, *Elohim* is used for God in the general sense of being the ruler of nature and source of all life, whereas in Genesis 2, *Yaweh* is used for the personal character of God in his direct redemption plan for humankind. Since it is not clear from Genesis 1:27-28 that the "male and female" in these verses refer to Adam and Eve, they could instead refer to male and female humankind being created along with the rest of the natural world, which is the specific topic of Genesis 1.

Genesis 2:4. We have already discussed in Chapter 2 how the "generations" (*tôlĕdôt*) of Genesis 2:4 link back to the Genesis 1:1-2:3 narrative and how this implies a longer period of creation than just seven literal days. In the context of our present discussion, the word "generations" could also imply a *genealogy* of humans before Adam. The discussion of two specific persons, Adam and Eve, doesn't start until Genesis 2:5, where the location for the garden of Eden is described.

In the Image of God

What does it mean to be created in God's image? From our modern worldview, some type of consensus opinion might include consciousness, self-awareness, creative intelligence, rationality, reasoning, language or abstraction ability, freedom of moral choice, dominion over animals, and capacity for fellowship with God. But from the worldview of the ancient world, people were considered to be "in the image of God" when they embodied his qualities and did his work; that is, they were the symbols of his presence, acting on his behalf as his representatives. The ancients also had different concepts of body, soul, and spirit than we do, and while their ideas and ours overlap, we have to be careful not to impose our modern theological concepts on the ancient text.

Notwithstanding, the biblical concept of "the image of God" seems to encompass the idea that humans can be distinguished from animals in having a body, soul, *and spirit*. To paraphrase theologian John Walton: It is this spirit nature that survives the death of the body. It is not something that can evolve, and it is not possessed by those other creatures in an evolutionary line of common descent. It is a special creative act of God, and it differentiates us from every other creature.

The Problem of Original Sin

The doctrine of "original sin" is one of the hardest theological issues to explain if Adam and Eve are considered not to be the parents of the whole human race. Not only is Adam supposedly responsible for sin and the physical death of humans, but according to tradition, the results of his sin have had even more far-reaching effects – it brought about a world of disease and disorder including the physical death of animals. However, in regard to this traditional view, it should be pointed out that the Old Testament never speaks of Adam's sin as bringing sin on all humankind. It also never uses the word "fall" as a term, and neither is there any mention of a "fall" in other ancient Near Eastern literature. This doctrine of sin being passed from generation to generation was first formulated by Augustine (A.D. 354-430), and, since then, has been traditionally regarded by the church as a depravity, or tendency to do evil, which was biologically transmitted to the entire human race as a consequence of Adam's fall (Fig 9-8). Therefore, since the Bible doesn't mention "the fall," and since Augustine had no knowledge of Origins as we now perceive them, it seems necessary that this doctrine – that sin was biologically transmitted to Adam's descendants – should be reinterpreted from a spiritual, not biological, standpoint.

Far-reaching effects of Adam's "fall." The traditional church view and Young-Earth Creationist position on "the fall" is that Adam's sin in Genesis 3 brought about the death of *all* living creatures in the natural world. This theology has sometimes generated amusing, nit-picking arguments such as "did Adam ever step on an ant?" Less amusing, it has alienated people who are educated with respect to Earth's fossil record (see Chapter 7). Many Young-Earth Creationists today believe that no animals died before the fall, although the Bible does not explicitly teach this doctrine. From a theological perspective, why is it that animals should die because of Adam and Eve's sin? They did not eat of the "tree of the knowledge of good and evil" (Gen. 2:17, KJV). It was humans who sinned, not animals (Rom. 5:12). Here also the Bible is not talking about physical death, it is talking about spiritual death, and about spiritual death after Adam's disobedience, not before it. Thus, the Earth's fossil record does not contradict Genesis because there is no theological reason why generations of animals and pre-Adamite humans could not have died before the so-called "fall" of Adam.

Spiritual or physical death? In order to understand the nature and extent of Adam's sin, we must closely examine the story of Genesis 2-3. Of particular importance to the subject of death are the verses of Genesis 2:17 and 3:2-4:

> But of the tree of the knowledge of good and evil, thou shall not eat of it; for in the day that thou eatest thereof thou shalt surely die.... And the woman said unto the serpent, We may eat of the fruit of the trees of the garden; but of the fruit of the tree which is in the midst of the garden, God hath said, Ye shall not eat of it, neither shall ye touch it, lest ye die. And the serpent said unto the woman, Ye shall not surely die. (KJV)

Did Adam and Eve die the exact day (*yôm*) that they ate the fruit of the tree of the knowledge of good and evil? No – they died many years later and not as the immediate physical consequence of eating the fruit. So who was the liar, God or Satan? John 8:44 says: "He [Satan] was a murderer from the beginning, not holding to the truth, for there is no truth in him" (NIV), while Titus 1:2 says that: "A faith and knowledge of the truth rest in the hope of eternal life, which God, who does not lie, promised before the beginning of time" (NIV). Christians must either have God the liar and Satan not the liar, or they have to assume that Adam and Eve's death was *spiritual*, not physical. First Corinthians 15:22 is key to a spiritual interpretation: "For as in Adam all die, so in Christ all will be made alive" (NIV). How are we made alive in Christ? We are "born again" – not physically but spiritually. We are born again to eternal life. Adam, as a human being, was "doomed to death" (the meaning of the Hebrew phrase in Gen. 2:17), but from the garden of Eden onward, a whole new kind of death and life enters into the picture – spiritual death and eternal life. This same idea of spiritual, rather than physical, inheritance is also expressed in Galatians 3:7 and 3:29: "Know ye, therefore, that they which are of faith, the same are the children of Abraham.... And if ye be Christ's, then are ye Abraham's seed,

Figure 9-8. *Adam and Eve* by Lucas Cranach the Elder, 1538, oil on lime. *Google Image.*

and heirs according to the promise" (KJV). Are all believers the biological (genetic) sons or offspring of Abraham? No, believers are the *spiritual* offspring of Abraham and heirs to the promise made by God way back in Genesis. Such an interpretation of Adam being the *spiritual* father of us all (instead of the biological father) greatly mitigates the role of Eve in Genesis 3:20.

Eve: mother of all living. "And Adam called his wife's name Eve because she was the mother of all living" (Gen. 3:20, KJV). If Adam's death was spiritual and if Eve was *not* the mother of all humans, as discussed earlier in this chapter, then perhaps the phrase "mother of all living" refers to Eve being the *spiritual* mother of the whole human race. This verse also begs the question: How did Adam know that Eve would be the mother of all living? This future tense is understandable from the point of view of authors/scribes of 2500-2000 B.C. who could see that the generations of *their* family lineage had come from Eve. Again, the important point is that the Bible traces a *specific* family lineage – the covenant line from Adam to Christ (Fig 1-4). Adam was the father of that chosen "living" line, and Eve was the "mother of all living" in that same family line.

In the next final chapter, we will attempt to pull together all the ideas presented in previous chapters into a coherent whole, from the perspective of both science and Scripture.

Moses with the Ten Commandments by Philppe de Champaigne, 1648.

Courtesy of The State Hermitage Museum, St. Petersburg; Photograph © The State Hermitage Museum/photo by Svetlana Suetova.

CHAPTER 10

PUTTING IT ALL TOGETHER

Colossians 1:16-17. For by him were all things created, that are in heaven, and that are in earth, visible and invisible, whether they be thrones, or dominions, or principalities, or powers: all things were created by him, and for him: And he is before all things, and by him all things consist. (KJV)

Revelation 1:8. I am Alpha and Omega, the beginning and the ending, saith the Lord, which is, and which was, and which is to come, the Almighty. (KJV)

We now return full-circle to Chapter 1 and our Figure 1-4 timeline (repeated below as Fig 10-1). In this final chapter, we will try to fit all the pieces together, from the creation of the universe to the present with respect to both the scientific evidence and a Worldview Approach to Scripture. The overall theme will be that the Bible emphatically claims that God/Christ is the creator of the universe and all that is in it, and by him all things consist (hold together). According to the Bible, the universe and life did not happen by *chance*, but was created, directed, and sustained by God.

The Timeline of Biblical History

In the timeline of Figure 10-1, God is considered infinite and the creator of time and space; he transcends time and, is thus, unlimited in being everywhere at every time (double open arrow). However, in the timeline related to the physical world and human history, time has a beginning and goes only in the forward direction (single open arrow). From his infinite time frame, God condescends to enter into the finite realm of human history as it is being played out at a certain time and place (vertical arrows), and he also accommodates all of the cultural trappings that go with it. In the context of biblical history, these "cultural trappings" were the customs and worldview of one people group (the line of Adam) in the Near East in the time frame of approximately 7,000 to 2,000 years ago – a special nation through whom God chose to spread his revelation and blessing to all humankind. And, since the Bible is a *historical* record of this one people group, the customs and worldview of this group became incorporated into the biblical text alongside God's revelation. In other words, the revelation is from God, but the concepts of the world and how it worked are commensurate

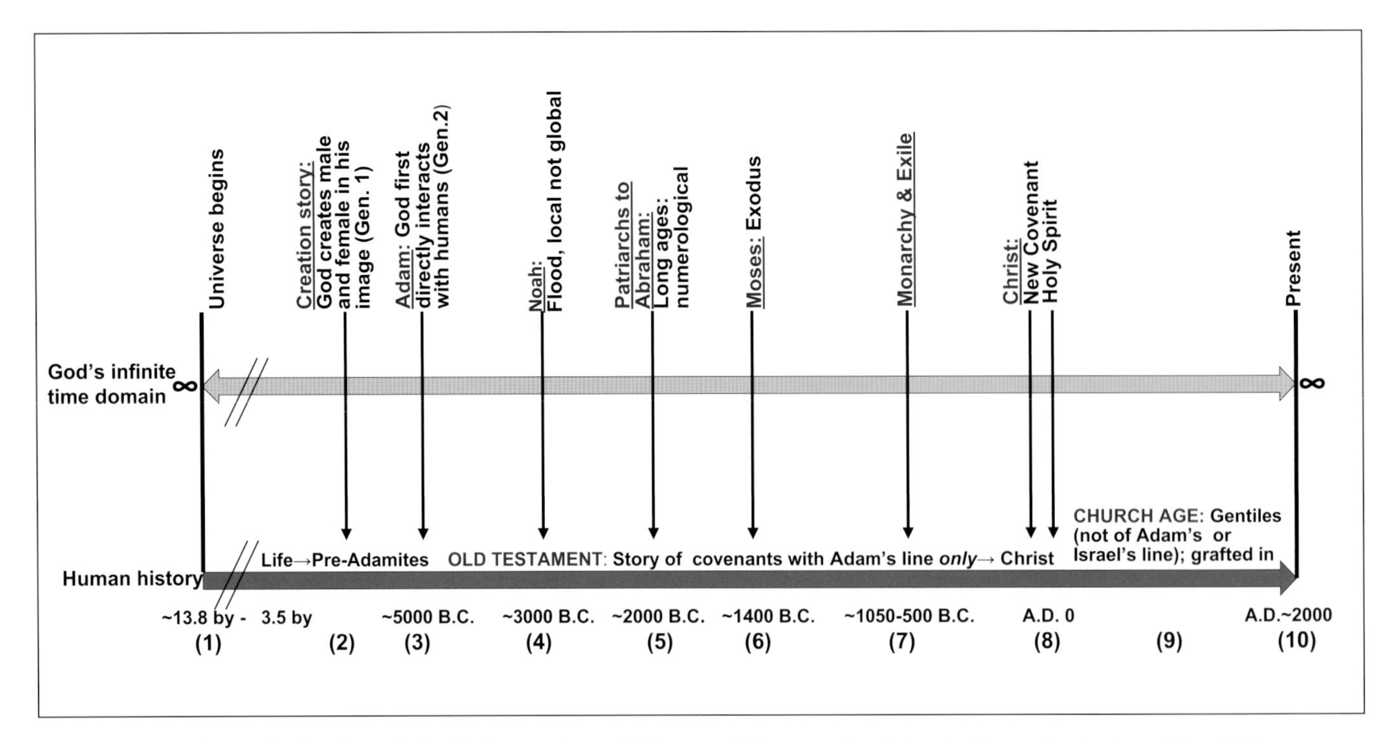

Figure 10-1. Schematic timeline of God's interaction with human history as it relates to the main stories of the Old Testament up to the time of Christ (by = billion years). The numbers (1) to (10) below the timeline are included in subject headings to make it easier for the reader to follow the flow of the text.

with the knowledge of this people group in time and place; not recognizing this distinction has been the major cause of conflict in the so-called "war" between science and Scripture.

In addition, two other basic misconceptions have caused problems in biblical interpretation. The first misconception is that biblical history is communicated the same way as modern history. Modern historians attempt to record "just the facts," but biblical history is colored by the worldview of the ancient biblical scribes, and that worldview must be stripped away to reveal the real history. The second basic misconception is that the Old Testament covers the entire human race. It does not: it is *primarily* concerned with the genealogical line from Adam to Christ, and it is only marginally concerned with non-Adamite people groups or the non-Israelite (Gentile) line of Adam. (In other words, it is Jewish history, not human history.) The main purpose of the biblical authors/scribes was to relay the story of God's interaction with, and revelation to, *their* ancestors. This concept is extremely important with respect to Origins: Adam and Eve were *not* the parents of the whole human race. They were only the parents of those in the *covenant* line of Adam leading to Christ (Fig 10-1). While contrary to many Old Testament scholars (both past and present), who have understood Genesis 1-11 as referring to the entire human race, this position is the *only* one that can account for the massive amount of anthropological and archeological evidence as it relates to Scripture.

The interpretation of a Worldview Approach is that the people, places, and events in the timeline of Figure 10-1 are historical. However, the *description* of these people, places, and events is modified by the worldview of the persons receiving God's revelation and/or by the worldview of the biblical author/scribes who wrote the text from oral accounts passed down to them. To beget someone is a *physical* act – either it happened or it didn't. But a descriptive interpretation of a historical event is a *cultural* act that stems from a particular worldview. Thus, the Genesis stories are dualistic in nature:

(1) *Historical*: they have a historical base, one that can be at least partly substantiated from archeological finds or historical documents, and

(2) *Worldview*: the worldview of the ancients is interwoven into this historical base. Sometimes this worldview involves figurative language or narrative style, and sometimes it involves primitive concepts of history or science. The ancients – like people of all cultures – saw the world not as it is, but through the eyes of that culture, and it is this worldview which has been passed down to us in the Bible.

Throughout the remainder of this chapter, we are going to try to see how this dualism "works" in practice along our timeline. The numbers in the section headings, for example (1), refer to the numbers in the timeline of Figure 10-1.

A Worldview Approach to Creation (1)

Creation and Genesis 1

We will start on the left side of Figure 10-1 when the universe began about 13.82 billion years ago in the Big Bang, and when the first life on Earth appeared in the fossil record about 3.5 to 4.0 billion years ago. Creation of the physical world of matter and energy could have begun by direct fiat of God, and then could have evolved under God's direction to its present form: from the creation of elements in supernovas, to DNA and life from organic molecules, to human consciousness and the ability to perceive a creator God. A Worldview Approach coincides with this God-directed, scientific view of creation.

A Worldview Approach does not interpret Genesis 1 to be a literal seven-day view or a temporal view of creation, but one that fits within the framework of the narrative literary style of the ancient Mesopotamians (Chapter 2). By taking this view of Genesis 1, the many problems encountered by Young-Earth Creationists and Progressive Creationists – when they try to force-fit modern science into Genesis – are circumvented.

A Worldview Approach to Evolution and Pre-Adamites (2)

Evolution and "After Its Kind"

A Worldview Approach to Scripture does not support the Young-Earth Creationist, Progressive Creationist, or Intelligent Design views on evolution where new species were created "full blown" in Genesis 1 or in subsequent creation acts throughout geologic history. "After its kind," as used in Genesis, is a folk classification system and is not equivalent to our modern conception of species: that would be forcing modern science onto the ancient biblical text (Chapter 8). Rather, a Worldview Approach favors an Evolutionary Creationist (theistic evolution) position based on the strong genetic evidence for evolution that has emerged over the past few decades. Life could have started by the direct interaction of God and then could have developed by a long process of God-directed evolution from simple organisms to humans. With respect to the timeline of human history (Fig 10-1), God let humans evolve physically (evolution), mentally, emotionally, and spiritually until such time as they were ready for his Progressive Revelation throughout a covenantal history that started with Adam and led to Christ.

Pre-Adamite Humans

God's revelatory program for humanity begins with pre-Adamite humans (Table 10-1). Throughout this discussion, it will be assumed that the "male and female" of Genesis 1:26-27 refers to modern humans who existed prior to, or contemporaneously with, Adam, but not specifically to populations of *Homo sapiens* or other *Homo* species that have existed before or since ~200,000 YBP because the biblical authors had no knowledge or conception of such populations. If we acknowledge a pre-Adamite status of "male and female," then we must also acknowledge that these modern humans were created in God's image, because such a relationship is stated in Genesis 1:26.

Time	<~50,000 YBP	~5000 B.C.	~3000 B.C.	~2000 B.C.	~1300 B.C.	~A.D. 30
People	**Pre-Adamite**	**Adam**	**Noah**	**Abraham**	**Moses**	**Christ**
Place	Entire Earth	Mesopotamia	Mesopotamia	Mesopotamia to Palestine	Egypt to Palestine	Palestine
Covenant	No covenant	Adamic Covenant	Noachian Covenant	Abrahamic Covenant	Sinaic Covenant (the Law)	New Covenant; church age starts
Religion	Animism, spirit world, shamans	Polytheism; monotheism starts with Adam, first of the chosen line	Flood sent to wipe out polytheism from Adam's chosen line	Monotheism; to be witness of one true God to the nations	Monotheism; the Law given to Israel to keep chosen line holy	Monotheism; Christ is the way to God
Holy Spirit	Holy Spirit, no	Adam is first person to encounter Holy Spirit	Holy Spirit of God protects Noah in the flood	Holy Spirit given to selected people in Old Testament	Holy Spirit given to the prophets + other selected persons	Holy Spirit comes on day of Pentecost to all believers
Humankind	Pre-Adamites (*Homo sapiens*)	Adam's chosen line + non-Adamites	Non-Adamites not wiped out by the flood; only Adam's chosen line	Line of Abraham + non-Adamites	Israel + Gentiles	Neither Jew nor Gentile in Christ; Gentiles grafted into Adam's line
Sin	Good & evil, yes; knowledge of sin, no; judgment of sin, no	Knowledge of good & evil; spiritual death by sin	Judgment of sin of Adam's line by the flood	Atonement for sin by the blood sacrifice of animals	Israel's sins judged by the Law; Gentiles by the glory of nature	Christ is the ultimate atonement of sin for all people

Table 10-1. Proposed chart of God's Progressive Revelation and covenantal history with humanity over time.

Did pre-Adamite humans have a spiritual nature? Yes, because to be created in God's image implies that they did, and because from the anthropological record, it is very likely that modern humans were practicing animistic/shamanistic religions by about 30,000 years ago (Table 9-1). Even animistic cultures have this innate recognition of God as a "Great Spirit," except that without the direct revelation of God to man (as God did with Adam), this spirit aspect is expressed by worshipping the created instead of the creator – that is, worshipping idols, ancestors, animals, the Sun and Moon, and so on. The "job" of Adam's line was to reveal the one true God to those who practiced animism and polytheism.

Progressive Revelation

The "spiritual nature" of humankind would have involved a *gradual* and *evolving* awakening of consciousness and ideas of morality – a long process of attaining the spiritual capacity and longing to seek and comprehend God. This spiritual awakening was universal to humankind with the geographical migration and expansion of the human race (Chapter 9), so that by the time of Adam, all humans had attained a "religious consciousness." Essentially, this process involves the idea that in the "fullness of [evolutionary] time," God decided to interact with the humans he created. It also involves the concept of Progressive Revelation, in that God did not reveal the knowledge of good and evil, atonement by the blood of animals, the Law, or the incarnate Christ as the ultimate atonement for sin until humans were spiritually ready to receive these progressive covenants (Table 10-1).

A New World Order Commences with Adam

So then, what changed with Adam? It wasn't because *humans* suddenly obtained a spiritual nature about 6,000-7,000 years ago as has been proposed by some. It was because *God* decided that this was the time in human history to impart his Holy Spirit into a specific human, Adam, and then to continue his intercession through Adam's biological line. Did God impart his Holy Spirit into that genetic line? Yes, he gave his Spirit to Noah, Abraham, Joseph, Gideon, and David, among others. But, before Christ, the Holy Spirit was imparted only to specific persons at specific times and for specific reasons, and this impartation was removable, as shown by the plea of David (Psalm 51:11): "Do not cast me from your presence or take your Holy Spirit from me" (NIV). After Christ, the Holy Spirit was poured out to all believers (Acts 2:17), and will not be taken from us (John 14:16; Eph. 4:30).

Did pre-Adamite humans sin? Yes, in the animalistic sense that committing violent and wrongful acts against others is the nature of *all* humanity, but not in the sense of a religious accountability to God. Even animistic societies have codes to live by, with some people being considered good or evil by these societies. However, it was not until Adam sinned that a *dispensational* state of sin was conferred on the entire human race, and judgment for that sin (Rom. 5:12): "Sin entered the world through one man [Adam], and death through sin, and in this way [spiritual] death came to all people, because all sinned" (NIV). Therefore, under this new "world order" set up by God, people groups such as the Australian aborigines or the European race (that I belong to) are accountable to God for their wrong doings, just as Adam's direct biological line is accountable.

PROGRESSIVE REVELATION

The "spiritual nature" of humankind would have involved a gradual and evolving awakening of consciousness and ideas of morality – a long process of attaining the spiritual capacity and longing to seek and comprehend God. This spiritual awakening was universal to humankind with the geographical migration and expansion of the human race (Chapter 9), so that by the time of Adam, all humans had attained a "religious consciousness." Essentially, this process involves the idea that in the "fullness of [evolutionary] time" God decided to interact with the humans he created. It also involves the concept of Progressive Revelation over time, in that God did not reveal his plan of salvation all at once. Instead, it was revealed in steps: the knowledge of good and evil, atonement by the blood of animals, the Law, and finally the incarnate Christ as humans became spiritually ready to receive these progressive covenants (Table 10-1).

Were pre-Adamites judged for their sins? No, because sin is not imputed where there is no (knowledge of the) law (Rom. 5:13). So while pre-Adamite humans had a religious nature, because they were ignorant of the true God, they were not held responsible for worshipping false gods. Although humans and sin were in the world before Adam, the manner of sin was in the form of offenses against nature, and all died a natural death. It was not until God imposed religious law, with Adam and Eve being the first to be subject to it, that humans were capable of "legal sin" – trespasses against God's law which leads to spiritual death.

A Worldview Approach to Adam and Eve (3)

We will now examine the dualistic nature of the Genesis text regarding Adam and Eve and the garden of Eden. A Worldview Approach to Scripture interprets Adam and Eve to have been historical persons, but not the first humans to exist on planet Earth based on the science of anthropology. They were merely the first in the *covenant* line leading to Christ (Table 10-1), which line was the only one that the biblical authors were concerned about. A Worldview Approach also does not consider Adam and Eve to be mythological, allegorical, or fictional persons, as is common among many Christians and non-Christians. The garden of Eden is also considered to have been a real historical place, with the four rivers of Eden (Gen. 2:10-14) pinpointing its location at the head of the Persian Gulf in today's Iraq (Chapters 3 and 5).

Were Adam and Eve Real Persons?

Why do many Christians believe that Adam and Eve were non-historical or mythological? Because if Adam and Eve were historical persons, then how does one explain the many fanciful aspects to the garden of Eden story, such as the creation of Eve from Adam's rib and the talking snake? This is where the concept of worldview comes in. In the case of Adam and Eve, the ancient Mesopotamian literary customs and motifs must be kept in mind, because until Abraham moved from Ur (in Mesopotamia) to Palestine, that is where the oral stories of creation and Adam and Eve originated, and it is also where the biblical authors/scribes first wrote down the early chapters of Genesis from these oral histories.

Remember in Chapter 2, we talked about how the literary conventions of the ancient Mesopotamians included analogy, carefully woven into language, and the use of repetition, which included not only words but also numbers, phrases, and structural elements. We also discussed in Chapter 2, how in the worldview of the Mesopotamians, language not only stated facts, but it could also establish them, and how the Mesopotamians loved a play on words. Then in Chapter 9, we elaborated on how figurative language is used in the Adam and Eve story, and how the use of such language was typical of the literary mindset of the ancient Mesopotamians. We also saw in Chapter 9 – in the matter of Adam being formed from the "dust of the ground" – how such a phrase could be a poetic figure of speech, one that conjured up the creation process in the minds of the people living at that time. These are all examples of how the worldview of the ancients affected the writing of the Genesis text.

As another example, consider the serpent motif of Genesis 3. In ancient Near East writings, serpents and serpentine creatures played prominent roles as adversaries of both humans and gods. The snake in the Sumerian epic of Gilgamesh is somewhat reminiscent of the serpent in Genesis 3 in that it deprives the hero Gilgamesh of immortality. Serpents similarly oppose humans and gods in other ancient Near Eastern literature, such as in the Mesopotamian *Enuma Elish* myth and in the Egyptian pyramid texts. The important point here is that: while the Adam and Eve/garden of Eden story involved real people residing in a real place, the writing of this story by the biblical authors was commensurate with the use of figurative images in narratives common to the ancient Near East. How else to describe the appearance of Satan in the garden of Eden except by using the snake motif, since that was the appropriate imagery in the minds of these ancient peoples (Fig 9-8)?

Other Creation Accounts Besides Genesis

The above Gilgamesh epic, *Enuma Elish*, and Egyptian examples are mentioned because they bring up an essential question: How has the cultural interaction between Adamite and non-Adamite lines influenced the writing of stories similar to the ones found in Genesis? This matter is of great importance in judging whether mixed-culture stories like the creation account, Adam and Eve, and Noah's flood are historical or mythological. When addressing the problem of mixed-culture stories, something that was briefly mentioned in Chapter 1 should be recalled: Our modern concept of history is that it be based on "just the facts," but that wasn't the concept of history for ancient peoples. They interpreted history from the standpoint of their various cultural and religious ideas. The basic worldview idea behind other creation accounts is this: the events behind these stories are historical, but the stories themselves were written down from the worldview of each people group that orally transmitted them from pre-literacy days.

A scenario is proposed in Figure 10-2 of how the story of Adam and Eve in Genesis 1 may relate to Mesopotamian creation myths and how these stories may have been passed down to a later generation of biblical authors. (For Egyptian creation myths, refer to the book *In the Beginning...We Misunderstood* by Miller and Soden.) The invention of cuneiform writing happened around 3000 B.C., but narrative writing didn't commence until ~2500-2000 B.C. (Chapter 5). Since Adam and Eve lived about 5000 B.C., their story must have been transmitted orally for about 2,000-3,000 years before it could have been written down. In other words, this was a prehistoric (pre-writing) story that involved a long transmittal, so it is understandable why the story is told in the figurative language and numerological style of the ancient Mesopotamians, according to how sacred texts were supposed to be written (Chapters 2 and 4). In other words, the Genesis author/scribes were simply using common literary motifs to convey truths about humanity that were familiar topics in their ancient world. This world-view input into the Genesis text does not imply, however, that God did not directly interact with a historical Adam and Eve or that his Holy Spirit did not inspire the Genesis authors.

Other creation stories, such as the Akkadian *Enuma Elish* or the Sumerian Enki myth, may have stemmed from an original Adamic story (Fig 10-2), but these Near Eastern creation myths provide a striking contrast to the Genesis creation story and may share no direct relationship to it. Rather, they are clearly the product of the myth-based, polytheistic Mesopotamian culture in which Adam's descendants lived as a sub-group. Peter Enns, in his book *Inspiration and Incarnation*, states that these ancient Near Eastern creation myths are almost certainly older than the creation accounts recorded for us in the Bible because the Hebrew language that we know from the Old Testament did not exist in the second millennium B.C. However, this does not

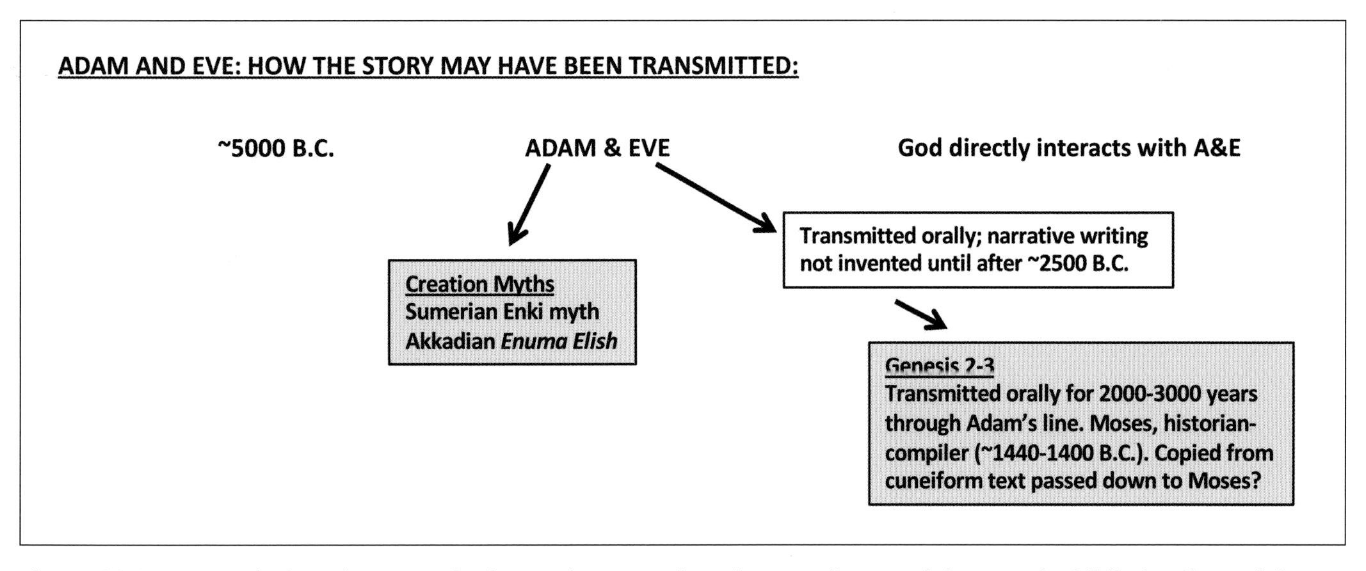

Figure 10-2. How and when the story of Adam and Eve may have been orally passed down to the biblical authors of the written text, and how it may have influenced other Mesopotamian creation stories. In oral tradition, one generation repeats the story by spoken word, to be memorized by the next generation. This is quite typical of preliterate societies: there is an almost ritualistic compulsion to keep track of ancestry because it establishes their identity as a people group. In the case of Adam and Eve, the story would have had to have been orally transmitted for about 2,000 to 3,000 years.

OTHER CREATION STORIES

The fact that the Hebrew version of the creation story came much later than the cuneiform creation myths of Akkad and Sumer does not preclude the possibility that the original (pre-writing) story of Adam was considerably older than all of these written versions, or that the original written version could have been in Sumerian rather than in the much later language of Hebrew. A Worldview Approach to these pre-writing stories considers it likely that the mythical versions of the creation of humans and the biblical version of Adam and Eve all derived from a much earlier common source – that is, from the original Adam and Eve story passed down orally to different people groups over a time span of 2,000 to 3,000 years. Why couldn't the Akkadian and Sumerian versions be borrowed from the original Adam and Eve story? Why couldn't these stories be mythological because they were written down by cultures that had polytheistic views? It should be expected that the Israelites held many concepts and perspectives in common with the rest of the ancient world, but this is far different from suggesting that their literature was borrowed or copied from other Near Eastern texts.

preclude the possibility that the original (pre-writing) story of Adam was considerably older than either the Akkadian or Sumerian written versions, or that the original written version was in Sumerian rather than in the much later language of Hebrew. A Worldview Approach to these pre-writing stories considers it likely that the mythical versions of the creation of humans and the biblical version of Adam and Eve *all* derived from a much earlier common source – that is, from the original Adam and Eve story passed down orally to different people groups (Fig 10-2). Thus, a Worldview Approach is decidedly different from that of (strict) Evolutionary Creationist Daniel Harlow:

> The parallels [of the Genesis and Mesopotamian texts] cited above should suffice to establish that virtually all of the narrative details in Genesis 2-8 are borrowed from Mesopotamian mythology but transformed to craft new stories with a decidedly different theology.

Why does the Genesis story have to be "borrowed" from Mesopotamian mythological stories? Why couldn't the Akkadian and Sumerian versions be borrowed from the original Adam and Eve story? Why couldn't these stories be mythological because they were written down by cultures that had polytheistic views? It should be expected that the Israelites held many concepts and perspectives in common with the rest of the ancient world, but this is far different from suggesting that their literature was borrowed or copied from other Near Eastern texts.

Adamic Covenant: Adam and Sin

The various points of the Adamic Covenant are summarized in Table 10-1. The Bible doesn't elucidate a specific Adamic covenant, except in Hosea 6:7 (NIV): "Like Adam they have broken the covenant," but the sentence in Genesis 4:26 (NIV) "at that time people began to call upon the name of the Lord" implies that Adam was in a covenantal relationship with God at this point in time. The first covenant set up with humanity by God in the garden of Eden initiated the following chain of cause and effect: sin, judgment of sin, blood-sacrifice atonement for sin, spiritual death or spiritual (eternal) life. That is why the story of the garden of Eden and Adam's sin is so crucial, because it heralds God's salvation plan for all humankind, leading directly to Christ, both in the physical sense that Adam was the first in a biological line to Christ and in the spiritual sense that Christ is the final atonement for sin. It was the first direct interaction of God with humanity, but all of the symbolism throughout the rest of the Bible hinges on this first covenant, even up to Revelation where it says, "To the one who is victorious, I will give the right to eat from the tree of life, which is in the paradise of God" (Rev. 2:7, NIV) (Fig 10-3).

Essentially, what we are talking about here is the institution of a new world *spiritual* order for humans. A standard of right and wrong is being set up, a way to be judged and forgiven for that wrong, with eternal life offered by the grace of God. This new order was to be played out through the history of one specific human lineage (Adam's) in the Old Testament, and would end in the New Testament with Christ when generic humans (the Gentiles) become grafted onto Adam's line (Rom. 11:17-24). The overall purpose of the Old Testament was to chronicle the history of God's covenants with this one people group.

Figure 10-3. *The Tree of Life* by Currier & Ives, 1892. The "Holy City, the new Jerusalem" of Rev. 21:2 (NIV) is shown in the middle right. *Image courtesy of the Library of Congress.*

A Worldview Approach to Noah's Flood (4)

We again return to our timeline and now to Noah's flood (Fig 10-1). A Worldview Approach interprets Noah to have been a historical person and the flood a historical event, but it does not consider the flood to have been global, as maintained by the Young-Earth Creationist position of flood geology. There are many reasons – both theological and scientific – for favoring a local flood in Mesopotamia. These have been covered in Chapters 6 and 7 and will not be repeated here. What will be discussed is the dualistic historical/figurative nature of the flood narrative.

The worldview of the biblical authors can be applied to the Noah story, just as we did for the Adam and Eve story. Again, the main problem with following a so-called "literal" account of Noah's flood is interpreting the text from a modern scientific worldview rather than from the pre-scientific worldview of the Mesopotamians. For example, as discussed in Chapter 6, when the Genesis flood story uses the word *earth,* it does not mean planet Earth (as we in the modern world use the word). It means the earth or ground, as is its sense in Genesis 1:10: "God called the dry land earth" (KJV). Similarly, the use of "universal language" in the text (e.g., "all," "every," "under heaven") was simply a type of emphatic speech used in that culture. So, while it seems that the narrative is describing a global flood, it is really only describing a local flood in the land of Mesopotamia. In their (the Mesopotamians') culture, Noah's flood was considered to be the "primitive flood catastrophe" (*abubu*) that wiped out the entire antediluvian world – not the whole world (planet Earth), but the only world the Mesopotamians knew about from their limited geographical knowledge base.

A Worldview Approach to Scripture disagrees with all three of the creationist positions on Noah and the flood. It disagrees with the (strict) Evolutionary Creationists who view both Noah and the biblical flood as being non-historical rather than historical. It disagrees with the Progressive Creationist view, where supposedly Noah's flood happened not long after the approximate 50,000 YBP date assigned to Adam and Eve; rather, it places Noah's flood in Mesopotamia in the time frame of about 2900 B.C. when large boats were being built using the tools then available for ship building (see Chapters 5 and 6). It severely disagrees with the Young-Earth Creationist/flood geologist so-called "literal" interpretation of the Genesis flood being global, which position changes the text from a historically based account into one totally unsupported by the findings of modern science.

Was Noah a Real Person?

From the 600-year-old age of Noah and from the supposed claim of a global flood, many biblical scholars discredit Noah as a historical person. However, *if* Noah was a real person, is there any evidence of his historicity? Remember in Chapter 9, in the Table of Nations section, that we talked about how ethnological studies have traced the linguistics and lineages of Noah's sons Shem, Ham, and Japheth to parts of the Near East (Fig 9-6). The fact that these studies are scientifically credible implies that Noah and his immediate descendants *were* non-fictional. With regard to lineage tracing, a present-day tribe

BOX 10-1.

"WE ARE THE PEOPLE OF 'AD"

"While investigating the well of the Oracle of 'Ad, we had visitors, tribesmen who drifted down from the mountains. Their bearing was elegant; their hair, done up in fine braids and tinted blue, had the fragrance of frankincense. Members of the Shahra tribe, they spoke in addition to Arabic, their own peculiar chirping, sing-song language, called by early explorers 'the language of the birds.' They confirmed that, indeed, the well was still known as a well of the People of 'Ad...and one of their number, speaking in crisp, Cambridge-accented English, matter-of-factly told us, 'You know, we *are* the people of 'Ad." Nicholas Clapp, *The Road to Ubar: Finding the Atlantis of the Sands*, p. 139. Photo on youtube.com; "Harun Yahya Perished Nations, The People of Ad Part 1."

linked to Noah are the Adites ("people of 'Ad"; Box 10-1) who claim to be descended from 'Ad, the great-great-grandson of Noah through the line of Shem (Shem-Aram-Uz-'Ad), and who, according to Islamic tradition, were the first (Semitic) inhabitants of the Dhofar region of southern Arabia (Fig 3-1) and the legendary "lost city of Ubar." These people still speak Shehri, an ancient dialect of the Semitic (meaning "from Shem") language.

Other Flood Accounts besides Genesis

Archeological and documental evidence exist both for a historical Noah and the flood. This documental evidence comes in the form of the Akkadian Atra-hasis epic, the Gilgamesh epic, and the Sumerian flood story, and King List (Fig 10-4). These ancient Mesopotamian texts attest to a great flood survived by Ziusudra (or Ut-napišhtim or Atra-hasis, alternate Babylonian names for Noah), who was supposedly "king" of the ancient city of Shuruppak in Sumer (southern Mesopotamia; Fig 3-3). These Mesopotamian cuneiform accounts of Noah's flood are known to predate the first written biblical accounts. According to Y. Chen in his book *The Primeval Flood Catastrophe*, the Mesopotamian stories of the flood told in the figurative and mythological tradition of the Mesopotamians, came down to us sometime between the Early Dynastic Period III and Old Babylonian Period (Table 9-2). The Mesopotamian account is known to be a late version of the Gilgamesh epic, which is itself an excerpt from the Atra-hasis epic, the oldest Babylonian version, dating to around 1646-1626 B.C. So if Moses was the historian author/compiler of Genesis, as is suggested later in this chapter, then the earliest confirmed date for a written Genesis flood story is around 1440-1400 B.C. – keeping in mind that this story could have been passed down in oral and/or written form long before Moses recorded it in Genesis.

The assumption that the primeval flood story might have existed as early as the Early Dynastic III Period (2600-2350 B.C.) is widely held among biblical scholars. However, Chen favored the position that while the different Mesopotamian flood stories must have started off with some basic factual information of the cities and rulers involved, these

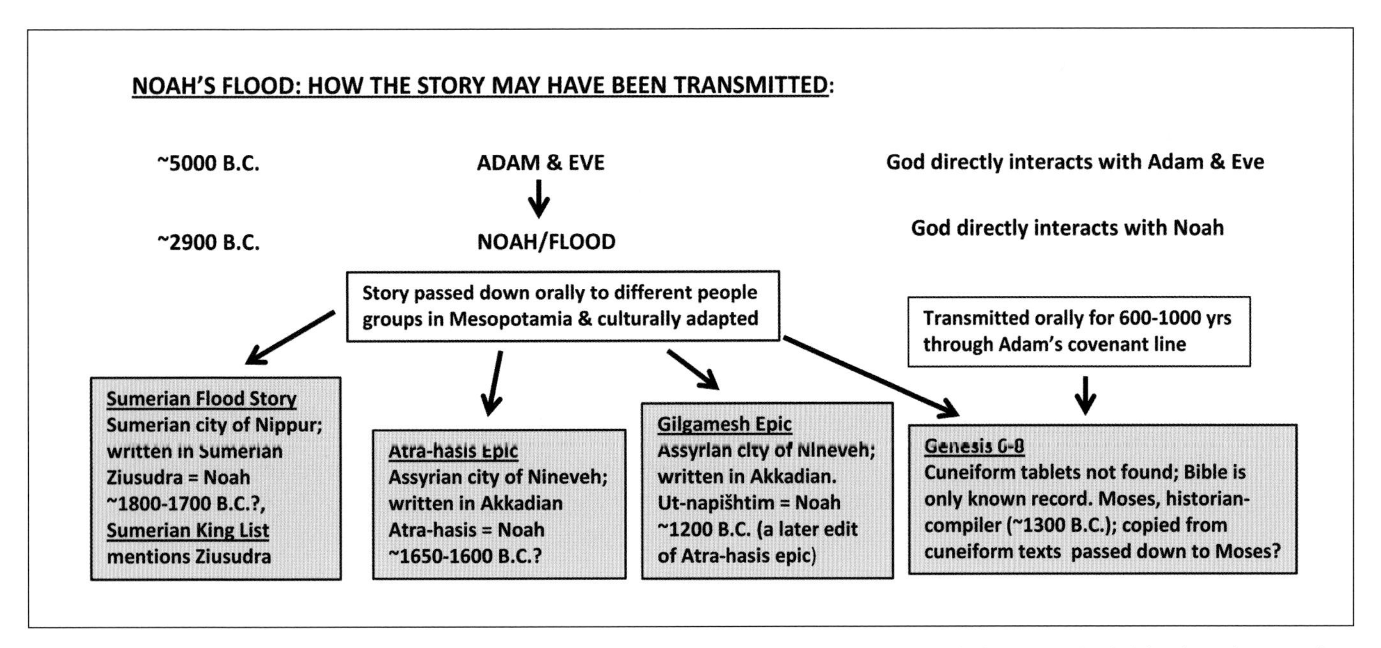

Figure 10-4. How and when the story of Noah and the flood could have been passed down to the biblical scribes, and how a real event could have spawned the biblical story and other Mesopotamian flood epics; that is, the Babylonian, Assyrian, and Hebrew narratives could have had a common historical source, which each narrative adapted to their own cultural context. The Sumerian flood story (Fig 5-1A) is the oldest reference to the deluge; the biblical account of the flood is the youngest.

non-biblical stories became assimilated into their present form only in the Old Babylonian Period (2000-1600 B.C.; Table 9-2). This assimilation process developed gradually, often by the accretion of different versions of the story based on the political ideology of each different Mesopotamian region; that is, these different flood traditions were the *cultural* products of their specific time and place. Or, as biblical archeologist Kenneth Kitchen stated about these different flood traditions:

> In detail the differences are so numerous as to preclude either the Mesopotamian or Genesis accounts having been copied directly from the other....So, an epochally important flood in far antiquity has come down in a tradition shared by both early Mesopotamian culture and Genesis 6-9, but which found clearly separate and distinct expression in the written forms left us by the two cultures.

Was the biblical account of Noah's flood only another cultural product of its time? Were the Mesopotamian texts taken from Genesis, was Genesis taken from them, or did they all come from a common literary source of an actual flood event? A Worldview Approach favors the last alternative: that a huge "1,000-year" flood inundated a substantial part of the Mesopotamian hydrologic basin (Chapter 5), a flood that would have disastrously affected a number of people groups in that area. It also favors Kitchen's view that all of these accounts came from a "historical memory" event, where these different people groups wrote down their own versions (including the Genesis version) of the story from their own religious/cultural worldviews, and that is why these versions differ from each other (Fig 10-4). It has not, however, been shown that the Hebrew version was borrowed, even indirectly, from the Babylonians. But the Genesis account would have had the advantage of being written by the covenant-line descendants of Noah, where the flood story would have been faithfully transmitted orally, and then in written cuneiform, to succeeding generations. However, this account would have still been written by the biblical authors/scribes from their limited scientific worldview (they had no concept of a global planet Earth) and from their narrative style of writing (use of sacred numbers and global-language expression).

Noachian Covenant: Noah and the Flood

Again, refer to Table 10-1 for the various elements of the Noachian covenant. By this time, generic modern humans (non-Adamites) existed outside of Mesopotamia in Europe, Australia, the Americas, and elsewhere (Fig 9-6). But what had happened to God's covenant line in Mesopotamia over the years since Adam? The descendants of Cain gloried in violence and vengeance, as illustrated by Lamech's boasting that if Cain shall be avenged seven-fold, then truly Lamech will be avenged seventy and sevenfold (Gen. 4:24). Genesis 6:11 says that "the earth also was corrupt before God, and the earth was filled with violence" (KJV). The line of Adam had become corrupted by the polytheism of non-Adamites with whom they had intermarried (Gen. 6:2).

But there was one man from Adam's line who was just and perfect in the sight of God (Gen. 6:9), one man who still retained Abel's kind of faith and who was not corrupted by the sin of the world around him, and that was Noah. So what should be done with this evil generation? To purge Adam's line from sin took a drastic action: the flood. The flood did not have to be global in scope because non-Adamite peoples outside of Mesopotamia were not of God's covenantal line. Thus, the "all flesh" of Genesis 7:21 included people in the line of Adam (except for Noah and his family), non-Adamites who also lived in southern Mesopotamia at that time, and animals that lived in the flooded area. In other words, God judged a faithless *covenant people* with the flood, not the human and animal populations of the entire Earth! Non-covenantal peoples, such as the Nephilim and others who lived outside the geographical limit of the flood or those who were flood survivors, returned after the flood to build over the ruins of pre-flood cities. That is why the Early Dynastic I culture immediately follows the Jemdet Nasr culture in places like the archeological site of Fara (i.e., Shuruppak, Noah's "home town;" Fig 3-3).

At the end of the flood, God established another covenant with Adam's chosen line: there would never be another flood that would wipe out this line, and the rainbow was to be a sign of that promise (Fig 10-5):

> I have set my rainbow in the clouds, and it will be the sign of the covenant between me and the earth. Whenever I bring clouds over the earth and the rainbow appears in the clouds, I will remember my covenant between me and you and all living creatures of every kind. (Gen. 9:13-15, NIV)

Note in Genesis 9:8-9 that God's promise was given to Noah, his sons, and descendants (the covenant line), *not* to all people then living on planet Earth! Therefore, the rainbow of Genesis 9:13 does *not* imply a global flood, as claimed by flood geologists (see Chapter 6). This so-called "literal" interpretation comes from assuming that "earth" (Gen. 9:11) means the whole planet Earth, and that Adam and Eve's descendants included all people living on Earth at the time of the flood. Numerous *local* floods have occurred on Earth since Noah's flood, but they haven't wiped out *Adam's entire line* or all living creatures of every kind, so God's promise has been kept to his covenant people.

But did Noah's flood stamp out evil? In just a few generations, the story of the Tower of Babel

Figure 10-5. *Noah Gives Thanks for Deliverance* (1901), by Domenico Morelli (1823-1901). According to the Young-Earth Creationist, global-flood model, this was the time when humans and animals spread over the entire planet Earth, since "all flesh" had died in the flood except for those on Noah's ark (i.e., the position of "anthropological universality" where all humans are descended from Noah). *Wikimedia Commons.*

shows that Noah's descendants were again building high places of worship (ziggurats) to pagan deities (page 130). But God kept his promise not to send another flood that would wipe out Adam's line. Instead, he began to unfold another part of his plan – one intended to eventually spread the news of the one true God to *all* nations and peoples of planet Earth.

A Worldview Approach to the Patriarchs (5)

A Worldview Approach to the scriptural and scientific evidence considers all of the patriarchs from Adam to Abraham to have been historical people, as documented by the genealogies of the Old and New Testaments. As already discussed in Chapter 9, the Bible goes to great lengths to establish the genealogy of Adam to Christ in Genesis, Numbers, Chronicles, Ezra-Nehemiah, Matthew, and Luke. Why such a pre-occupation with detailed descent records if it wasn't theologically important that Adam be genealogically related to Christ, the "second Adam?" If these genealogies are not comprised of historical people, then where do the historical people start and the non-historical people end? Furthermore, what impact does this kind of thinking have on the integrity of the Old Testament record – which Jesus quoted often in the New Testament?

Were the Patriarchs Real Persons?

If the patriarchs were real persons, is there any evidence for their historicity? There is very little evidence outside of the Genesis account, which is not surprising considering the antiquity of these stories. However, there is some evidence for the existence of Terah (Abraham's father), Nahor (Abraham's grandfather), and Serug (Abraham's great-grandfather) because all three of these names have survived from antiquity as names of towns in the vicinity of Haran where these patriarchs supposedly resided (Gen. 11:31-32). Also, there has been DNA testing of Arabs and Jews linking both groups to a common male ancestor (Abraham?) several thousand years ago.

If these patriarchal genealogies do comprise a historical record, then why do most people dismiss the patriarchs so readily? One reason is because the patriarchal ages are of unbelievably long duration and that automatically makes the patriarchs suspect as historical persons. What people (including some theologians) are missing in their rejection of these patriarchal genealogical records is the numerological worldview of the biblical authors (Chapter 4). The numbers dealing with patriarchal ages are numerological (sacred numbers) rather than numerical (real numbers), and this tradition of exaggerated "long reigns" for gods and kings seems to have been a common religious tradition for the peoples of the ancient Near East. *They* (the biblical authors/scribes plus the people they wrote for) knew that these numbers were exaggerated, but this did not concern them because their worldview included a dual concept of numbers. *We* have no such dual conception of numbers in our modern worldview, and so the exaggerated ages in Genesis make the patriarchs unbelievable to us.

The Critical Scholars

Young-Earth Creationists and Progressive Creationists believe that the patriarchs were historical people, while some (strict) Evolutionary Creationists consider all of the patriarchs from Adam down to, but not including, Abraham to be non-historical persons. In addition, there is another set of theologians who deny the genealogical record of Adam, including all the patriarchs from Adam to Abraham, Isaac, Jacob, and Joseph, and up to and including Moses and the Exodus. These are the "critical scholars" (also called "minimalists," or followers of "higher criticism" or "historical criticism") who were briefly mentioned in Chapters 2 and 3. These biblical scholars take the position that the Genesis stories are merely a literary reflection of conditions under the Hebrew Monarchy or later during the Exile (Fig 10-1). The minimalist movement started in the 1870s with Julius Wellhausen and others, and continues up until today in a number of different forms and schools. Wellhausen proposed – based mainly on word studies – that a number of different biblical authors (J, E, D, P) had actually written Genesis much later than the patriarchial period – during the Monarchy/Exile, and even as late as the fourth to third centuries B.C.

There are two claims of critical scholarship that will now be discussed in relationship to the historicity of the patriarchs: (1) the scientific evidence for placing the patriarchs in an Old Testament time frame rather than during the Monarchy/Exile, and (2) different literary authors or sources for different sections of Genesis.

(1) As mentioned at the beginning of Chapter 1, there are problems with basing biblical interpretation on word studies alone, such as was done by Wellhausen, because they do not account for the scientific or historical evidence, which in this case involves almost 150 years of Near Eastern archeological and literary findings since Wellhausen. We have just mentioned the DNA and dating evidence for the antiquity of the patriarchs. Also, in a few chapters, we mentioned "historical memories" that attest to the antiquity of the text, one prime example in Chapter 3 being the Pishon River: How could an author who lived in the eighth to fifth centuries B.C. have known about a river that had dried up some 1,000 to 3,000 years previously? In Chapter 9 we discussed the anthropological and archeological evidence for placing Adam in a Neolithic-Chalcolithic time frame and concluded from this evidence that the biblical authors/scribes seemed to have had an approximate idea of how long before them Adam and his descendants lived. That is, based on these various lines of evidence, the patriarchs seem to *fit* within the time frame and culture of the Old Testament story, and for these reasons, a Worldview Approach agrees with archeologist Kenneth Kitchen concerning the antiquity of the Genesis text:

> The content of the Hebrew Bible is not "invented history" and never was (p. 462). [The Wellhausen] scheme, pumped into generations of students, both future and practicing biblicists, is and (alas!) always was pure, unadulterated fantasy. It clashes horribly with real-life historical profiles for the cultures that we can test (p. 487). The world of Gen. 12-50 is certainly *not* that of the monarchy period...it fits only the period before that, the twentieth to seventeenth centuries [ca. 2000-1600 B.C].... (p. 495)

(2) However, the claim of critical scholars that different sources contributed to different sections of the Genesis text seems valid – in fact, it should be *expected* if different authors/scribes were writing down versions of different oral stories, and then trying to piece them together into a coherent whole. For example, in Chapter 9 it was mentioned how there is a break between the "first creation" and "second creation" in the middle of Genesis 2:4, with "These are the generations of the heavens and of the earth when they were created" (KJV) being an account written by a Chapter 1 source, and "in the day that the LORD God made the earth and the heavens" (KJV) being another source's account of Chapter 2. Most biblical scholars agree that these two accounts came from different sources because of their dissimilarity in style and key words (such as *Elohim* for God in Chapter 1, and *Yahweh* in Chapter 2). However, this does not necessarily imply that Genesis 2 is a completely separate story from Genesis 1, only that they are but two renditions that focus on different aspects of the story. Furthermore, different sources definitely do *not* imply that these stories were made up during the much-later Monarchy/Exile period!

Another possible multiple-source example in Genesis involves the parallel lines of Cain (Gen. 4:17-18) and Seth (Gen. 5:6-25) (Box 10-2). While the agreement is not perfect, there is enough similarity to suspect two source versions of these same genealogies. Although some scholars have attributed these name similarities to Cain's line and Seth's line living in close geographical proximity to each other, it seems odd that Jared and Enoch (an especially important person in the story; Gen. 5:24) are included in the Seth list but not in the Cain list. This suggests that different scribes may

Box 10-2. Two different versions of the genealogies of Cain and Seth							
Cain	Enoch	Irad	Mehujael			Methushael	Lamech
Seth	Enosh	Kenan	Mahalalel	Jared	Enoch	Methuselah	Lamech

have had access to different sources for the same Adam line. In the flood story, the author/source of Genesis 6 mentions that Noah should bring *two* of every sort of creature onto the ark (Gen. 6:19), while in Genesis 7 the source requests pairs of *clean* animals be brought by *sevens* into the ark (Gen. 7:2). This variation in the story line is no different from the New Testament gospels where Mark may have heard a slightly different version of the story of Christ's life than Matthew, Luke, or John, but it does suggest that different parts of Genesis may have been written by different authors, with the Genesis story later compiled by Moses into one amalgamated version.

Abrahamic Covenant: Abraham, Man of Faith

With Abraham, God began his plan to include generic humans outside of the line of Adam. In Abraham, all of the families (nations) of the world were to be blessed (Gen. 12:3), not because Abraham was the *biological* father of all peoples, but because he was to *become* their *spiritual* father by *faith*. Abraham had a test of faith, and, as with Abel, it involved a blood sacrifice. God asked Abraham to sacrifice Isaac, the promised son of faith, and Abraham was obedient unto God by faith: "Consider Abraham: he believed God, and it was credited to him as righteousness" (Gal. 3:6, NIV). The purpose of this covenant with Abraham was for Abraham's specific descendant line to take the news of the one God to all nations (Gal. 3:8): "Scripture foresaw that God would justify the Gentiles by faith, and announced the gospel in advance to Abraham: 'All nations will be blessed through you'" (NIV).

A Worldview Approach to Moses (6) and the Monarchy/Exile (7)

A Worldview Approach disagrees with the critical scholars and considers the Exodus to have been a real event that happened around 1300-1200 B.C., and that Moses and Aaron were real people who led the Israelites out of Egypt. This position comes from both the scientific evidence and from the New Testament. DNA evidence has confirmed a distinct paternal genealogy for Jewish priests as claimed in Exodus 40:12-16, and although this evidence doesn't prove that Aaron was a real person, it is supportive of that claim. In the New Testament, the historical personage of Moses was affirmed by Jesus and the apostles when Moses appeared and talked to Jesus at the time of his transfiguration (Matt. 17:3; Mark 9:4; Luke 9:30).

A Worldview Approach also considers Moses to have been the author of Genesis – albeit the *historian* author. While Moses is the traditional author of Genesis, this authorship is only assumed. Various Old Testament books ascribe the "books of the Law" (Exodus, Leviticus, Numbers, Deuteronomy) to Moses, but nowhere in the Bible is Moses specified as being the author of Genesis. However, it is not unreasonable that Moses transcribed Genesis since he was well educated in Egypt (Acts 7:22) and was probably able to read, write, and translate many ancient languages, including ancient cuneiform texts that may have been passed down to him by his Levite ancestors. The *NIV Archaeological Study Bible* describes this view of Mosaic authorship thusly:

> We might view Moses as an editor/historian who, in addition to receiving God's direct and supernatural communication, drew

together details of the family histories of Abraham and his descendants, as they existed in the Israelite community in Egypt, into a single text.

In addition to the family histories of Abraham and his descendants, a Worldview Approach also attributes the compilation of the Genesis 1-11 stories to Moses. In other words, Moses could have *compiled* Genesis, but he did *not* write it from his own fifteenth century B.C. perspective because the worldview implicit in these ancient stories was already set long before the time of Moses. Sometime after narrative writing was invented (~2500 to 2000 B.C.), these ancient stories could have become inscribed onto cuneiform tablets by scribes from the line of Adam who faithfully transmitted the oral and written versions passed down to them. For the history of the written book of Genesis, consider the following proposed scenario (also shown in Fig 10-7):

The Genesis stories could have originated from different ancient sources and then passed down to Abraham who lived about 2000 B.C. Then Abraham could have taken this early Genesis account with him when he left Ur (southern Mesopotamia; Fig 3-3) for Haran – taken either in oral form or perhaps in written form since the colophon "This is the book of the generations of Adam" (Gen. 5:1, KJV) could imply a written genealogy. Abraham passed on family tradition to Isaac; Isaac to Jacob; Jacob to Joseph and his brothers. They in turn maintained an ongoing tradition (written, oral, or both) of updating family archives that were eventually passed down the patriarchal line to Moses, who translated and compiled these stories into a single chronological narrative sometime between the fifteenth and thirteenth centuries B.C. Historical "memories" and geographical information, as well as old words that had disappeared from the living language before the time of Moses, attest to these stories being handed down from more ancient times.

Later, during the time of the Israelite Monarchy/Exile (~800-400 B.C.), redactor scribes could have edited the Genesis text in order to put it into a smooth, effective, and understandable literary form suitable for that generation. This editing may have included "tidbits" of information that were known by this later time, superimposed over the more basic ancient text. But since redactor scribes would have regarded the texts as sacred, they

WAS MOSES THE AUTHOR OF GENESIS?

A Worldview Approach considers Moses to have been the author of Genesis – albeit the historian author. While Moses is the traditional author of Genesis, this authorship is only assumed. Various Old Testament books ascribe the "books of the law" (Exodus, Leviticus, Numbers, Deuteronomy) to Moses, but nowhere in the Bible is Moses specified as being the author of Genesis. However, it is not unreasonable that Moses transcribed Genesis from ancient texts since he was well educated in Egypt (Acts 7:22) and was probably able to read, write, and translate many ancient languages. In other words, Moses could have compiled Genesis, but he did not write it from his own fifteenth century B.C. perspective because the worldview implicit in these ancient stories was already set long before the time of Moses. Sometime after narrative writing was invented (~2500 to 2000 B.C.), these ancient stories could have become inscribed onto cuneiform tablets by scribes from the line of Adam who faithfully transmitted the oral and written versions passed down to them; then these stories could have been passed down to Moses.

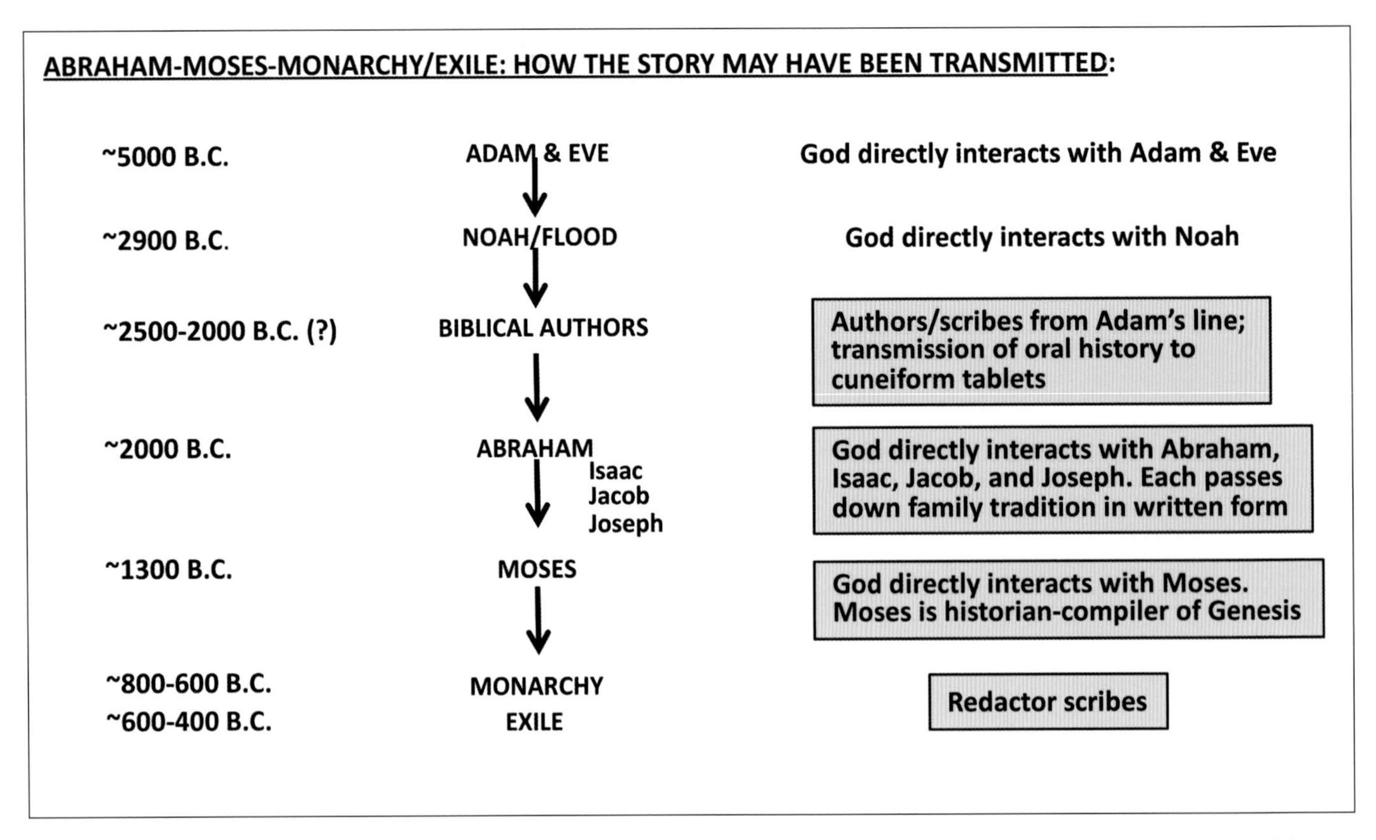

Figure 10-7. How the Genesis stories could have been passed down through Moses to redactor scribes at the time of the Monarchy/Exile.

would have tried to preserve the literary integrity of these ancient manuscripts. Also at this time, the Genesis text could have been put together in a logical chronological sequence and converted by author/scribes into classical Masoretic Hebrew – a form of the Hebrew language that did not even exist in the time of Moses. Thus, all of these elements could figure into the authorship of Genesis.

Was Moses a Real Person and the Exodus a Real Event?

Although most recent archeologists that have worked in Egypt and other parts of the Near East have been critical of the Old Testament claim that there was an Israelite presence in Egypt and an Exodus to Canaan from Egypt, Jewish and Christian faith is strongly dependent on the historicity of these events. Part of the problem of finding archeological evidence for the Exodus is that biblical scholars cannot agree on the date or the route of the Exodus. The "early date" view of biblicists is that the Exodus occurred around 1445 B.C. during the reign of pharaoh Amenhotep II, and the "late date" view is that it occurred around 1260 B.C. during the reign of Ramesses II (Ramesses the Great). Since there is only scanty archeological evidence for either date or place, critical scholarly consensus has been that there was no Exodus as claimed in the Bible. However, it should be remembered that *the absence of evidence is not evidence of absence*. Biblical archeologist Kenneth Kitchen, who favors a late date for the Exodus, states this absence in his book *The Reliability of the Old Testament*:

> The Delta [Egypt's East Delta where the Hebrews resided] is an alluvial fan of mud deposited through many millennia by the annual flooding of the Nile; it has no source of stone within it. Mud, mud and wattle, and mud-brick structures were of limited duration and use, and were repeatedly leveled

> and replaced, and very largely merged once more with the mud of the fields. So those who squawk intermittently, "No trace of the Hebrews has ever been found" (so, of course, no exodus!), are wasting their breath.... On these matters, once and for all, biblicists must shed their naïve attitudes and cease demanding "evidence" that *cannot* exist.

So where is the evidence of Hebrew inhabitation in the Nile Delta? It is probably buried beneath, or contained in, over 3,000 years of erosion and clay build up of soil, which has been tilled and farmed from then up until today. Also, with the advent of satellite imagery, large-scale surface features and patterns that are difficult to recognize on the ground are now being used to trace the most probable route of the Exodus out of Egypt and into Sinai.

In addition to archeological evidence (or the lack of it), the model of Humphreys and Waddington of a known solar eclipse that occurred in 1207 B.C., possibly in conjunction with Joshua's victory at Gibeon (Fig 1-3), gives credence to the late-date view of the Exodus. This date of 1207 B.C. fits with a late-date view given that the Israelites wandered 40 years in the wilderness plus Joshua's entrance into Canaan and his subsequent military campaign, and it also fits with Ramesses the Great being the pharaoh of the Exodus. It does not fit with the early 1445 B.C. view for the Exodus.

Sinaic Covenant: Moses and the Law

Four hundred and thirty years after Abraham, the descendant of Abraham, Moses, was called upon by God to institute another covenant with the chosen line of Adam (by now the descendant line was called "Israel"). This was a covenant of law (see page 158, Moses), which did not negate the earlier covenants that God had made with Adam, Noah, or Abraham. It simply was added on to these earlier covenants.

Why a law covenant? There were a number of reasons why the law was added. One practical reason was that the Israelites were – for the first time – becoming a nation state and the new nation needed a law code. Another reason was that the righteousness of the law was intended to set Israel apart (Lev. 20:22-26) so Israel might be holy as God is holy (Lev. 11:44), in order that God's "chosen people" would be a witness to other nations (Isa. 42:6): "I, the LORD, have called you in righteousness; I will take hold of your hand. I will keep you and will make you to be a covenant for the people and a light for the Gentiles" (NIV). Still another reason was that through the law, the Israelites would become more conscious of sin. The law very specifically spelled out what sin was, so the Israelites would be aware of why they were being judged (Rom. 3:20): "For by the law is the knowledge of sin" (KJV). The law brought conviction of sin; it showed that redemption was needed.

A Worldview Approach to Christ (8), the Church Age (9), and Present (10)

We now come to the end of our Old Testament timeline and to the focal point of its entire 5000-year history: Christ. Adam, Noah, the patriarchs, Abraham, Moses, the prophets – the whole story of God's covenants with Adam's line – all lead up to Christ and his atonement for the sins of humanity. It is Paul, in Romans 5:12-18 and in 1 Corinthians 15:21-22, 45-47, who relates Christ to Adam and thus back to the beginning of our timeline (Fig 10-1).

Paul compares Christ to Adam in the matters of sin, death, and grace. Paul connects sin with death in Rom. 5:12 where he says: "Sin entered the world through one man [Adam], and death through sin, and in this way death came to all people, because all sinned" (NIV). As previously discussed, the "death through sin" that Paul is talking about is *spiritual* death, not physical death. Next, comes grace and how it ties in with Adam's sin (trespass). Romans 5:15 says: "For if the many [spiritually] died by the trespass of the one man [Adam], how much more did God's grace and the gift [of righteousness, v. 17] that came by the grace of the one man, Jesus Christ, overflow to the many" (NIV). So we now have the following logic: Adam's sin brought spiritual death upon all humankind – upon Adamites and non-Adamites alike. But grace to cover that sin comes through Jesus Christ – both to the line of Adam (Israelites) and to the "Gentiles" because Paul tells us in Romans 11 that Gentiles are "grafted" into Adam's line by faith (Fig 10-1). It is this covenant of grace that has been the foundation of the church age up until today, a covenant that extends to all people groups on planet Earth and not just to the line of Adam.

Continuing further with the parallel construction of Paul's logic in Romans 5:15: Since Christ is assumed by Paul to have been a historical person through whom grace was given to all humanity (a foundation on which our faith lies), then it must also be assumed that Adam was a historical person through whom sin was conferred on humanity. Romans 5:18 also implies the same thing: it says that Adam's "one trespass" brought condemnation to all men. How can a person who is not real commit a trespass? Conversely, using Paul's logic: If one assumes that Adam was *not* a historical person, and that he and his trespass are to be taken symbolically, then from the parallel construction of Romans 5:15 shouldn't Christ and his grace also be taken symbolically?

A number of (strict) Evolutionary Creationists, such as Daniel Harlow, consider Adam to have been a symbolic-literary figure and contend that we are not obliged to consider Adam as a historical person based on the Apostle Paul's statements in Romans:

> This paper explains why most biblical scholars regard Adam and Eve as purely symbolic figures....The paper also examines Paul's interpretation of Genesis in his typology of Adam and Christ, arguing that though Paul probably did regard Adam as a historical figure, we are not obliged to. Paul was chiefly interested in Adam as a representative counterpart to Christ, and the role he assigned Adam in the entry of sin into the world was more temporal than causal. The doctrine of original sin does not require that Adam and Eve be historical figures.

First, *most* biblical scholars do not regard Adam and Eve as purely symbolic figures – "most" applies only to those that Harlow feels are worthy of the title "biblical scholar." Second, how can it be stated that Paul was "chiefly interested in Adam as a representative counterpart to Christ?" This sounds like a modern idea, and not the first-century worldview of Paul and what he would have been interested in. From the above verses and other passages in Romans, and from others of Paul's books, it seems like Paul's "interest" was in showing Christ to be the fulfillment of God's interaction with the line of Adam, starting with Adam himself. Third, why aren't we "obliged" to regard Adam as a historical figure if Paul regarded him as such? What about Paul's receiving God's supernatural revelation on the important doctrines of sin and grace? Is this factor to be ignored? The two persons – Adam and Christ – are historically (through genealogy) and theologically (through Paul) linked, and it seems that a denial of the historicity of Adam *does* impact the historicity of Christ and the foundational doctrines of the Christian faith.

Covenant of Grace: Christ, the Final Atonement

The focal point of the whole story, and the historical climax of God's interaction with humankind from Adam onward, was the atonement that Christ accomplished on the cross of Calvary (pages 215, 225, 230). Seven hundred years earlier, Isaiah prophesied of this final atonement, this ultimate blood sacrifice (Is. 53:5-6): "But he was wounded for our transgressions, he was bruised for our iniquities; the chastisement of our peace was upon him; and with his stripes we are healed. All we like sheep have gone astray, we have turned everyone to his own way; and the LORD hath laid on him the iniquity of us all" (KJV). This was the "new covenant" that Jeremiah talked about (Jer. 31:31-33): "Behold the days come, saith the LORD, that I will make a new covenant with the house of Israel, and with the house of Judah....I will put my law in their inward parts, and write it in their hearts; and will be their God and they shall be my people" (KJV). This was God's final new covenant, one that would change the history of the world.

This covenant of grace was given to us in Jesus Christ before the beginning of time (2 Tim. 1:9), but was only revealed through his appearance (Fig 10-1). In the "fullness of *time*" (eternal, evolutionary, and historical time), God sent forth his Son to redeem those under the law and to adopt believing Gentiles into the line of Adam as the "sons of God" (Gal. 4:4-5). Through Christ, Gentiles became grafted into the line of Adam by faith, Abraham became the "father" of the faithful, and the words "he bled for Adam's helpless race" attained a universal meaning. There was no further reason for the blood-sacrifice of animals because Christ became the final atonement for sin. There was no further reason to be in bondage under the law because his grace sets us free. Faith in the incarnate Christ is all that is required for the sins of the whole human race:

> You are worthy to take the scroll and to open its seals, because you were slain, and with your blood you purchased for God persons from every tribe and language and people and nation. (Rev. 5:9, NIV)

Final Thoughts

Why should there be a Worldview Approach to biblical interpretation in the science-Scripture debate? Aren't the old positions of Young-Earth Creationism, Progressive Creationism, and Evolutionary Creationism adequate? The Young-Earth and Progressive Creationist positions are not adequate because they do not resolve the conflict between the biblical text and scientific evidence. The Progressive Creationist position does not incorporate all the anthropological, archeological, and biological (genetic) data into its theological framework and the Young-Earth Creationist position ignores practically all of science or twists it into a pseudoscience that fits its own particular ideology. The (strict) Evolutionary Creationist position correctly recognizes the scientific issues but denies some of the fundamental doctrines of the Christian faith. A Worldview Approach tries to resolve the conflict between science and Scripture by proposing that the basic problem with compatibility involves understanding the pre-scientific worldview of the biblical authors. It is a mistake to try and impose our twenty-first century scientific worldview on the ancient biblical text, which is the main reason why there is so much confusion and contention in the science-Scripture debate.

In applying the Worldview Approach to biblical interpretation it should never be forgotten that the biblical text is God's revelation to humanity and thus foundational, whereas the worldview of the biblical authors only imparts a historical slant to its expression. Once this fundamental premise is lost, the truth and power of Scripture are severely compromised. Not only were the biblical authors inspired by God through his Holy Spirit in the writing of the text, but they also faithfully transmitted the stories handed down to them from more ancient times, even though these stories may be scientifically inaccurate by today's standards. Despite the limited knowledge of the ancients, we should not discard the concept of God's revelation being given to them in the timeline of human history (Fig 10-1). And, even though the biblical authors wrote down this revelation from their own worldview perspective, this worldview itself is part of real history and thus important for biblical interpretation. *Most important: the worldview of the ancients in the biblical text does not obscure the heart of the message of God's love, his forgiveness, and his promise of eternal life.*

REFERENCES

General Reading

Collins, F. S., *The Language of God* (New York, NY: Free Press, 2006) 294 pp.

Davidson, G. R., *Friend of Science, Friend of Faith: Listening to God in His Works and Word* (Grand Rapids, MI; Kregel Publications (2019) 296 pp.

Fischer, R. J., *Historical Genesis: From Adam to Abraham* (Lanham, MD: University Press of America) 227 pp.

Hill, C., Davidson, G., Helble, T., and Ranney, W., eds., *The Grand Canyon, Monument to an Ancient Earth: Can Noah's Flood Explain the Grand Canyon?* (Grand Rapids, MI: Kregel Publications, 2016) 240 pp.

Hoffmeier, J. K., *Israel in Egypt, The Evidence for the Authenticity of the Exodus Tradition* (Oxford, U.K.: Oxford University Press, 1996) 244 pp.

Kitchen, K. A., *On the Reliability of the Old Testament* (Grand Rapids, MI: Eerdmans, 2006) 662 pp.

Lamoureux, D. O., *Evolutionary Creation: A Christian Approach to Evolution* (Eugene, OR: Wipf & Stock, 2008) 493 pp.

Miller, J. V., and Soden, J. M., *In the Beginning...We Misunderstood: Interpreting Genesis 1 in Its Original Context* (Grand Rapids, MI: Kregel Publications, 2012) 220 pp.

Morris, S. C., *Life's Solution: Inevitable Humans in a Lonely Universe* (Cambridge, U.K.: Cambridge University Press, 2003) 464 pp.

Ramm, B. *The Christian View of Science and Scripture* (Grand Rapids, MI: Eerdmans, 1954) 253 pp.

Walton, J. H., and Sandy, D. B., *The Lost World of Scripture: Ancient Literary Culture and Biblical Authority* (Downers Grove, IL: InterVarsity Press Academic, 2013) 320 pp.

Chapter 1: A Worldview Approach

Page 3

What is hermeneutics?: R. McQuilkin, *Understanding and Applying the Bible* (Chicago, IL: Moody Press, 1992) 339 pp.

Page 4

The "go the extra mile principle": H. W. Robinson, *The Christian Salt and Light Company: A Contemporary Study of the Sermon on the Mount* (Grand Rapids, MI: Discovery House Publishers, 1988) 168-169.

Salt and gypsum mined from the Dead Sea: H. W. Robinson, *The Christian Salt and Light Company: A Contemporary Study of the Sermon on the Mount* (Grand Rapids, MI: Discovery House Publishers, 1988) 101.

Page 6

Big Bang Theory: H. Ross, *Hidden Treasures in the Book of Job: How the Oldest Book in the Bible Answers Today's Scientific Questions* (Grand Rapids, MI: Baker Books, 2011) 240 pp.

Reading science in between lines of Scripture: J. H. Walton and D. B. Sandy, *The Lost World of Scripture: Ancient Literary Culture and Biblical Authority* (Downers Grove, IL: InterVarsity Press Academic, 2013) 53.

Topic of Accommodation: P. Schaff, "Accommodation," *in* L. A. Loetscher, ed., *New Schaff-Herzog Encyclopedia of Religious Knowledge*, v. 1 (Grand Rapids, MI: Baker Book House, 1955) 22-24. One of the first to remark on the subject of general accommodation was St. Gregory Nyssa (A.D. 355-394): "...in the various manifestations of God to humanity, God both adapts to humanity and speaks in human language" (*Eun*, 2.419); quoted in *Christianity and Classical Culture*, J. Pelikan (New Haven, CT: Yale University Press, 1993) 88.

Divine accommodation: P. H. Seely, "The Date of the Tower of Babel and Some Theological Implications," *Westminster Theological Journal* 63

(2001) 1; P. H. Seely, "Noah's Flood: Its Date, Extent, and Divine Accommodation," *Westminster Theological Journal* 66 (2004) 291-311.

Evolutionary Creationist position: D. O. Lamoureux, *Evolutionary Creation: A Christian Approach to Evolution* (Eugene, OR: Wipf & Stock, 2008) 493 pp.

For an excellent book on the range and different perspectives of people who subscribe to "evolutionary creation," refer to K. Applegate and J. B. Stump, eds., *How I Changed My Mind About Evolution: Evangelicals Reflect on Faith and Science* (Downers Grove, IL: InterVarsity Press Academic) 196 pp.

Evolutionary Creationist position of real history beginning with Abraham: D. O. Lamoureux, *Evolution: Scripture and Nature Say YES!* (Grand Rapids, MI: Zondervan, 2016) 120.

Page 7

A general perspective of worldview: J. W. Sire, *Naming the Elephant: Worldview as a Concept* (Downers Grove, IL: InterVarsity Press Academic, 2015) 2nd ed., 198 pp.

A worldview approach to Genesis 1: J. V. Miller and J. M. Soden, *In the Beginning…We Misunderstood: Interpreting Genesis 1 in Its Original Context* (Grand Rapids, MI: Kregel, 2012) 220 pp.

Page 8

Understanding the Arab mind: R. McQuilkin, *Understanding and Applying the Bible* (Chicago, IL: Moody Press, 1992) 178.

Shamanistic cave art: M. Aldhouse-Green and S. Aldhouse-Green, *The Quest for the Shaman* (London, U.K.: Thames and Hudson, 2005) 19-64.

Many of the ideas presented herein on worldview came from two Sunday School classes taught by Terry and Pam Moore at Heights Cumberland Presbyterian Church.

Page 9

The ancients' concept of "firmament:" P. H. Seely, "The Firmament and the Water Above, Part 2: The Meaning of 'The Water above the Firmament' in Gen. 1:6-8," *Westminster Theological Journal* 53 (1991) 37.

The ancients' concept of mountains as the "pillars of the sky:" J. H. Walton, "Flood," T. D. Alexander and D. W. Baker, eds., *Dictionary of the Old Testament Pentateuch* (Downers Grove, IL: InterVarsity Press, 2003) 320.

Page 10

The ancients' view of the cosmos: N. M. Sarna, *Understanding Genesis: The Heritage of Biblical Israel* (New York, NY: Schocken Books, 1966) 5.

Page 11

The ancient three-tiered concept of cosmology: P. H. Seely, "The Geographical Meaning of 'Earth' and 'Seas' in Genesis 1:10," *Westminster Theological Journal* 59 (1997) 231-234; for an excellent book on ancient Near East views of cosmology, the reader is referred to K. Greenwood, *Scripture and Cosmology: Reading the Bible Between the Ancient World and Modern Science* (Downers Grove, IL: InterVarsity Press Academic, 2013) 250 pp.

Joshua's long day as a poetic literary genre: W. C. Kaiser, senior ed., *The NIV Archaeological Study Bible* (Grand Rapids, MI: Zondervan, 2005) 319.

Joshua's long day interpreted from the world of omens, not physics: J. H. Walton, *Ancient Near Eastern Thought and the Old Testament* (Grand Rapids, MI: Baker Academic, 2006) 262-263.

Joshua 10:12-15 translated figuratively: J. H. Walton, *Ancient Near Eastern Thought and the Old Testament* (Grand Rapids, MI: Baker Academic, 2006) 263.

Joshua's long day interpreted from the viewpoint of a solar eclipse: C. J. Humphreys and W. G. Waddington, "Solar Eclipse of 1207 B.C. Helps to Date the Pharaohs," *A&G (Astronomy and Geophysics)*, published by the Royal Astronomical Society, 58 (2017) 5.39-5.42.

Page 13

The solar eclipse of 1207, viewed from the land of Canaan: F. R. Stephenson, L. V. Morrison, and C. Y. Hohenkerk, "Measurement of the Earth's rotation: 720 B.C. to A.D. 2015," *Proceedings of the Royal Society of Astronomy*, 472 (2016) 20160404, http://dx.doi.org/10.1098/rspa.2016.0404.

The definition of worldview as presented in this book: C. A. Hill, "A Third Alternative to Concordism and Divine Accommodation: The Worldview

Approach," *Perspectives on Science and Christian Faith* 59, no. 2 (2007) 129-134.

The Hebrew Masoretic text is a later translation of an original text. J. H. Walton and D. B. Sandy, *The Lost World of Scripture: Ancient Literary Culture and Biblical Authority* (Downers Grove, IL: InterVarsity Press, 2013) 320 pp.

Worldview is a fundamental orientation of the heart: J. W. Sire, *Naming the Elephant: Worldview as a Concept* (Downers Grove, IL: InterVarsity Press Academic, 2015) 2nd ed., 14.

Page 14

Genesis comes from the cultural and worldview perspectives of the ancient authors/scribes: W. C. Kaiser, senior ed., *The NIV Archeological Study Bible* (Grand Rapids, MI: Zondervan, 2005), 2,306 pp.

Page 15

Entanglement of worldview with the divine: G. L. Murphy, "Kenosis and the Biblical Picture of the World," *Perspectives on Science and Christian Faith* 64, no. 3 (2012) 157-165.

Jesus entered history at a specific time and place and lived within the pre-scientific ideas of that time: D. O. Lamoureux, *Evolution: Scripture and Nature Say YES!* (Grand Rapids, MI: Zondervan, 2016) 196 pp.

Chapter 2: The Six Days of Creation

Page 17

A young-earth faith requirement for membership in some churches: D. F. Siemens, "Considering the Probabilities of Creation and Evolution," *Perspectives on Science and Christian Faith* 52, no. 3 (2000) 194-199.

Page 18

Books that supports a very old universe: H. Ross, *The Fingerprint of God* (Orange, CA: Promise Publishing, 1989) 233 pp.; H. Ross, *The Creator and the Cosmos* (Colorado Springs, CO: NavPress, 1993) 185 pp.; H. Ross, *Creation and Time* (Colorado Springs, CO: NavPress, 1994) 187 pp.

Page 19

Plank satellite evidence that the universe is 13.82 billion years old: C. Crockett, "A More Precise Picture of the Cosmos," *Science News* 186, no. 13 (2014) 11.

The length of the seventh day of Genesis 1: G. L. Archer, *Encyclopedia of Bible Difficulties* (Grand Rapids, MI: Zondervan, 1982) 62.

If the seventh day is long, why not the other six? J. V. Miller and J. M. Soden, *In the Beginning...We Misunderstood: Interpreting Genesis 1 in Its Original Context* (Grand Rapids, MI: Kregel Publications, 2012) 52-53.

The word "generations" in Genesis 2:4 pertains to descent over a long period of time: P. J. Wiseman, *Ancient Records and the Structure of Genesis – A Case for Literary Unity* (New York, NY: Thomas Nelson Publishers, 1985) 17; H. F. Blank, "On the Structure of Genesis," *Perspectives on Science and Christian Faith* 56, no. 1 (2004) 74; W. C. Kaiser, senior ed., *The NIV Archaeology Bible* (Grand Rapids, MI: Zondervan, 2005) 6-7.

Page 20

If "generations" is taken as an extended time period, six literal days for Genesis 1 is a questionable interpretation: D. Fischer, "Young-Earth Creationism: a Literal Mistake," *Perspectives on Science and Christian Faith* 55, no. 4 (2003) 224; R. K. Harrison, "From Adam to Noah: a Reconsideration of the Antediluvian Patriarchs Ages," *Journal of the Evangelical Theological Society* 37 (1994) 161-168.

The Gap Theory was in favor in the early 20th Century: P. P. T. Pun, *Evolution: Nature and Scripture in Conflict?* (Grand Rapids, MI: Zondervan, 1982) 312.

Hebrew syntax of Genesis 1:1-3 does not allow a "gap" as per the Gap Theory: Alan Millard, personal communication, January, 2019.

The Gap Theory today is almost extinct: For a historical perspective on the gap theory, go to R. L. Numbers, *The Creationists: The Evolution of Scientific Creationism* (Berkeley, CA: University of California Press, 1992) 458 pp; D. A. Young, *The Bible, Rocks and Time: Geological Evidence for the Age of the Earth* (Downers Grove, IL: InterVarsity Press, 2008) 120-127.

Page 21

A "double creation" is because the Sun, Moon, and stars were not visible until the fourth day: H. Ross, *Creation and Time* (Colorado Springs, CO: NavPress, 1994) 148-151.

Page 22

Birds did not appear until the middle Jurassic: R. T. Bakker, *The Dinosaur Heresies* (New York, NY: Zebra Books, 1986) 304.

Making ancient cosmology into modern cosmology changes the meaning of the text: J. A. Walton, *The Lost World of Genesis One: Ancient Cosmology and the Origins Debate* (Downers Grove, IL: InterVarsity Press Academic, 2009) 17.

The Literary View of Genesis is gaining in popularity: D. G. Hagopian, ed., *The G3n3sis Debate: Three Views on the Days of Genesis* (Mission Viejo, CA: CruXpress) 2001, 319 pp.

Page 23

The Literary View of Genesis 1 has only been recently introduced to English-speaking, non-Jewish readers: N. H. Ridderbos, *Is There a Conflict Between Genesis 1 and Science?* (Grand Rapids: Eerdmans Pathway Books, 1957) 88 pp. Translated by John Vriend from a 1924 Dutch edition entitled *Beschouwingen over Genesis I* by Ridderbos, who was a professor of Old Testament at the Free University of Amsterdam. However, Ridderbos (1957) attributes his knowledge of the Literary (framework) view to an *A. Noordtzij* (p. 37), *God's Woord en der eeuwen getuigenis, pp. 116ff* (no date or publication given). Wikipedia identifies A. Noordtzij as Arie Noordtzij, who wrote *De Filistijnen* in 1905, so his *God's Woord* book could have been written ca. 1905. It is also possible that some of Noordtzij's views could have influenced later Hebrew scholar Umberto Cassuto (1883-1951), or that both Noordtzij and Cassuto could have been influenced by previous rabbinic scholarship tradition.

The whole chapter of Genesis 1 is based on a system of numerical harmony: U. Cassuto, *A Commentary on the Book of Genesis*, Pt. 1, translated from Hebrew by Israel Abrahams (Jerusalem: Magnes Press, 1972) 12-17; C. A. Hill, "Making Sense of the Numbers of Genesis," *Perspectives on Science and Christian Faith* 55, no. 4 (2003) 246-247.

It was customary to divide six days of work into three pairs: U. Cassuto, *A Commentary on the Book of Genesis*, Pt. 1, translated from Hebrew by Israel Abrahams (Jerusalem: Magnes Press, 1972) 13.

A completely harmonious account of creation would be the parallel form of symmetry found in Genesis 1: U. Cassuto, *A Commentary on the Book of Genesis*, Pt. 1, translated from Hebrew by Israel Abrahams (Jerusalem: Magnes Press, 1972) 10-11, 17.

The Genesis author wrote in the customary prose-narrative style of that day: C. Hyers, "The Narrative Form of Genesis 1: Cosmogenic, Yes; Scientific, No," *Journal of the American Scientific Affiliation* 36, no. 4 (1984) 212; P. H. Seely, "The First Four Days of Genesis in Concordist Theory and in Biblical Context," *Perspectives on Science and Christian Faith* 49, no. 2 (1997) 85-95.

Page 24

In the worldview of the Mesopotamians, language not only stated facts, it could establish them: J. M. Sasson, ed., *Civilizations of the Ancient Near East* (New York, NY: Charles Scribner's, 1995) 1818; W. C. Kaiser, senior ed., *The NIV Archaeology Bible* (Grand Rapid, MI: Zondervan, 2005) 4.

A being or thing only came into existence once it was given a name: J. H. Walton, *Ancient Near Eastern Thought and the Old Testament* (Grand Rapids, MI: Baker Academic, 2006) 88; D. Schmandt-Besserat, *When Writing Met Art: From Symbol to Story* (Austin, TX: University of Texas Press, 2007) 64.

A "play on words" was the very basis of intellectual thought: J. M. Sasson, ed., *Civilizations of the Ancient Near East* (New York, NY: Scribner's, 1995) 1819.

Page 25

A prime example of literary and numerical symmetry is the first chapter of Genesis: Alan Millard, Emeritus Rankin Professor of Hebrew and Ancient Semitic Languages, University of Liverpool, provided the Hebrew text for Gen. 1:1-2.

Page 26

Repetitions were introduced for the sake of parallelism: U. Cassuto, *A Commentary on the Book of Genesis*, Pt. 2, translated from Hebrew by Israel Abrahams (Jerusalem: Magnes Press, 1972) 38.

Numerical repetition and symmetry also applies to the garden of Eden, Cain and Abel and Noah's flood: U. Cassuto, *A Commentary on the Book of Genesis*, Pt. 1, translated from Hebrew by Israel Abrahams (Jerusalem: Magnes Press, 1972) 94, 192; Pt. 2, 32-33.

This numerical symmetry is proof of antiquity of Genesis: U. Cassuto, *A Commentary on the Book of Genesis*, Pt. 1, translated from Hebrew by Israel Abrahams (Jerusalem: Magnes Press, 1972) 15.

The narrative writing displayed in Genesis 1-11 is evidence of its antiquity: K. A. Kitchen, *On the Reliability of the Old Testament* (Grand Rapids, MI: Eerdmans, 2001) 422.

To faithfully interpret Genesis is to faithfully follow how it was originally written: C. Hyers, "The Narrative Form of Genesis 1: Cosmogenic, Yes; Scientific, No," *Journal of the American Scientific Affiliation* 36, no. 4 (1984) 209, 212.

Chapter 3: The Garden of Eden

Page 29

The land of Havilah is Arabia: C. F. Keil and F. Delitzsch, *Commentary on the Old Testament* 1, The Penteteuch (Grand Rapids, MI: Eerdmans, 1975) 171; R. L. Harris, "The Mist, the Canopy and the Rivers of Eden," *Bulletin of the Evangelical Theological Society* 11, no. 4 (1968) 179.

The climate in Saudi Arabia was much wetter in the past: H. A. McClure, "Late Quaternary Palaeogeography and Landscape Evolution of the Rub' Al Khali," D. T. Potts, ed., *Araby the Blest*, Carsten Niebuhr Institute of Ancient Near Eastern Studies (Copenhagen: Tusculanum Press, 1988) 9-13. For an excellent synthesis of recent work on the paleo-climate, sea-level changes, the ancient Karun and Wadi Batin Rivers, and prehistoric human populations of the Arabian-Persian Gulf region refer to J. I. Rose, "New Light on Human Prehistory in the Arabo-Persian Gulf Oasis," *Current Anthropology* 51, no. 6 (2010) 849-883.

Page 30

The Pishon River incises into sedimentary rock: U. S. Geological Survey-Arabian American Oil Company, *Geologic Map of Saudi Arabia* (1963), scale 1:2,000,000.

The Pishon River dried up sometime between 3500 and 2000 B.C.: J. A. Sauer, "The River Runs Dry - Creation Story Preserves Historical Memory," *Biblical Archaeology Review* 22, no. 4 (1996) 52-64; E. Uchupi, S. A. Swift, and D. A. Ross, "Gas Venting and Late Quaternary Sedimentation in the Persian (Abrabian) Gulf," *Marine Geology* 129 (1996) 237-269; C. A. Hill, "The River Pishon Flows Again?" *Perspectives on Science and Christian Faith* 61, no. 2 (2009), 135, a note on rare modern flow in the Wadi Rimah-Batin (the ancient Pishon River); B. Bower, "Ancient Deluges Drenched Sahara," *Science News* 191, no. 3 (2017) 18.

Map of the ancient incense rotes in Arabia: C. A. Hill, "Garden of Eden: A Modern Landscape," *Perspectives on Science and Christian Faith* 52, no. 1 (2000) 33.

Page 31

A flowing Pishon River can be thought of as a "historical memory" and very rarely as a recent event. Regarding the river Pishon, the author received an interesting e-mail from a Saudi Arabian who read my article "Garden of Eden: A Modern Landscape" on the *PSCF* website. Here is what his e-mail said:

"I read your article on the Pishon River – this totally amazed me as something interesting happened recently. Just in November 2008 there were very heavy rains in northern Saudi Arabia – the heaviest in 70 years. There was so much water that the desert turned into lakes (still there, and people are jet-skiing in these waters!). The flow cleared a lot of dust and sand from an ancient riverbed that nobody cared much for. This is Wadi Rumma (or Rimah as per the map in your article). I did go there a week later and saw the water was still flowing. Unfortunately my camera conked out on me but I do have pictures taken by others." This e-mail supports the idea that the Wadi Rimah-Wadi al Batin was the ancient Pishon

River of Gen. 2:11-12, and if climatic conditions are right, it will flow again.

Gold deposits in the Mahd adh Dhahab mining district: A. M. Afife, "Geology of the Mahd adh Dhahab District, Kingdom of Saudi Arabia," *U. S. Geological Survey*, Open File Report 90-315 (1990) 49 pp.

In antiquity, all forms of chalcedony were referred to as "onyx": C. A. Hill, "The Garden of Eden: A Modern Landscape," *Perspectives on Science and Christian Faith* 52, no. 1 (2000) 31-46.

Onyx stone has been found in archeological levels from about 4000-3200 B.C.: S. Ratnager, *Encounters - the Westerly Trade of the Harappa Civilization* (Delhi: Oxford University Press, 1981) 201.

Frankincense, myrrh, and bdellium are all different kinds of gum resins: E. M. Meyers, ed., *The Oxford Encyclopedia of Archaeology in the Near East* (New York, NY: Oxford University Press, 1997) 375.

The Gihon River is the ancient Gyudes: R. J. Fischer, *Historical Genesis: From Adam to Abraham* (New York, NY: University Press of America, 2008) 14.

The "land of Cush" is the land of the Kassites: E. A. Speiser, *Anchor Bible Commentary* 1 (Garden City, NY: Doubleday, 1981) 17.

The land of "Kush" is the land of the Kassites (Kashshu): K. A. Kitchen, *On the Reliability of the Old Testament* (Grand Rapids: Eerdmans, 2003) 429.

Page 32

Susa was the primary city of ancient Persia: R. J. Fischer, *Historical Genesis: From Adam to Abraham* (New York, NY: University Press of America, 2008) 173-174.

Page 33

The route of the Euphrates River in Noah's time: M. C. DeGraeve, *The Ships of the Ancient Near East (c. 2000-500 B.C.)* (Lewen, U.K.: Dept. Orientalisticli, 1981) 225.

The limit of the Persian Gulf in Noah's time: A. Parrot, *The Flood and Noah's Ark* (London: SCM Press Ltd., 1955) 33; translated by Edwin Hudson from the French book *Déluge et Arche de Noé*, 1953.

Page 34

The dates (radiocarbon ^{14}C, calibrated) come from a variety of sources. F. A. Hassan and S. W. Robinson, "High Precision Radiocarbon Chronometry," *Antiquity* 61 (1987) 119-135, and R. M. Adams, *Heartland of Cities: Surveys of Ancient Settlements and Land Use on the Central Floodplain of the Euphrates* (Chicago, IL: Univ. of Chicago Press, 1981) 60, are two older sources, with Hassan and Robinson dating the Jemdet Nasr/Early Dynasty 1 boundary at 2960 ± 167 cal B.C. (time of Noah's flood?). The most recent dates are from C. Hritz et al., "Mid-Holocene Dates for Organic-Rich Sediment, Palustrine Shell, and Charcoal from Southern Iraq," *Radiocarbon* 54 (2012), 65-79, who dated some of the earliest archeological sites in Southern Mesopotamia and who tried to shed light on the extent of marshlands and the rise and fall of sea level in conjunction with the occupation of early cities. Two occupation site dates related to key biblical persons are: Eridu = 5837-5644 cal B.C., base of mound (first city, according to cuneiform texts; possible abode of Adam & Eve?); and Jemdet Nasr = 2917-2898 cal B.C. (Noah's "home town"). Therefore, absolute dates can be roughly estimated for Adam and Eve at about 5500 B.C., and for Noah at about 2900 B.C.

The seasonal flow rate of the Euphrates and Tigris Rivers: A. I. Bağiş, "Turkey's Hydropolitics of the Euphrates-Tigris Basin," *International Journal of Water Resources Development* 13, no. 4 (1997) 567-581.

Page 35

The cobbles and pebbles from the ancient Pishon River once extended as far to the vicinity of Ur: U. S. Geological Survey-Arabian American Oil Company, *Geologic Map of Saudi Arabia* (1963), scale 1:2,000,000.

By around 3500 B.C., a marine transgression had inundated the southern part of the Mesopotamian plain: J. I. Rose, "New Light on Human Prehistory in the Arabo-Persian Gulf Oasis," *Current Anthropology* 51, no. 6 (2010) 849-883.

Ur was described as having landing docks where oceangoing vessels changed their cargos: T. Jacobsen, "The Waters of Ur," *Iraq* 22 (1960)

174-85; S. Lloyd, *The Archaeology of Mesopotamia* (London: Thames and Hudson, 1978) 43-57.

When the marine transgression retreated the Karun River appears to have shifted to the southeast: J. I. Rose, "New Light on Human Prehistory in the Arabo-Persian Gulf Oasis," *Current Anthropology* 51, no. 6 (2010) 850.

No archeological mounds exist where the land was submerged by the marine transgression: S. Lloyd, *The Archaeology of Mesopotamia* (London: Thames & Hudson, 1978) 15.

Cuneiform inscriptions indicate that Eridu was located near a garden, a "holy place" containing a sacred palm tree: J. C. Munday, "Eden's Geography Erodes Flood Geology," *Westminister Theological Journal* 58 (1996) 144.

Eridu is one of the oldest settlements in southern Mesopotamia: C. C. Lamberg-Karlovsky and J. A. Sabloff, *Ancient Civilizations - the Near East and Mesopotamia* (Menlo Park, CA: Benjamin Cummings Publications, 1979) 110. C-14 ages of Eridu date to ca 5800 B.C.: G. A. Cooke, "Reconstruction of the Holocene Coastline of Mesopotamia," *Geoarchaeology* 2, no. 1 (1987) 15-28; C. Hritz et al., "Mid-Holocene Dates for Organic-Rich Sediment, Palustrine Shell, and Charcoal from Southern Iraq," *Radiocarbon* 54 (2012) 65-79. Hritz et al.'s ^{14}C calibrated dates for the occupation of Eridu actually spans the time frame of ca. 5800 B.C. (basal section) to ca. 5000 B.C. (center of main mound) to ca. 2500 B.C. (top of ziggurat).

Page 36

The kingship was lowered from heaven; the kinship was in Eridu: T. Jacobsen, *The Sumerian King List* (Chicago, IL: Oriental Institute, 1996) Assyriological Studies no. 1, 59.

Possible name variations for Adam in Babylonian, Akkadian, Sumerian, and Egyptian: R. J. Fischer, *Historical Genesis: From Adam to Abraham* (New York, NY: University Press of America, 2008) 35; for a video of Fischer's talk on a Historical Adam at the American Scientific Affiliation's 2015 conference in Tulsa, go to http://youtu.be/XMPW8PrBPsU.

Some scholars have placed the location of the garden of Eden in Armenia near the source of the Tigris and Euphrates: G. J. Wenham, "Genesis 1-15," *Word Biblical Commentary* 1 (Waco, TX: Word Books, 1987) 66.

Very strictly, it is not "the Garden of Eden" at all, but "*a* garden *in* Eden:" K. A. Kitchen, *On the Reliability of the Old Testament* (Grand Rapids, MI: Eerdmans, 2003) 428-429.

A river rises *in* Eden to water the garden: E. A. Speiser, "Genesis," *Anchor Bible Commentary* (Garden City, NY: Doubleday, 1981) 16-17.

Page 37

Subterranean freshwater upwells through karstic limestone beneath the Persian Gulf: J. I. Rose, "New Light on Human Prehistory in the Arabo-Persian Gulf Oasis," *Current Anthropology* 51, no. 6 (2010) 853.

The Damman Formation crops out only a few miles southwest of Eridu: H. E. Wright, "The Southern Margins of Sumer: Archaeological Survey of the Area of Eridu and Ur:" *in* R. M. Adam, ed., *Heartland of Cities* (Chicago, IL: University of Chicago Press, 1981) 300.

Typical of Mesopotamian literature is the motif of four streams flowing from the temple to water the garden and from there to water the four corners of the earth: J. A. Walton, *Ancient Near Eastern Thought and the Old Testament* (Grand Rapids, MI: Baker Academic, 2006) 122.

The presence of God was the key to the garden: J. A. Walton, *Ancient Near Eastern Thought and the Old Testament* (Grand Rapids, MI: Baker Academic, 2006) 124-125.

Page 38

Some authors have even identified Jerusalem with Eden: L. E. Stager, "Jerusalem as Eden," *Biblical Archeology Review* 26, no. 3 (2000) 36-47.

The garden of Eden was located on a *modern landscape*: C. C. Lamberg-Karlovsky and J. A. Sabloff, *Ancient Civilizations – The Near East and Mesopotamia* (Menlo Park, CA: Benjamin Cummings Publications, 1979) 110.

Page 39

Six miles of sedimentary rock exists below the garden of Eden before Precambrian rock is encountered: Z. R. Beydoun, "Arabian Plate Hydrocarbon

Geology and Potential - a Plate Tectonics Approach," *American Association of Petroleum Geologists*, Studies in Geology 33 (1991) 77 pp.

Bitumen was used extensively by the Mesopotamians for every type of adhesive-construction need: C. A. Hill, "The Garden of Eden: A Modern Landscape," *Perspectives on Science and Christian Faith* 52, no. 1 (2000) 31-46.

Bitumen is known to have been used in the construction of reed boats even as early as the Ubaid Period: J. I. Rose, "New Light on Human Prehistory in the Arabo-Persian Gulf Oasis," *Current Anthropology* 51, no. 6 (2010) 864.

The Hit bitumen occurs in "lakes" where a line of hot springs are upwelling along deep faults: S. A. al-Sinawi and D. S. Mahood, "Geothermal Measurements in the Upper Euphrates Valley, Western Iraq," *Iraqi Journal of Science* 23, no. 1 (1982) 94-106.

In southern Iraq, oil and gas are produced from Jurassic, Cretaceous, and Miocene formations: B. P. Tissot and D. H. Welte, *Petroleum Formation and Occurrence* (New York, NY: Springer-Verlag, 1984) 699 pp.

Chapter 4: The Numbers and Chronologies of Genesis

Page 41

One of the first attempts at calculating a literal chronology from Genesis was Jewish scholar Jose Ben Halafta: E. G. Richards, *Mapping Time: The Calendar and History* (Oxford: Oxford University Press, 1998) 224-225.

Page 42

The long lifespans in Genesis make it seem fictional or legendary: H. Ross, *The Genesis Question* (Colorado Springs, CO: NavPress, 1998) 115.

Page 43

The Mesopotamians were the first accomplished mathematicians and astronomers: H. W. Saggs, *The Greatness That Was Babylon: A Survey of the Ancient Civilization of the Tigris-Euphrates Valley*, Ch. 13, Mathematics and Astronomy (New York, NY: Hawthorn, 1962) 445-453.

The Babylonians were regularly using a sexagesimal (base 60) system by Uruk time: J. Friberg, "Numbers and Measures in the Earliest Written Records," *Scientific American* 250, no. 2 (1984) 110-118.

There are influences of a decimal system that operated together with the sexagesimal system: R. C. Archibald, "History of Mathematics Before the Seventeenth Century," *American Mathematical Monthly* 56, issue 1, pt. 2 (1949) 7; G. Sarton, "Decimal Systems Early and Late," *Osiris* 9 (1950) 582; J. F. Scott, *History of Mathematics: From Antiquity to the Beginning of the Nineteenth Century* (London: Taylor and Francis, 1969) 9; K. A. Kitchen, *On the Reliability of the Old Testament* (Grand Rapids, MI: Eerdmans, 2001) 445.

A sexagesimal 360 day-year also needed to be intercalated with seven lunar months every 19 years to keep to a solar year: Alan Millard, personal communication, January, 2019.

Numbers could be used cryptographically where a name could have a corresponding numerical value: D. C. Harlow, "The Genesis Creation Accounts," paper given at the *Origins Symposium*, Calvin College, Grand Rapids, MI, March 13-14, 2006.

Barley was the currency of ancient Sumaria and allotments of barley were recorded on cuneiform tablets: K. R. Nemet-Nejal, "Mathematics," *Daily Life in Ancient Mesopotamia* (Westport, CT: Greenwood Press, 1998) 83.

Page 44

The purpose of numbers in ancient religious texts could be *numerological* rather than *numerical*: C. Hyers, "The Narrative Form of Genesis 1: Cosmogenic, Yes; Scientific, No," *Journal of the American Scientific Affiliation* 36, no. 4 (1984) 212.

"Seven stanza template" parallelisms in the Old and New Testaments is another example of the predominance of the use of number seven by the ancient Hebrews: K. E. Bailey, *Jesus through Middle Eastern Eyes* (Downers Grove, IL: InterVarsity Press, 2008) 443 pp.

Sometimes a Genesis author would selectively penetrate a descendant line to achieve a final number of 70: W. C. Kaiser, senior ed., *NIV*

Archaeological Study Bible (Grand Rapids, MI: Zondervan, 2005) 18.

The number 120 was regarded as a maximal and *ideal* age that could be reached by extremely virtuous individuals: J. Klein, "The 'Bane' of Humanity: A Lifespan of One Hundred and Twenty Years," *Acta Sumerologica* 12 (1990) 62.

In ancient Egypt, "he died at 110" was actually an epitaph commemorating an outstanding life: R. K. Harrison, "Reinvestigating the Antediluvian Sumerian King List," *Journal of the Evangelical Theological Society* 36, no. 1 (1993) 3-8; J. K. Hoffmeier, *Israel in Egypt* (Oxford: Oxford University Press, 1996) 95; K. A. Kitchen, *On the Reliability of the Old Testament* (Grand Rapids, MI: Eerdmans, 2001) 351.

Which numbers should be taken symbolically or figuratively is a question that scholars have been struggling with for centuries: Two of the earliest "modern" attempts to understand biblical numbers were by W. H. Green, "Primeval Chronology" *Bibliotheca Sacra*, Ch. 7 (Andover, MA: Draper, 1890) 285-303 [reprinted in W. C. Kaiser, ed., *Classical Evangelical Essays in Old Testament Interpretation* (Grand Rapids, MI: Baker Books, 1972) 13-28]; and by E. W. Bullinger, *Number in Scripture: Its Supernatural Design and Spiritual Significance* (London: Eyre and Spottiswoode, 1913) 3rd ed., Kessinger Publishing Rare Reprint Series) 303 pp.

The 70 years of captivity for the Israelites may or may not be literal: W. C. Kaiser, senior editor, *NIV Archaeological Study Bible* (Grand Rapids, MI: Zondervan, 2005) 1,234.

Exaggerated ages seems to have been a common religious tradition for peoples of the ancient Near East: J. K. Hoffmeier, *Israel in Egypt* (Oxford: Oxford University Press, 1996) 41; K. A. Kitchen, *On the Reliability of the Old Testament* (Grand Rapids, MI: Eerdmans, 2001) 443.

The long lifespans in the Sumerian King Lists and long ages of the patriarchs in Genesis have only a superficial similarity: T. C. Hartman, "Some Thoughts on the Sumerian King List and Genesis 5 and 11," *Journal of Biblical Literature* 91 (1972) 26.

The sexagesimal numbering system was only gradually replaced by the decimal system after Abraham left Ur and the Israelites met other Semitic people groups: G. Sarton, "Decimal Systems Early and Late," *Osiris* 9 (1950) 581-601.

A historical chronology had become established in Israel by the time of David: J. Oppert, "Chronology," J. Singer, ed., *The Jewish Encyclopedia* (New York, NY: Funk and Wagnalls, 1903) 68.

Before and during the Middle Ages, the concept of "sacred" numbers was lost: K. Greenwood, *Scripture and Cosmology: Reading the Bible Between the Ancient World and Modern Science* (Downers Grove, IL: InterVarsity Press Academic, 2015) 250 pp.

Page 47

A year could have marked the orbit of the Moon (a month) or a season (3 months): Diodorus Siculus, *Diodorus on Egypt* (London: McFarland, 1985, translated by Edwin Murphy) 32.

The astronomical hypothesis of a supernova decreasing the age of humans: H. Ross, *The Genesis Question* (Colorado Springs, CO: NavPress, 1998) 119.

Page 48

The canopy theory hypothesis of a vapor canopy shielding the Earth from harmful radiation, allowing people to live to long ages: J. C. Whitcomb and H. M. Morris, *The Genesis Flood* (Philadelphia, PA: Presbyterian and Reformed Publishing, 1966) 518 pp.

Humans had an average life span of about 40 years in the Bronze Age: F. Kendig and R. Hutton, *Life Spans – Or How Long Things Last* (New York, NY: Holt, Rinehart and Winston, 1979) 265 pp.

The ancients living in Mesopotamia had a dual numerical-numerological numbering system: C. A. Hill, "Making Sense of the Numbers of Genesis," *Perspectives on Science and Christian Faith* 55, no. 4 (2003) 239-251.

The sexagesimal numbers in the Genesis chronologies can be placed in two groups: multiples of *five* (5 or 0) or multiples of five with the addition of *seven*: U. Cassuto, *A Commentary on the Book of Genesis*,

Pt. 1, translated from Hebrew by Israel Abrahams (Jerusalem: Magnes Press, 1972) 258-259.

Page 50

Mathematical improbabilities continue for the 26 generations between Adam and Moses: R. Johnson, "Patriarchal Ages in Genesis," *Perspectives on Science and Christian Faith* 56, no. 2 (2004) 152-153.

Page 51

Love for symmetry and order also applies to numbers and to early Mesopotamian numbering on tablets: D. Schmandt-Besserat, *When Writing Met Art: From Symbol to Story* (Austin, TX: University of Texas Press, 2007) 8-9, 21.

To break a text into a ten-generational pattern was common for many Near Eastern people groups of that time: U. Cassuto, *A Commentary on the Book of Genesis*, Pt. 1, translated from Hebrew by Israel Abrahams (Jerusalem: Magnes Press, 1972) 254; D. J. Hamilton, *The Book of Genesis*, Chapters 1-17 (Grand Rapids, MI: Eerdmans, 1990) 254.

Abridgment was a general rule for biblical authors who did not want to encumber their texts with more names than necessary: W. H. Green, "Primeval Chronology," *Bibliotheca*, Ch. 7 (Andover, MA: Draper, 1890) 286.

Biblical father-to-son sequences can be a condensation from an originally longer series of generations: K. A. Kitchen, *On the Reliability of the Old Testament* (Grand Rapids, MI: Eerdmans, 2001) 440.

Page 52

The "overlap" of patriarchal lifespans is also evidence that there could be gaps in the Genesis chronologies: W. H. Green, "Primeval Chronology," *Bibliotheca*, Ch. 7 (Andover, MA: Draper, 1890) 300; J. H. Raven, *Old Testament Introduction: General and Special* (New York, NY: Fleming H. Revell, 1910) 134; P. P. T. Pun, *Evolution: Nature and Scripture in Conflict?* (Grand Rapids, MI: Zondervan, 1982) 259.

There are three different numbering systems: the Masoretic, Septuagint, and Samaritan versions: W. H. Green, "Primeval Chronology," *Bibliotheca*, Ch. 7 (Andover, MA: Draper, 1890) 299-302; P. P. T. Pun, *Evolution: Nature and Scripture in Conflict?* (Grand Rapids, MI: Zondervan, 1982) 259.

Page 53

The Septuagint was the first translation, leading up to the work of the Masoretes: J. H. Walton and D. B. Sandy, *The Lost World of Scripture: Ancient Literary Culture and Biblical Authority* (Downers Grove, IL: InterVarsity Press, 2013) 190.

Hebrew scholars such as Cassuto argue that the Masorite text was the autograph copy: U. Cassuto, *A Commentary on the Book of Genesis*, Pt. 1, translation by Israel Abrahams (Jerusalem: Magnes Press, 1972) 264-265.

Biblical chronologies were intended to confirm a specific "line of Adam" descent for the Jews in the Old Testament and for Jesus in the New Testament: W. H. Green, *Bibliotheca Sacra*, Ch. 7 "Primeval Chronology" (Andover, MA: Draper, 1890) 303.

Chapter 5: Noah's Flood: Historical or Mythological?

Page 56

The ancients did not invent spurious history, but were content to interpret real history in accord with their views: K. A. Kitchen, *On the Reliability of the Old Testament* (Grand Rapids, MI: Eerdmans, 2003) 63-64.

The Genesis account of Noah's age (600 years) is not nearly as exaggerated as the Sumerian account (36,000 years): Y. S. Chen, *The Primeval Flood Catastophe: Origins and Early Development in Mesopotamian Traditions* (Oxford: Oxford University Press, 2013) 141.

Page 58

More than one million cuneiform tablets have now been excavated: J. A. Walton, *The Lost World of Adam and Eve* (Downers Grove, IL: InterVarsity Press, 2015) 23.

Abraham is known to have lived about 2000-1900 B.C.: M. C. Tenney, ed., *The Zondervan Pictorial Encyclopedia of the Bible*, 1 (Grand Rapids, MI: Zondervan, 1975) 23.

An average length of a "generation" in the Bible is assumed to be 40 years. K. A. Kitchen, *On the Reliability of the Old Testament* (Grand Rapids, MI: Eerdmans, 2003) 307.

Page 59

The word *father* can also be translated as "ancestor" or "instructor": E. A. Speiser, *Genesis: Anchor Bible Commentary* 1, (Garden City, NY: Doubleday, 1981) 34.

Uruk cuneiform text mentions the occupations of farmer, shepherd, cowherd, fisherman, metal smith, weaver, and potter: H. J. Nissen, "The Emergence of Writing in the Ancient Near East," *Interdisciplinary Science Reviews* 10, no. 4 (1985) 360.

The domestication of cattle, sheep, and goats occurred around 10,000 years before present: J. Zarins, "Early Pastoral Nomadism and the Settlement of Lower Mesopotamia," *American School of Oriental Research* 280 (1990) 31-35.

Page 60

It is known that nomads occupied the Negev by the Early Bronze Age: S. A. Rosen, "Finding Evidence of Ancient Nomads," *Biblical Archaeology Review* 14, no. 5 (1988) 50.

Sumerian characters on stone tablets representing harps have been found as early as the Uruk Period: S. Sadie, *The New Grove Dictionary of Music and Musicians* 12 (London: McMillian, 1980) 196.

Harps have been found in the Royal Cemetery of Ur: S. Gaidos, "More Than a Feeling," *Science News* 178, no. 4 (2010) 29.

The beginning of the Bronze Age is around 3200 B.C.: E. M. Meyers, ed., *The Oxford Encyclopedia of Archaeology in the Near East* 4 (Oxford: Oxford University Press, 1997) 269.

Ziusudra, Ut-napišhtim, or Atra-hasis are alternate names for the biblical Noah on Mesopotamian tablets: W. G. Lambert and A. R. Millard, *Atra-Hasis: The Babylonian Story of the Flood* (Oxford: Clarendon Press, 1969) 1-28.

Gilgamesh was the fifth king of the first dynasty of Uruk following the great flood: M. E. Mallowan, "Noah's Flood Reconsidered," *Iraq* 26 (1964) 68.

The city of Shuruppak (supposedly Noah's "home town") was founded in the Jemdet Nasr Period: H. P. Martin, *Fara: A Reconstruction of the Ancient Mesopotamian City of Shuruppak* (Birmingham: Martin Associates, 1988) 113.

Page 61

Noah most likely lived somewhere around 2900 B.C. (±200 years): W. W. Hallo also argued from archeological evidence that the absolute date for the flood was approximately 2900 B.C.; see W. W. Hallo, "The Early Bronze Age, ca 3100-2100 B. C." *in* W. W. Hallo and W. K. Simpson, eds., *The Ancient Near East – a History* (New York, NY: Harcort-Brace, 1971) 36.

Noah's flood is not the same flood as proposed for the Black Sea around 5600 B.C.: W. Ryan and W. Pitman, *Noah's Flood: New Scientific Discoveries about the Event that Changed History* (New York, NY: Simon & Schuster, 1999) 319 pp.

Archeological discoveries of a Neolithic stone-tool site show people who inhabited a freshwater lake before the Black Sea rose: C. A. Reed, "Noah's Village," *Geotimes* 45, no. 11 (2000) 15.

Geologic evidence supports a complex and progressive reconnection of the Black and Mediterranean Seas over the past 12,000 years: A. E. Aksu, P. J. Mudie, A. Rochon, M. A. Kaminiski, T. Abrajano, and D. Yasar, "Persistent Holocene Outflow from the Black Sea to the Eastern Mediterranean Contradicts Noah's Flood Hypothesis," *GSA Today* 12, no. 5 (2002) 4-10.

The land where Noah lived was Mesopotamia, "land between the rivers": C. A. Hill, "A Time and Place for Noah," *Perspectives on Science and Christian Faith* 53 (2001) 24-40.

The low gradient of the four rivers of Mesopotamia sometimes causes flow to back up and breach levees: A. Dickin, "New Historical and Geological Constraints on the Date of Noah's Flood," *Perspectives on Science and Christian Faith* 70, no. 3 (2018) 181.

Page 62

Even as early as the Ubaid Period, rivers of southern Mesopotamia were diverted into canals: J. N. Postgate, *Early Mesopotamia – Society and Economy at the Dawn of History* (London: Routledge, 1992) 174.

By Noah's time, land trade routes had become established with neighboring regions: S. Pollock, "Bureaucrats and Managers, Peasants and Pastoralists, Imperialists and Traders: Research on the Uruk and Jemdet Nasr Periods in Mesopotamia," *Journal of World Prehistory* 6, no. 3 (1992) 297-336; G. Algaze, "The Uruk Expansion – Cross Cultural Exchange in Early Mesopotamian Civilization," *Current Anthropology* 30, no. 5 (1989) 571-608; J. Oates, "Trade and Power: Evidence from Northern Mesopotamia," *World Archeology* 24, no. 3 (1993) 413; C. A. Hill, "The Garden of Eden: a Modern Landscape," *Perspectives on Science and Christian Faith* 52, no. 1 (2000) 31-46; C. A. Hill, "A Time and Place for Noah," *Perspectives on Science and Christian Faith* 53, no. 1 (2001) 24-40.

By Noah's time, precious minerals were imported into southern Mesopotamia from Afghanistan, India, Iran, Turkey, and Arabia, while timber was dragged down the Amanus Mountains (in the modern Nur range) where the logs were tied into rafts and floated down the Euphrates River: M. C. Astour, *Overland Trade Routes in Ancient Western Asia* 3 (New York, NY: Charles Scribners, 1995) 1402-1408.

Sea routes were established between Mesopotamia and India and Egypt by the end of the fourth millennium B.C.: G. F. Bass, "The Earliest Seafarers in the Mediterranean and the Near East," G. F. Bass, ed., *A History of Seafaring Based on Underwater Archeology* (New York, NY: Walker and Company, 1972) 14-15.

Page 63

Representation of a ziggurat on an Assyrian bas-relief found at Nineveh: A. Parrot, *The Tower of Babel* (New York, NY: Philosophical Library, 1955) 29, 33-34, translated by Edwin Hudson from the French *La Tour de Babel*, 1954.

The Mesopotamians invented board games to play: W. C. Kaiser, senior ed., *The NIV Archaeological Study Bible* (Grand Rapids, MI: Zondervan, 2005) 10.

Page 64

Writing evolved from clay tokens, to markings on envelopes enclosing these tokens, to impressed signs on tablets to pictographic script: D. Schmandt-Besserat, *How Writing Came About* (Austin, TX: University of Texas Press, 1996) 193 pp; D. Schmandt-Besserat, *When Writing Met Art* (Austin, TX: University of Texas Press, 2007) 134 pp.

It wasn't until about 2500-2400 B.C. that narrative and religious writings were being recorded on cuneiform tablets: H. J. Nissen, "The Emergence of Writing in the Ancient Near East," *Interdisciplinary Science Reviews* 10, no. 4 (1985) 349-361.

Noah would have spoken the archaic Sumerian language, which is known from during or just before the Jemdet Nasr Period: C. A. Rollston, "Writing the Literacy in the World of Ancient Israel: Epigraphic Evidence from the Iron Age," *SBL Archaeology and Biblical Studies* 11 (Atlanta, GA: Society of Biblical Literature, 2010) 2; Wikipedia, "Sumerian language: Archaic Sumerian" (~3100-2600 B.C.) is the earliest stage of inscriptions with linguistic content, beginning with the Jemdet Nasr period. It succeeds the proto-literate period, which spans roughly the 35th to 30th centuries.

Noah could have been engaged in importing cedar wood from the Amanus Mountains and bitumen from Hit, both necessary items for building the ark: C. A. Hill, "A Time and Place for Noah," *Perspectives on Science and Christian Faith* 3, no. 1 (2001) 37.

Page 65

The *known world* of Noah would have consisted basically of the Mesopotamian alluvial plain and surrounding area: P. Enns, *Inspiration and Incarnation* (Grand Rapids, MI: Baker Academic, 2005) 40.

Page 66

The "land of the five seas" refers to the lands encompassed by the Mediterranean, Black, Caspian, Red, and Arabian Seas: U. S. Weather Bureau, *Climate of Southwestern Asia*, Report no. 40 (Washington, D.C.: U. S. Army Air Forces Weather Division, 1944) 1.

The eastern tropical maritime air masses originate in the Indian Ocean and can travel northwestward via the Arabian Sea and Persian Gulf to as far as the Mosul area: A. H. Shalash, *The Climate of Iraq* (M. S. Thesis: University of Maryland, 1957) 22-23.

Long-duration downpours are caused by the stalling or blocking of a Mediterranean frontal system: The term "100-year flood" (or "1,000-year" or "5000-year" flood) is a statistical designation meaning that there is a 1 in 100 (or 1 in 1,000 or 5,000) chance that a flood this size will happen during any one year.

Cyclonic circulation patterns over the Jordan basin brought an average of 75 inches of rain – the highest amount for a 150-year period: M. Inbar, "Effects of a High Magnitude Flood in a Mediterranean Climate: A Case Study in the Jordan River," L. Mayer and D. Nash, eds., *Catastrophic Floods* (Boston, MA: Allen and Unwin, 1987) 337.

Southern Iraq has an average rainfall of less than 4 inches: K. Takahashi and H. Arakawa, *Climates of Southern and Western Asia* 9 (New York, NY: Elsevier, 1981) 221.

Page 67

The very dry air brought by the shamal wind causes intense heating and evaporation of the land surface: U. S. Weather Bureau, *Climate of Southwestern Asia*, Report no. 40 (Washington, D.C.: U. S. Army Air Forces Weather Division, 1944) 122; H. C. Metz, *Iraq: A Country Study* (Washington, D. C.: Library of Congress Federal Research Division, 1988) 78.

The "second month, seventeenth day of the month" is when the flood started; it is not an extension of Noah's age: U. Cassuto, *A Commentary on the Book of Genesis*, Pt. 2, translated from Hebrew by Israel Abrahams (Jerusalem: Magnes Press, 1972) 83-84.

When the ancient Mesopotamian sidereal calendar is compared with today's tropical calendar, this puts the "second month, seventeenth day of the month" in about the middle of March when meteorological conditions bring the most rain to the Mesopotamian region: E. M. Plunket, *Ancient Calendars and Constellations* (London: John Murray, 1903) 2-3; E. G. Richards, *Mapping Time: The Calendar and its History* (Oxford: Oxford University Press, 1994) 147.

Genesis 7:12 implies that it was a long ("forty days and forty nights"), heavy rain, and this is the type of continuous downpours that are characteristic of this season: E. A. Speiser, *Anchor Bible Commentary: Genesis* 1 (Garden City, NY: Doubleday, 1981) 53.

Then in summer, the wind howls southward and the drying process begins: A. F. Aveni, *Ancient Astronomers* (Washington, D. C.: Smithsonian Books, 1993) 52.

Page 68

The destructive action of violent flooding of the Euphrates and Tigris can easily sweep away human habitations in an area that is so flat: A. Parrot, *The Flood and Noah's Ark* (London: SCM Press Ltd., 1955) 52, translated by Edwin Hudson from the French book *Déluge et Arche de Noé*, 1953.

When deep snow melts quickly, it can produce immediate flooding, but if melting is slow, water will be released over a long period of time: K. Smith and R. Ward, *Floods: Physical Processes and Human Impacts* (New York, NY: John Wiley, 1998) 10.

Prolonged flooding is known to cause profound flooding in the upper Mesopotamian basin, such as the 1954 flood along the upper Tigris River: S. N. Kramer, "Reflections on the Mesopotamian Flood: The Cuneiform Data New and Old," *Expedition* 9, no. 4 (1967) 16.

Ras-el-ain near the border of Syria and Turkey is one of the largest karst springs in the world: D. J. Burdon and C. Safadi, "Ras-el-ain: The Great Karst Spring in Mesopotamia," M. M. Sweeting, ed., *Karst Geomorphology* (London: Academic Press, 1963) 244, 258.

Page 69

Many springs could have contributed water to the basin during Noah's flood: Iraq Ministry of Development, *Groundwater Resources of Iraq* 7 (Baghdad: Government of Iraq, 1956) 28, 38; S. A. Al Sinawi and D. S. Mahmood, "Geothermal Measurements in the Upper Euphrates Valley, Western Iraq," *Iraqi Journal of Science* 23, no. 1

(1982) 94; United Nations, "Groundwater in Eastern Mediterranean and Western Asia," *Natural Resources/Water Series* 9 (New York: Department of Technical Cooperation for Development, 1982) 65, 69.

Tributaries to the Tigris River also merge from large caves along the foothills of the Zagros Mountains: A. S. Issar, *Water Shall Flow from the Rock: Hydrology and Climate in the Lands of the Bible* (New York, NY: Springer-Verlag, 1990) 43.

One of the most important of these springs emerged from Shalmaneser's Cave; an Assyrian bronze panel in the British Museum commemorates this visit by Shalmaneser: T. R. Shaw, "Historical Introduction," C. A. Hill and P. Forti, eds., *Cave Minerals of the World*, 2nd ed. (Huntsville, AL: National Speleological Society, 1997) 29.

Sargon II learned the secret of tapping water from subterranean strata: R. J. Forbes, *Studies in Ancient Technology* 2 (Leiden: Brill, 1965) 22.

The best known springs of the Cudi Limestone of the Cudi Dag (Jabel Judi Mountains) is in the Beytişebab area: E. I. Altini, *Geology of Eastern and Southeastern Anatolia: Cizre Sheet* 76, 92.

The possibility of a tidal wave due to a storm surge cannot be ruled out: A. Parrot, *The Flood and Noah's Ark* (London: SCM Press Ltd., 1955) 52-53, translated by Edwin Hudson from the French book *Déluge et Arche de Noé*, 1953.

The technical word for flood in Sumerian, Akkadian, and Hebrew are all applicable to a wind-driven rainstorm: A. S. Yahuda, *The Accuracy of the Bible* (New York, NY: Dutton, 1935) 191.

Abubu indicates moving water caused by a rainstorm or storm that drives seawater landward: R. E. Simoons-Vermeer, "The Mesopotamian Flood Stories: A Comparison and Interpretation," *Numen* 21 (1974) 18-19.

Abubu can also describe the destructive winds and gales that accompany the rainstorm: U. Cassuto, *A Commentary on the Book of Genesis*, Pt. 2, translated from Hebrew by Israel Abrahams (Jerusalem: Magnes Press, 1972) 11.

The Sumerian cuneiform tablets found at Nippur describes the Noachian deluge: J. H. Walton, "Flood," *in* T. D. Alexander and D. W. Baker, eds., *Dictionary of the Old Testament Pentateuch* (Downers Grove, IL: InterVarsity Press, 2003) 317; Y. S. Chen, "The Primeval Flood Catastrophe: Origins and Early Development in Mesopotamian Traditions" (Oxford, Oxford University Press, 2013) 213.

In the Gilgamesh epic, the flood is recorded as being a "southstorm": M. Rice, *The Archaeology of the Arabian Gulf* (London: Routledge, 1994) 306-307; S. N. Kramer, "Reflections on the Mesopotamian flood: The Cuneiform Data New and Old," *Expedition* 9, no. 5 (1967) 3.

Chapter 6: Noah's Flood: Global or Local?

Page 72

Land is a better translation for *eretz* because it extends to the "face of the ground" we can see around us (the horizon): J. H. Sailhammer, *Genesis Unbound – A Provocative New Look at the Creation Account* (Multnomah, OR: Sisters Publications, 1996) 45.

The Mesopotamian concept of the land seems to have meant the entire alluvial plain: J. N. Postgate, *Early Mesopotamia – Society and Economy at the Dawn of History* (London: Routledge, 1992) 34.

In Genesis 1:10, "God called the dry land *eretz,*" and if God defined *earth* as dry land then so should we: D. Fischer, *The Origins Solution* (Lima, OH: Fairway Press, 1996) 172.

Page 73

Noah's eyewitness account was passed down orally to biblical scribes, who wrote the story down from their worldview perspective: J. H. Walton and D. B. Sandy, *The Lost World of Scripture: Ancient Literary Culture and Biblical Authority* (Downers Grove, IL: InterVarsity Press, 2013) 320 pp.

It was not a universal deluge; it was a vast flood in Mesopotamia, which for the people who lived there was *all the world*: L. Woolley, *Excavations at Ur* (London: Ernest Benn, 1955) 36.

The ark didn't need to land on a mountain for Genesis 8:5 to also be true for "foothills:" R. J. Fischer, *Historical Genesis: From Adam to Abraham* (Lanham, MD: University Free Press of America, 2008) 111-112.

The Sumerians considered their ziggurats to be "mountains:" M. A. Beek, *Atlas of Mesopotamia* (London: Nelson, 1962), Map 8; M. Roaf, "Palaces and Temples in Ancient Mesopotamia," *in* J. M. Sasson, ed., *Civilizations of the Ancient Near East* (New York, NY: Charles Scribners, 1995) 425.

"Fifteen cubits upward" in Genesis 7:20 could refer to how deep its 30-cubit depth was submerged in the water: B. Ramm, *The Christian View of Science and Scripture* (Grand Rapids, MI: Eerdmans, 1974) 164.

Page 74

A global flood interpretation is not a "literal" reading of Genesis 7:20: D. Snoke, *A Biblical Case for an Old Earth* (Grand Rapids, MI: Baker, 2006) 165.

Page 75

The "gopher wood" of Genesis 6:14, while difficult to translate, is probably cedar wood: U. Cassuto, *A Commentary on the Book of Genesis: From Noah to Abraham*, Pt. 2, translated from Hebrew by Israel Abrahams (Jerusalem: Magnes Press, 1972) 61.

Cedar trunks up to one hundred feet long were being imported by Jemdet Nasr (Noah's) time: M. C. DeGraeve, *The Ships of the Ancient Near East* (Lewen: Department Orientalistich, 1981) 94, 151-154, 158-159.

Lumber and other goods were hauled upriver either by rowing or towing: S. Dalley, *Mari and Karana, Two Old Babylonian Cities* (London: Longman, 1984) 6; M. B. Rowton, "The Woodlands of Ancient Western Asia," *Journal of Near Eastern Studies* 26 (1967) 272; C. A. Hill, "A Time and Place for Noah," *Perspectives on Science and Christian Faith* 53, no. 1 (2001) 37.

According to Cassuto the word for "rooms" literally means "nests:" U. Cassuto, *A Commentary on the Book of Genesis: From Noah to Abraham*, Pt. 2, translated from Hebrew by Israel Abrahams (Jerusalem: Magnes Press, 1972) 62.

Speiser translated Genesis 6:14 as "make it an ark with compartments (cells)": E. A. Speiser, *Anchor Bible Commentary: Genesis* 1 (Garden City, NY: Doubleday, 1981) 52.

Yahua thought the correct translation was "fiber-tight shalt thou make the ark": A. S. Yahuda, *The Accuracy of the Bible* (New York, NY: Dutton, 1935) 192.

Page 76

Large quantities of bitumen were used for the coating and caulking of boats, large or small: R. J. Forbes, "Bitumen and Petroleum in Antiquity," *Studies in Ancient Technology* 8 (Leiden: Brill, 1964) 92; M. C. DeGraeve, *The Ships of the Ancient Near East* (Lewen: Department Orientalistich, 1981) 105. Sir Austen Layard compared the *tiradas*, or "black boats" of the Tigris and Euphrates of his day, with those shown in the bas-reliefs of Nineveh: P. Johnsone, *The Sea-Craft of Prehistory* (Cambridge, MA: Harvard University Press, 1980) 11.

A Mesopotamian cubit was 20 inches in proto-Sumerian time: J. Friberg, "Numbers and Measures in the Earliest Written Records," *Scientific American* 250, no. 2, 111-112.

The stated dimensions of the ark would have made it about the size of a modern ocean liner: G. Frame, "Some Neo-Babylonian and Persian Documents Involving Boats," *Oriens Antiquus* 25 (1986) 38; E. Robson, "The Uses of Mathematics in Ancient Iraq 6000-600 B.C.," *in* H. Selin and U. D'Smbrosio, eds., *Mathematics Across Cultures: The History of Non-Western Mathematics* 2 (Dondrecht: Kluwer, 2000) 107.

The size of wooden ships is limited to about 500 feet due to their inherent strength instability above this size: F. D. Hobbs, "Transportation," *Encyclopedia Britannica* 28, 15th ed. (1987) 770.

The stated dimensions of the ark may be numerological numbers: U. Cassuto, *A Commentary on the Book of Genesis: From Noah to Abraham*, Pt. 2, translated from Hebrew by Israel Abrahams (Jerusalem: Magnes Press, 1972) 62-63.

The stated 6:1 length-width ratio of the ark may involve numerological numbers: J. W. Montgomery, *The Quest for Noah's Ark* (Minneapolis, MN: Bethany Fellowship, 1972), 48.

The Hebrew word for "ark" comes from the Egyptian word for "box:" U. Cassuto, *A Commentary on the Book of Genesis: From Noah to Abraham,* Pt. 2, translated from Hebrew by Israel Abrahams (Jerusalem: Magnes Press, 1972) 59; A. S. Yahuda, *The Accuracy of the Bible* (New York, NY: Dutton, 1935) 192.

"The ark was simply a large box that would float on the surface of the water and not need to be steered:" Alan Millard, personal communication, January, 2019.

Page 77

Make a *skylight* for the ark, terminating it within a cubit of the top: M. C. DeGraeve, *The Ships of the Ancient Near East* (Lewen: Department Orientalistich, 1981) 19 + figure 54.

Skylights were a feature characteristic of Egyptian houses and temples in ancient times: L. Casson, *Ships and Seafaring in Ancient Times* (Austin, TX: University of Texas Press, 1994) 15.

Skylights were located in a ship's roof in order for it to be lighted and ventilated from a spot where no water could penetrate: E. A. Speiser, *Anchor Bible Commentary: Genesis* 1 (Garden City, NY: Doubleday, 1981) 47.

Boats in the fourth millennium B.C. are known to have had a cabin (or two cabins) placed at midship: A. S. Yahuda, *The Accuracy of the Bible* (New York, NY: Dutton, 1935) 193.

The Hebrew specifies that there were many rooms on each of three decks in the ark: S. Yahuda, *The Accuracy of the Bible* (New York, NY: Dutton, 1935) 194.

There is only one account of a Mesopotamian cargo boat containing decks: G. F. Bass, "The Earliest Seafarers in the Mediterranean and Near East," in G. F. Bass, ed., *A History of Seafaring Based on Underwater Archaeology* (New York, NY: Walker and Company, 1972) 13; L. Casson, *Ships and Seafaring in Ancient Times* (Austin, TX: University of Texas Press, 1994) 14.

Egyptian ships are known to have had decks: U. Cassuto, *A Commentary on the Book of Genesis: From Noah to Abraham*, Pt. 2, translated from Hebrew by Israel Abrahams (Jerusalem: Magnes Press, 1972) 65; P. Johnstone, *The Sea-Craft of Prehistory* (Cambridge, MA: Harvard University Press, 1980) 71.

Page 78

Sails are known to have been used on both sea and river craft from the fourth millennium B.C. onward: M. C. DeGraeve, *The Ships of the Ancient Near East* (Lewen: Department Orientalistich, 1981) 185.

Direct evidence exists for the use of sails dating to as early as the Ubaid Period: P. R. Moorey, *Ancient Mesopotamian Materials and Industries* (Oxford: Clarendon, 1994) 6; M. C. DeGraeve, *The Ships of the Ancient Near East* (Lewen: Department Orientalistich, 1981) 177.

A summary of the Gilgamesh epic by Berossus mentions a "sail" and "sailing" with regards to the flood ship: U. Cassuto, *A Commentary on the Book of Genesis: From Noah to Abraham*, Pt. 2, translated from Hebrew by Israel Abrahams (Jerusalem: Magnes Press, 1972) 14; S. Burstein, *The Babyloniaca of Berossus* (Malibu: Undena, 1978) 20.

The ark has been assigned to at least eight different landing places: L. R. Bailey, "Wood from 'Mount Ararat': Noah's Ark," *Biblical Archaeologist* 40, no. 4 (1977) 137.

Page 79

"Ararat" is the biblical name for the mountainous region of Urartu, as this region was known to the Assyrians: W. H. Stiebling, "A Futile Quest: The Search for Noah's Ark," *Biblical Archaeology Review* 2, no. 2 (1976) 16.

Mt. Ararat was probably not part of Urartu in Noah's time: E. M. Yamauchi, "Urartians and Manneans," *Foes from the Northern Frontier – Invading Hordes from the Russian Steppes*, Ch. 2 (Grand Rapids, MI: Baker Book House, 1982) 31.

Page 80

The name Urartu was transformed into Ararat by later vocalizations imposed on the Hebrew Bible: B. B. Piotrovsky, *The Ancient Civilization of Urartu* (New York, NY: Cowles, 1969), translated from Russian by James Hogarth, 13.

The first modern search for Noah's ark on Mt. Ararat was by Fernand Navara in 1955 and 1969: F. Navarra, *Noah's Ark: I Touched It* (Plainfield, NJ: Logos International, 1974) 137 pp.

Radiocarbon dates for wood collected by Navara were A.D. 720-790 (1955) and A.D. 620-640 (1969): L. R. Bailey, "Wood from 'Mount Ararat:' Noah's Ark," *Biblical Archaeologist* 40, no. 4 (1977) 138, 142.

A two-hour TV special on Noah's ark was seen by an estimated 20 million people: R. S. Dietz, "Ark-Eology: A Frightening Example of Pseudoscience," *Geotimes* 38, no. 9 (1993) 4.

The so-called "ark" of Doğubayazit, Turkey is a natural geologic formation: L. G. Collins and D. F. Fasold, "'Noah's Ark' from Turkey Exposed as a Common Geologic Structure," *Journal of Geoscience Education* 44 (1996) 439-444; L. G. Collins, "Noah's Ark near Doğubayazit, Turkey?," *Perspectives on Science and Christian Faith* 68, no. 4 (2016) 218-228; http://www.csun.edu/vcgeo005/bogus.html; L. G. Collins, "A Supposed Cast of Noah's Ark in Eastern Turkey" (2011), http://www.csun.edu/~vcgeo005/Sutton%20Hoo%2014.pdf.

Mt. Ararat as the landing place of the ark became a tradition only in the eleventh and twelfth centuries A.D.: D. Young, *The Biblical Flood – A Case Study of the Church's Response to Extrabiblical Evidence* (Grand Rapids, MI: Eerdmans, 1995) 34; B. Crouse, "Noah's Ark: Its Final Berth," *Archaeology and Biblical Research* 5, no. 3 (1992) 67; B. Crouse and G. Franz, "Mount Cudi – True Mountain of Noah's Ark," *Bible and Spade* 19, no. 4 (2006) 99-109.

Page 81

Wine grapes became domesticated in the area between the Black and Caspian seas around 6000-5000 B.C.: G. Algaze, "Fourth Millenium B.C. Trade in Greater Mesopotamia: Did it Include Wine?," P. E. McGovern, S. J. Reming, and S. H. Katz, eds., *The Origins and Ancient History of Wine* (Luxembourg: Gordon and Breach, 1995) 95.

Viticulture was practiced and wine was made in Northern Mesopotamia sometime before 3000 B.C.: T. Unwin, *Wine and the Vine – An Historical Geography of Viticulture and the Wine Trade* (New York, NY: Routledge, 1991) 63-64; H. P. Olmo, "The Origin and Domestication of the *Vinifera* Grape," *in* P. E. McGovern, S. J. Flemings, and S. H. Katz, eds., *The Origins and Ancient History of Wine* (Luxembourg: Gordon and Breach, 1995) 36.

Grapevines grow where the winters are not too severe, and the climate is not too hot and dry: E. Isaac, *Geography of Domestication* (Englewood Cliffs, NJ: Prentice Hall, 1970) 69.

Olive trees need an elevated, well-drained soil to survive: E. C. Semple, "The Regional Geography of Turkey: A Review of Banse's Work," *Geographical Review* 6 (1921) 338-350; R. L. Gorney, "Viticulture and Ancient Anatolia," P. E. McGovern, S. J. Flemings, and S. H. Katz, eds., *The Origins and Ancient History of Wine* (Luxembourg: Gordon and Breach, 1995) 139.

The return of the olive leaf by the dove suggests the survival of relatively unharmed trees outside of the flooded area: T. H. Everett, *Encyclopedia of Horticulture* 7 (New York, NY: Garland, 1981) 2,380.

Doves were part of the ancient Mesopotamians diet: J. Bottéro, "The Cuisine of Ancient Mesopotamia," *Biblical Archaeologist* 48, no. 1 (1985) 42.

Doves were probably domesticated in Mesopotamia by Ubaid time: W. M. Levi, *The Pigeon* (Sumter, SC: Levi Publishing Co., 1969) 1-2.

The homing instinct of doves was recognized and exploited since early times: J. Hansell, *The Pigeon in History or The Dove's Tail* (Bath, U.K.: Millstream Books, 1998) 15-16; R. A. Caras, *A Perfect Harmony: The Intertwining of Lives of Animals and Humans Throughout History* (New York, NY: Simon and Schuster, 1997) 219.

Page 82

How far an ancient breed of dove could have flown from the ark and back again in one day was probably less than 100 miles: S. J. Bodie, *Aloft: A Meditation on Pigeons and Pigeon-Flying* (New York, NY: Lyons and Burford, 1990) 23.

Jabel Judi is located only about 80 miles from Nineveh, a garden region renowned in ancient times for its grapevines and olive trees: D. J. Wiseman, "Mesopotamian Gardens," *Anatolian Studies* 33 (1983) 138.

If the flood was global, why wouldn't the ark have floated to somewhere outside of the boundaries of Mesopotamia? D. Young, *The Biblical Flood - A Case Study of the Church's Response to Extrabiblical Evidence* (Grand Rapids, MI: Eerdmans, 1995) 52.

Page 83

At the start of the eighteenth century, an agonizing battle over the history of the Earth began between scriptural chronology and the newly founded science of geology: A. Hallam, *Great Geological Controversies* (Oxford: Oxford University Press, 1989) 30-64.

Before 1810, most geologists had abandoned the idea that all sedimentary rock had formed in Noah's flood: C. C. Gillispie, *Genesis and Geology* (New York, NY: Harper-Row, 1959) 98-120.

After about 1810, only fossils of a recognizable modern type were attributed to changes in climate brought about by Noah's flood: L. M. Davies, "Scientific Discoveries and Their Bearing on the Biblical Account of the Noachian Deluge," *Journal of the Transactions of the Victoria Institute* 62 (London: Philosophical Society of Great Britain, 1930) 64-70; J. Imbrie and K. P. Imbrie, *Ice Ages: Solving the Mystery* (Cambridge, MA: Harvard University Press, 1979) 33-46.

The "Ice Ages" revelation then left *no* deposits that could be attributed to Noah's flood: For the history of geology, including the Ice-Age controversy, see A. Hallam, *Great Geological Controversies* (Oxford: Oxford University Press, 1992) 244 pp.

Page 84

The Mesopotamian hydrologic basin includes the entire region drained by the four rivers of Eden: C. A. Hill, "The Garden of Eden: A Modern Landscape," *Perspectives on Science and Christian Faith* 52, no. 1 (2000) 31-46; C. A. Hill, "Qualitative Hydrology of Noah's Flood," *Perspectives on Science and Christian Faith* 58, no. 2 (2006) 120-129.

Archeologist Leonard Woolley reported eight to eleven feet of mud at the Royal Cemetery of Ur and pronounced it the result of Noah's flood: L. Woolley, *Excavations at Ur* (London: Ernest Benn, 1955) 27.

The flood deposits at Nineveh seem too early to correlate with Noah's flood: M. E. Mallowan, "Noah's Flood Reconsidered," *Iraq* 26 (1964) 78-79.

Water-lain clay is nearly five feet thick at Uruk and two feet thick at Shuruppak: P. Carleton, *Buried Empires: The Earliest Civilizations of the Middle East* (London: Edward Arnold, 1939) 64; H. Crawford, *Sumer and the Sumerians* (London: Cambridge University Press, 1991) 19; A. Parrot, *The Tower of Babel* (New York, NY: Philosophical Library Publishers, 1955) 50.

Record of flood deposits for the most important archeological sites in Mesopotamia: R. J. Fischer, *Historical Genesis: From Adam to Abraham* (Lanham: University Press of America, 2014) 227 pp.

The flood event that occurred at about 2800-2900 B.C. is the one most likely to have been Noah's flood: A. Parrot, *The Flood and Noah's Ark*, translated from the 1953 French book *Déluge et Arche de Noé* (London: SCM Press Ltd., 1955) 52.

Page 85

Radiocarbon dates give approximate ages of the archeological periods of Mesopotamia, but the absolute dates of the flood deposits is less certain: A. Hassan and S. W. Robinson, "High Precision Radiocarbon Chronometry," *Antiquity* 61 (1987) 119-135; C. Hritz, J. Pournelle, S. Smith, B. Albadran, B. M. Issa, and A. Al-Handal, "Mid-Holocene Dates for Organic-Rich Sediment, Palastrine Shell, and Charcoal from Southern Iraq," *Radiocarbon* 54, no. 1 (2012) 65.

The Mississippi flood of 1973 was out of its banks for two or three months, and the average sediment thickness was 12 feet in back-swamp areas: H. Kesel, K. C. Dunne, R. C. McDonald, and K. R. Allison, "Lateral Erosion and Overbank Deposition on the Mississippi River in Louisiana Caused by 1973 Flooding," *Geology* 2, no. 9 (1974) 461; K. K. Hirschboeck, "Catastrophic Flooding and Atmospheric Circulation Anomalies," *in* L. Mayer and D. Nash, eds., *Catastrophic Floods* (Boston, MA: Allen and Unwin, 1987) 46.

Computer models show that a strong wind could have blown the ark northward from southern

Mesopotamia to the mountains of Ararat: A. E. Hill, "Quantitative Hydrology of Noah's Flood," *Perspectives on Science and Christian Faith* 58, no. 2 (2006) 130-141.

Page 86

A marine transgression could have helped back up river flow, preventing flood water from rapidly draining away: A. Dickin, "New Historical and Geological Constraints on the Date of Noah's Flood," *Perspectives on Science and Christian Faith* 70, no. 3 (2018) 186.

Divine action can be understood as higher-order laws working seamlessly with lower-order laws: S. Choi, "Knowledge of the Unseen: A New Vision for Science and Religion Dialogue," *Perspectives on Science and Christian Faith* 53, no. 2 (2001) 100.

The nature of "nature miracles" is to have the *timely* action of God into natural processes: W. F. Tanner, "How Many Trees Did Noah Take on the Ark?" *Perspectives on Science and Christian Faith* 47, no. 4 (1995) 262.

Records show that in 1267, 1906, and 1927, landslides upstream from Jericho dammed the river for up to twenty-one hours: K. A. Kitchen, *On the Reliability of the Old Testament* (Grand Rapids, MI: Eerdmans, 2003) 306.

Any theory, no matter how feeble, can be "proved" by recourse to the miraculous: B. Ramm, *The Christian View of Science and Scripture* (Grand Rapids, MI: Eerdmans, 1974) 177.

It is a weak interpretation that has to invent all sorts of miracles that the text says nothing about in order to compensate for logistical problems: J. H. Walton, "Flood," *in* T. D. Alexander and D. W. Baker, eds., *Dictionary of the Old Testament Pentateuch* (Downers Grove, IL: InterVarsity Press, 2003) 321.

Chapter 7: Flood Geology

Page 89

So popular has the flood geology position become in the conservative Christian church that it is held by almost half of the Christian public: Gallup, 2015, "Evolution, Creationism, Intelligent Design:" http://www.gallup.com/poll/21814/evolution-creationism-intelligent-design.aspx.

Page 91

Traditional flood geologists envision a thick vapor canopy ("mist") enshrouding planet Earth before the flood: J. C. Whitcomb and H. M. Morris, *The Genesis Flood: The Biblical Record and Its Scientific Implications* (Philadelphia, PA: Presbyterian and Reformed Publishing Company, 1966) 121, 254-257, 305-306.

"Mist" is better translated as "flow" in the sense of an underground swell or spring: E. A. Speiser, "Anchor Bible Commentary," *Genesis* 1 (Garden City, NY: Doubleday, 1981) 16-17.

Archeological mounds such as Jericho, which exhibit fairly continuous occupation for the last 10,000 years or so, show no signs of a great flood: J. T. Ator, *The Return of Credibility* (New York, NY: Vantage Press, 1998) 37.

The flood does not extend even to the land of Israel: W. H. Stiebling, "A Futile Quest: The Search for Noah's Ark," *Biblical Archaeology Review* 2, no. 2 (1976) 16.

Other flood stories around the world are essentially different from the biblical narrative and have only a few indeterminate elements in common with it: U. Cassuto, *A Commentary on the Book of Genesis*, Pt. 2, translated from Hebrew by Israel Abrahams (Jerusalem: Magnes Press, 1972) 26, 46.

The controversies regarding Noah's flood and modern geology were resolved in the 1800s and 1900s: A. Hallam, *Great Geological Controversies* (Oxford: Oxford Scientific Publications, 1989) 244 pp.

Page 94

Radiometric dating by radioactive decay (also called "absolute dating") has been proven to be reliable: For a simple discussion of radiometric dating, see R. Wiens, "So Just How Old is That Rock?," *in* C. Hill, G. Davidson, T. Helble, and W. Ranney, eds., *Grand Canyon: Monument to an Ancient Earth – Can Noah's Flood Explain the Grand Canyon?*, Ch. 9 (Grand Rapids, MI: Kregel Publications, 2016) 131-143.

Page 95

Increasingly complex life forms are found in successively younger strata: For a simple discussion of the fossil record in the Grand Canyon and planet Earth, see R. Stearley, "Fossils of the Grand Canyon and Grand Staircase," *in* C. Hill, G. Davidson, T. Helble, and W. Ranney, eds., *Grand Canyon: Monument to an Ancient Earth – Can Noah's Flood Explain the Grand Canyon?*, Ch. 13 (Grand Rapids, MI: Kregel Publications, 2016) 88-97.

The first flowering plants appeared late in the Cretaceous Period: R. J. Duff, "Flood Geology's Abominable Mystery," *Perspective on Science and Christian Faith* 60, no. 3 (2008), 166-176; R. J. Duff, "Tiny Plants - Big Impact; Pollen, Spores, and Plant Fossils;" *in* C. Hill, G. Davidson, T. Helble, and W. Ranney, eds., *Grand Canyon: Monument to an Ancient Earth – Can Noah's Flood Explain the Grand Canyon?*, Ch. 14 (Grand Rapids, MI: Kregel Publications, 2016) 145-151.

The Paluxy "man tracks" were made by the dinosaur genus *Irenesauripus*, not by humans: W. Langston, "Nonmammalian Comanchean Tetrapods," *Geoscience and Man* 8 (1974) 77-102.

The Whitcomb and Morris hypothesis of order in the fossil record: J. C. Whitcomb and H. M. Morris, *The Genesis Flood* (Philadelphia, PA: The Presbyterian and Reformed Publishing Company, 1966) 265-266.

Page 97

If dinosaurs and other classes of animals were not carnivores, then what were their sharp teeth and claws used for?: An excellent book on the modern concepts of paleontology is R. T. Bakker, *The Dinosaur Heresies* (New York, NY: Zebra Books, 1986) 481 pp.

Page 98

Mount Ararat is a 17,000 foot-high volcano that is still intermittently active: F. Navarra, *Noah's Ark: I Touched It* (Plainfield, NJ: Logos International, 1974) 121.

Mount Ararat rises above the high plateau of eastern Turkey, which was formed by the uplift of the Taurus and Zagros mountain systems: E. C. Semple, "The Regional Geography of Turkey: A Review of Banse's Work," *Geographical Review* 6 (1921) 344.

The volcanic rock of Mount Ararat is intruded by sedimentary rock: I. E. A'tinli, *Geologic Map of Turkey, Van Sheet* (with map notes in English), 1:500,000 (1961) 50-62; I. E. A'tinli, "Geology of Eastern and Southeastern Anatolia," *Bulletin of the Mineral Research and Exploration Institute of Turkey* 66 (1966) 42-46.

The point of view of Grand Canyon geology according to the Young-Earth Creationist book *Grand Canyon: A Different View*: T. Vail, compiler, *Grand Canyon: A Different View* (Green Forest, AK: Master Books, 2003) 104 pp.

The point of view of Grand Canyon geology according to modern geology: C. Hill, G. Davidson, T. Helble, and W. Ranney, eds., *Grand Canyon: Monument to an Ancient Earth – Why Noah's Flood Cannot Explain the Grand Canyon?* (Grand Rapids, MI: Kregel Publications, 2016) 240 pp.

A modern-geology diagram of the stratigraphic sequence of Grand Canyon rocks: C. A. Hill and S. O. Moshier, "Flood Geology and the Grand Canyon: A Critique," *Perspectives on Science and Christian Faith* 61, no. 2 (2009) 106.

Page 101

Roughly 75 percent of the once-present sedimentary rock record in the Grand Canyon is missing, primarily due to having been eroded away over time during periods when the land rose above sea level: S. Moshier and C. Hill, "Missing Time: Gaps in the Rock Record," *in* C. Hill, G. Davidson, T. Helble, and W. Ranney, eds., *Grand Canyon: Monument to an Ancient Earth – Can Noah's Flood Explain the Grand Canyon?* (Grand Rapids, MI: Kregel Publications, 2016) 99-101.

Page 103

Small geologic features in the sedimentary rock of the Grand Canyon attest to their non-flood origin: C. A. Hill and S. O. Moshier, "Flood Geology and the Grand Canyon: A Critique," *Perspectives on Science and Christian Faith* 61, no. 2 (2009) 99-115.

The sedimentary rock sequence in the Grand Canyon is due to the alternating coming (transgression) and going (regression) of the sea over the region, leaving behind the fossils, rocks, and sedimentary structures in the canyon's rocks that attest to its long geologic history: For an excellent book showing the transgressions and regressions of the sea over time in the Grand Canyon-Colorado Plateau area, refer to the book by R. Blakey and W. Ranney, *Ancient Landscapes of the Colorado Plateau* (Grand Canyon, AZ: Grand Canyon Association, 2008) 156 pp.

Chapter 8: Evolution and the New Genetics

Page 109

A number of pastors and lay Christians accept Young-Earth Creationism, are anti-evolutionists, and hold to flood geology: D. A. Young, and R. F. Stearley, *The Bible, Rocks and Time* (Grand Rapids, MI: InterVarsity Press, 2008) 162; An oft-cited Gallup poll illustrates the lack of progress in promoting evolution among the American public. In 1982, a question was posed regarding beliefs about human evolution. At the time, 44% believed God made humans in their present form. After a quarter century of improved educational materials, upgraded K-12 science standards, and several successful court battles to curb anti-science influences, that number has remained essentially unchanged. The last poll in 2014 pegged the number at 42% (Gallup, 2015).

Page 110

"Microevolution" is a term that has been used in publications by a number of Christian biologists and authors, but it is not a term recognized by most biologists: P. P. T. Pun, *Evolution: Nature and Scripture in Conflict?* (Grand Rapids, MI: Zondervan, 1982) 49, 50, 53, 176, 228, 264, 295; P. E. Johnson, *Darwin on Trial* (Grand Rapids, MI: InterVarsity Press, 1991) 40, 68-69; M. J. Behe, *Darwin's Black Box* (New York, NY: Touchstone, 1996) 13-15, 202; C. J. Collins, *Science and Faith: Friends or Foes?* (Wheaton, IL: Crossway, 2003) 277-278.

There is no universal agreement among biologists as to what constitutes a species: C. Zimmer, "What is a Species?" *Scientific American* 298, no. 6 (2008) 72-79.

Page 111

An example of natural selection is the beaks of finch populations on the Galápagos Islands: J. Weiner, *The Beak of a Finch* (New York, NY: Random House, 1995) 332 pp.

Page 112

The original word *kind* (as in "after its kind") referred to subdivisions within the type of life described: J. B. Payne, "The Concept of 'Kinds' in Scripture," *Journal of the American Scientific Affiliation* 10 (1958) 17-20.

Scholars holding various points of view have equated "kind" with species, genus, family, and even higher orders in the Linnaean classification scheme: P. H. Seeley, "The Meaning of Mîn, 'Kind':" *Science and Christian Belief* 9, no. 1 (1997) 47-56.

Can a "splitting off" support the splitting of species in divergent evolution?: A. Held and P. Rüst, "Genesis Reconsidered," *Perspectives on Science and Christian Faith* 51, no. 4 (1999) 235.

Page 112

"Folk" classification systems are based on the most distinctive species of a local habitat and on the characteristics of plants and animals that are readily observed: B. Berlin, *Ethnobiological Classification: Principles of Categorization of Plants and Animals in Traditional Societies* (Princeton, NJ: Princeton University Press, 1992) 31-35.

"Native" taxonomies characteristically break down the classification of animal life forms into five categories or less: C. H. Brown, "Folk Zoological Life-Forms, Their Universality and Growth," *American Anthropologist* 81, no. 4 (1979) 791-817.

The four life-form categories mentioned in Genesis 1 are typical for a proto-literate society, both in number and their non-scientific nature: For example, the King James Version of the Bible translates the Hebrew word *tannîym* as "whale," but other versions translate this word as "sea creature," "sea monster," or "dragon," implying some kind of dinosaur-like reptile. Strong's concordance translates it either way. For the purpose of this discussion, it doesn't matter, as neither a sea mammal nor sea reptile fit within the Linnaean category of fish.

The meaning of "kind" can relate to any of the taxonomic divisions of the Linnaean system: P. H. Seely, "The Meaning of Mîn, 'Kind':" *Science and Christian Belief* 9, no. 1 (1997) 49.

With owls, each of Palestine's eight species of owls in mentioned: E. Hunn, "The Abominations of Leviticus Revisited," *in* R. F. Ellen and D. Reason, eds., *Classifications in Their Social Context* (New York, NY: Academic Press, 1979) 103-116.

Page 113

Even as early as Uruk time, archaic cuneiform texts from Uruk document the categorizing of sheep into breeding bulls and rams: M. W. Green, "Animal Husbandry at Uruk in the Archaic Period," *Journal of Near Eastern Studies* 39, no. 1 (1980) 4.

Genesis 1:11 breaks the general class of "vegetation" into separate categories of plant-bearing seed and fruit-bearing seed: U. Umberto, *La Questione della Genesi* (Firenze: F. LeMonnier, 1934) 262. Translated from Italian for author by Arrigo Cigna.

Page 114

The distortion of the meaning of "kind" comes from forcing modern scientific knowledge onto the ancient biblical text: P. H. Seely, "The Meaning of Mîn, 'Kind':" *Science and Christian Belief* 9, no. 1 (1997) 55.

Much of the evidence which eventually convinced scientists that organisms change over time was collected *before* Darwin lived and proposed his theory of evolution: D. A. Young, and R. F. Stearley, *The Bible, Rocks and Time* (Grand Rapids, MI: InterVarsity Press, 2008) 108, 242.

Simple, single-celled organisms are found in the lowest and oldest sedimentary rock, while more complex organisms are found in the topmost and youngest layers: R. Stearley, "Fossils of the Grand Canyon and Grand Staircase," *in* C. Hill, G. Davidson, T. Helble, and W. Ranney, eds., *The Grand Canyon: Monument to an Ancient Earth* (Grand Rapids, MI: Kregel Publications, 2016) 134.

Homologous bone structure suggests a common descent connection between all of life: G. R. Davidson, *When Faith and Science Collide* (Oxford, MS: Malius Press, 2009) 120.

Whales were once legged animals that walked on land fifty million years ago: K. Wong, "The Mammals that Conquered the Sea," *Scientific American* 286 (2002) 70-79; T. Mueller, "Valley of the Whales," *National Geographic* 218, no. 2, 118-137.

On rare occasions a modern whale is found with a partially formed hind leg: R. C. Andrews, 1921, "A Remarkable Case of External Hind Limbs in a Humpback Whale," American Museum Novitates, no. 9, June 3; R. Ogawa and T. A. Kamiya, "Case of the Cachalot (Sperm Whale) with protruded rudimentary hind limb," *Scientific Reports of the Whales Research Institute*, no. 12 (1957) 197-208; R. T. Bakker, *The Dinosaur Heresies* (New York, NY: Zebra Books, 1986) 317; G. R. Davidson, *When Faith and Science Collide* (Oxford, MI: Malius Press, 2009)126-127.

Human embryos form a yolk sac during the early stages of their development: K. R. Miller, *Finding Darwin's God* (New York, NY: Harper-Collins, 1999) 338 pp.

Genetic comparisons between animal species are now being made for all kinds of plants and animals: S. Milius, "Should We Junk Linnaeus?" *Science News* 146 (1999) 268-269.

If one considers the overall genetic sequence identity between humans and other animals, then differences can be observed; percentages vary somewhat depending on the genetic parameters chosen: see *The Language of God* by Francis Collins (New York, NY: Free Press, 2006) 126-129.

The tiny one percent portion of DNA between humans and chimps makes a world of difference between their behavior and intelligence: K. Wong, "The One Percent Difference," *Scientific American* 311, no. 3 (2014) 100.

Page 116

It is extraordinary to view large segments of chimp and human genes and pseudogenes aligned side by side: G. Finlay, B. Choong, J. Flenley, N. Karunasinghe, G. O'Brien, R. Prestidge, C. Print, A. Shelly, and M. West, "Creation versus Creationism," *Perspectives on Science and Christian Faith* 58, no. 3 (2006) 236-239.

This genetic evidence is in strong support of evolutionary theory: F. S. Collins, "Faith and the Human Genome," *Perspectives on Science and Christian Faith* 55 (2003) 148.

The interested reader of evolutionary theory is referred to Francis Collins' book *The Language of God*: (New York, NY: Free Press, 2006) 294 pp.; D. R. Venema, "Genesis and the Genome: Genomics Evidence for Human-Ape Common Ancestry to Ancestral Hominid Population Sizes," *Perspectives on Science and Christian Faith* 62, no. 3 (2010) 166-178.

Genetic terms used in the following discussion come mainly from the work of geneticists James Shapiro and Sy Garte: S. A. Shapiro, *Evolution: A View From the 21st Century* (Upper Saddle River, NJ: FT Press, 2011) 253 pp.; S. Garte, "New Ideas in Evolutionary Biology: from NDMS to EES," *Perspectives on Science and Christian Faith* 68, no. 1 (2016) 3-11.

Transpoons are mobile segments of DNA that can "jump" into new positions in the genome: G. B. Kolata, "Jumping Genes: A Common Occurrence in Cells," *Science* 193 (1976) 392-394; F. H. Gage and A. R. Muotri, "What Makes Each Brain Unique?" *Scientific American* 306, no. 3 (2012) 26-31.

Page 117

The perceived need to reject evolution is still the prevailing and widely accepted view of biologists: S. A. Shapiro, *Evolution: A View From the 21st Century* (Upper Saddle River, NJ: FT Press, 2011) 253 pp.

Evolution does not allow for anything and everything: if it isn't in the tool kit, it doesn't happen: S. Garte, "New Ideas in Evolutionary Biology: From NDMS to EES," *Perspectives on Science and Christian Faith* 68, no. 1 (2016) 7-8; S. Garte, "Teleology and the Origin of Evolution," *Perspectives on Science and Christian Faith* 69, no. 1 (2017) 42-40.

The oldest known fossils have been found in Archean rock and date to about 3.7 billion years ago: A. M. Rosen, "Life's Early Traces," *Science News* 85, no. 2 (2014) 16-19.

The probability of life evolving by pure chance is astronomical: F. Hoyle and C. Wiekramasinghe, *Evolution from Space: A Theory of Cosmic Creation* (New York: Simon and Schuster, 1981) 24; http://en.Wikipedia.org/wiki.FredHoyle, p. 5.

Page 118

Besides the time problem, there is also the serious biochemical problem of how to bridge the gap between complex organic chemistry and a replicating *living* system: S. C. Morris, *Life's Solution: Inevitable Humans in a Lonely Universe* (New York, NY: Cambridge University Press, 2008) 464 pp.

The term "Cambrian Explosion" refers to the so-called "explosion" of animal life during the Cambrian Period: J. S. Levinton, "The Big Bang of Animal Evolution," *Scientific American* 267 (1992) 84-91.

Within about 50 million years, nearly all the main phyla known today appear in the fossil record: A. Parker, *In the Blink of an Eye* (Cambridge, MA: Perseus Publishing, 2003) 299 pp.

Early Cambrian life is best represented in Canada's Burgess Shale and China's Qingjiang biota: C. Gramling, "Fossils expand variety of early life: site in China offers a new look at the Cambrian explosion," *Science News*, 195, no. 8 (2019) 14, 32.

The lobopod *Hallucigenca*: fossil oddball of the Burgess Shale: J. Yang, J. Ortega-Hernández, S. Gerber, N. J. Butterfield, J. Hou, T. Lan, and X. Zhang, "A superarmored lobopodian from the Cambrian of China and early disparity in the evolution of Onychophora," *Proc. Natl. Acad. Sciences*, 112, no. 25 (2015) 8,678-8,683.

Page 119

Ongoing studies by paleontologists of Ediacaran-Cambrian fossils all over the world are helping to substantiate this early record of animal life and why and how it happened: R. Stearley, "The Cambrian Explosion: How Much Bang for the Buck?" *Perspectives on Science and Christian Faith* 65, no. 4 (2013) 245-257; K. Miller, "The Fossil Record of the Cambrian 'Explosion': Resolving the Tree of Life," *Perspectives on Science and Christian Faith* 66, no. 2 (2014) 67-82; J. D. Schiffbauer, J. W. Huntley, G. R. O'Neil, S. A. Darroch, M. Laflamme, and Y. Cai, "The Latest Ediacaran Wormworld Fauna: Setting the Ecological Stage for the Cambrian Explosion," *GSA Today* 26, no. 11 (2016) 4-11; R. A. Wood, "The Rise of Animals: New Fossils and Analyses of Ancient Ocean Chemistry Reveal the Surprisingly Deep Roots of the Cambrian Explosion," *Scientific American* 320, no. 6 (2019) 24-31.

There are numerous examples of fossils with transitional morphologies at all taxonomic levels: S. Perkins, "Step-By-Step Evolution," *Science News* 175, no. 3 (2009) 33; G. R. Davidson, *When Faith and Science Collide* (Oxford, MI: Malius Press, 2009) 143-147.

The famous *Archaeopteryx* fossils display both a reptile skeleton with a full set of teeth and a long bony tail and also wings and avian feathers: R. T. Bakker, *The Dinosaur Heresies*, (New York, NY: Zebra Books, part of the William Morrow and Company, 1986) 481 pp.; P. Wellnhofer, "*Archaeopteryx*," *Scientific American* 262, no. 5 (1990) 71-72; K. Padian and L. M. Chiappe, "The Origin of Birds and Their Flight," *Scientific American* 278, no. 2 (1998) 38-47; S. Brusatte, "Taking Wing: A Remarkable Fossil Record of the Dinosaurs that Led to Birds Reveals How Evolution Produces Entirely New Kinds of Organisms," *Scientific American* 316, no. 1 (2017) 48-55.

Page 120

A fossil form can remain unchanged in layer upon layer of rock for millions of years, examples of static species being blue-green algae or stromatolites: E. Pennisi, "Static Evolution," *Science News* 145 (1994) 168-169.

"Punctuated equilibrium" characterizes evolution as consisting of long periods of little change punctuated by relatively brief periods of rapid change: S. J. Gould and N. Eldredge, "Punctuated Equilibria: The Tempo and Mode of Evolution Considered," *Paleobiology* 3 (1977) 115-151.

Page 122

Different genes can mutate to converge on the same result: R. Borowsky, "Restoring Sight in Blind Cave Fish," *Current Biology* 18 (2008) R23-R24.

This type of *convergent evolution* is ubiquitous in the natural world: S. C. Morris, *Life's Solution: Inevitable Humans in a Lonely Universe* (New York, NY: Cambridge University Press, 2008) 469 pp.

The family *Amblyopsidae* is comprised of seven species in five genera and the whole range of cave adaption is represented by this one small family: T. L. Poulson and W. B. White, "The Cave Environment," *Science* 165 (1969) 971-980.

Chart shows the representatives of the family Amblyopsidae: A. Vandel, *Biospeleology: The Biology of Cavernicolous Animals* (Oxford: Pergamon Press, 1965) 524 pp.; Vandel modified it from Woods & Inger, "The Cave, Spring, and Swamp Fishes of the Family *Amblyosidae* of Central and Eastern United States," *American Midland Naturalist* 58, copyright permission granted.

Page 123

The freshwater cave fish *Astyanax mexicanus* may have lost their pigment in 10,000 years or less: R. W. Mitchell, W. H. Russell, and W. R. Elliot, "Mexican Eyeless Characin Fishes, Genus *Astyanax*: Environment, Distribution, and Evolution," *Special Publications of the Museum of Texas Tech University* SP-012 (1977) 89 pp.

Surface relatives of cave-adapted *Oliarus* planthoppers are characterized by large compound eyes, long wings, and dark pigmentation: H. Hoch and F. G. Howarth, "Evolutionary Dynamics of Behavioral Divergence Among Populations of Hawaiian Cave-Dwelling Planthopper *Oliarus polyphemus* (Homoptera: Fulgoroidea: Cixiidae)," *Pacific Science* 47, no. 4 (1993) 303-315.

Oliarus polyphemus has completely regressed from a surface planthopper species in less than 500 years – the highest recorded speciation rate of any taxon known: A. Wessel, H. Hoch, M. Asche, T. Rintelen, B. Stelbrink, V. Heck, F. D. Stone, and F. G. Howarth, "Founder Effects Initiated Rapid Species Radiation in Hawaiian Cave Planthoppers," *Proceeding of National Academy Sciences* 110, no. 23 (2013) 9,391-9,396.

The problem of regressive evolution puzzled Darwin, who ascribed the loss of eyes in cave creatures as "wholly to disuse." C. Darwin, *On the Origin of Species By Means of Natural Selection* (London: John Murray, 1859) Ch. V, Laws of Variation: Effects of Use and Disuse, 168-170.

Page 125

Division among Christians over the topic of evolution has segregated different church denominations into separate factions: Examples in this "principal battlefield" are the opposite positions of the Young

Earth Creationist organization *Answers in Genesis* and that of the online Christian organization *BioLogos*. *BioLogos'* constituents consider themselves to be "evolutionary creationists" who subscribe to the unifying theme of God having used evolution and common descent as his tool to carry out the creation of all life. *Answers in Genesis'* response to Christians who are theistic evolutionists is, "The thrust of BioLogos is not in accord with the biblical doctrines of Christianity; thus it is in reality from the spirit of anti-Christ. Church be warned." ("Is BioLogos Promoting Heresy?" Feb. 29, 2016, *Answers in Genesis* website.) Other positions are intermediate variations of these two extremes, but these also generate separate factions within the church.

In a teleological sense, a "creationist" is someone who believes that the world and everything in it was *designed* and *exists* for a purpose: P. E. Johnson, *Darwin on Trial* (Downers Grove, IL: InterVarsity Press, 1993) 220 pp.

With regards to evolution, some Young-Earth Creationists allow for limited evolutionary adaption within kinds: K. Ham, "Did God Create Poodles?" *Creation Magazine* 25, no. 4 (2003) 19-22; M. Christian "Purring Cats and Roaring Tigers," *Answers Magazine* 4, no. 4 (2007) 20-22.

A somewhat different version of Progressive Creationism is that God has divinely introduced or manipulated genetic material intermittently through time to bring about new life forms: G. C. Mills, "A Theory of Theistic Evolution as an Alternative to Naturalistic Theory," *Perspectives on Science and Christian Faith* 47 (1995) 112-122.

The Progressive Creationist view sounds like a "god of the gaps" argument, but is this unexplainable gap a *gap in our knowledge* or a *true gap in nature*?: D. O. Lamoureux, "Beyond the 'Evolution vs. Creation' Debate," *Canyon Institute for Advanced Studies Newsletter* 6, no. 2 (2006) 8-14.

Over and over in the history of theology, what was once attributed to God has been explained by natural processes: J. H. Brooke, *Science and Religion: Some Historical Perspectives* (Cambridge: Cambridge University Press, 2006) 422 pp.

The Intelligent Design view challenges the Darwinian view that valid science must appeal *only* to natural causes: C. J. Collins, *Science and Faith: Friends or Foes?* (Wheaton, IL: Crossway, 2003) 448 pp.

The constants of physics are so finely tuned as to make the universe appear to be designed to support life: H. Ross, *The Fingerprint of God* (Orange, CA: Promise Publishing, 1989) 119-138.

If these physics constants were even a tiny bit larger or smaller, it would preclude life from even being here: M. Rees, *Just Six Numbers* (London: Weidenfeld and Nicholson, 1999) 195 pp.

Physics constants would seem to suggest a universe that has been carefully crafted for our benefit: W. L. Bradley, "The Fine Tuning of the Universe: Evidence for the Existence of God?" *Perspectives on Science and Christian Faith* 70, no. 3 (2018) 147.

The premise of Intelligent Design is that God could have followed the same blueprint and created all life to only *appear* as if it is connected: G. R. Davidson, *When Faith and Science Collide: A Biblical Approach to Evaluating Evolution and the Age of the Earth* (Oxford, MS: Malius Press, 2009) 288 pp.

What geneticists observe time and time again is that genetic sequences in organisms thought to be close evolutionary relatives match at all genomic levels: D. R. Venema, "Genesis and the Genome: Genomics Evidence for Human-Ape Common Ancestry to Ancestral Hominid Population Sizes," *Perspectives on Science and Christian Faith* 62, no. 3 (2010) 172.

Page 126

Intelligent design is a belief that the world's beauty, complexity, and functionality point toward an Intelligent Designer: D. O. Lamoureux, *Evolution: Scripture and Nature Say YES!* (Grand Rapids, MI: Zondervan, 2016) 196 pp.

In some studies, acknowledging the student's religiosity actually increases their acceptance of evolution: R. Lloyd, "Dissent with Modification: Acknowledging Students' Religiosity Could Increase Acceptance of Evolution," *Scientific American* 318, no. 5 (2017) 14.

The human genome was written in the DNA language by which God spoke life into being: F. S. Collins,

The Language of God (New York, NY: Free Press, 2006) 143.

In the view of theistic evolutionists, natural processes and divine action are not in competition with each other but are complementary: H. J. Van Till, *The Fourth Day: What the Bible and the Heavens are Telling Us about the Creation* (Grand Rapids, MI: Eerdmans, 1986) 223; D. F. Siemens, "Considering the Probability of Creation and Evolution," *Perspectives on Science and Christian Faith* 52 (2000) 194-199.

All living things have slowly evolved under the sovereign control of God's Holy Spirit: G. L. Murphy, "The Third Article in the Science-Theology Dialogue," *Perspectives on Science and Christian Faith* 45, no. 3 (1993) 162-168.

Miracles involve the *spiritual dimension*, which falls outside the laws of nature: J. C. Lennox, *Seven Days That Divide the World: The Beginning According to Genesis and Science* (Grand Rapids, MI: Zondervan, 2011) 167-171.

Theistic evolution has the weakness of straining Genesis 2 and 3 in terms of Adam and Eve being historical figures: P. P. T. Pun, "Integration and Confrontation of Contemporary Worldviews: Evolution and Intelligent Design," *Perspectives on Science and Christian Faith* 59, no. 2 (2007) 102-109.

Has God manipulated genetic material in some way impossible for science to detect?: M. J. Behe, *The Edge of Evolution: The Search for the Limits of Darwinism* (New York, NY: Free Press, 2007) 320 pp.

Could evolution be a self-organizing process that God directs from within? N. Gregerson, "The Idea of Creation and the Autopoietic Process," *Zygon* 33 (1988) 333-367; T. H. Saey, "Gene Tweak Led to Human Big Toe," *Science News* 189, no. 3 (2016) 15.

Is the evolutionary process an emergent property that is wired into the biosphere: S. C. Morris, *Life's Solution: Inevitable Humans in a Lonely Universe* (New York, NY: Cambridge University Press, 2008) 148.

Chapter 9: Adam and Eve and Origins

Page 132

It seems like the church is afraid to look into paleoanthropology: M. A. Noll, *The Scandal of the Evangelical Mind* (Grand Rapids, MI: Eerdmans, 1994) 232.

For an excellent recent summary of the anthropological evidence for human origins see D. L. Wilcox, "Updating Human Origins," *Perspectives on Science and Christian Faith*, v. 71, no. 2 (2019) 37-49.

"Peking man" is thought to have lived in China from about 780,000 to 400,000 years ago: A. D. Aczel, *The Jesuit and the Skull* (New York: Riverhead Books, 2007) 302 pp; Shen, G., Gao, X., Gao, B., and Granger, D. E., "Age of Zhoukoudian *Homo erectus* Determined With ^{26}Al/^{10}Be Burial Dating," *Nature* 458 (2009) 198-200.

The interpretation of *Homo floresienus* descending from *Homo erectus* instead of *Homo sapiens* is highly controversial: K. Wong, "The Littlest Human," *Scientific American* 292, no. 2 (2005) 56-65; K. Wong, "Rethinking the Hobbits of Indonesia," *Scientific American* 301, no. 5 (2009) 66-73; B. Bower, "Hobbit History Gets New Beginning: Fossils of Potential Ancestor Fuel Debate over *Homo floresiensis*," *Science News* 190, no. 1 (2016) 6.

Page 134

Relationships of hominids to *Homo* are not clear, and "family trees" always seem to be changing: B. Bower, "Tangled Roots: Mingling Among Stone-Age Peoples Muddies Human's Evolutionary Story," *Science News* 182, no. 4 (2012) 22-25; For a popular discussion of different hominids and *Homo* species living in the Pre-Paleolithic, the reader is referred to *Lucy's Legacy: The Quest for Human Origins* by D. C. Johanson and K. Wong (New York, NY: Three Rivers Press, 2010) 321 pp.

Neanderthals remains are almost always found in caves because the cave environment is conducive to preserving human remains and tracks: B. P. Onac et al., "U-Th Ages Constraining the Neanderthal Footprint at Vârtop Cave, Romania," *Quaternary Science Review* 24 (2005) 1,151-1,157.

Homo sapiens DNA has also been retrieved from Neanderthal bones: R. E. Green et al., "A Draft Sequence of the Neandertal Genome," *Science* 328 (2010) 710-722; B. Bower, "Oldest Known Human DNA Analyzed, *Science News* 186, no. 11 (2014) 8-9; T. S. Saey, "Human DNA Found in Neandertal Bone," *Science News* 189, no. 6 (2016) 6.

For an excellent synthesis of recent work on the paleoclimate, sea-level changes, and human populations of the Arabia-Persian Gulf region, refer to J. I. Rose, "New Light on Human Prehistory in the Arabo-Persian Gulf Oasis," *Current Anthropology* 51, no. 6 (2010) 849-883.

The fantastic cave paintings in France and Spain date from about 35,000 to 12,000 years ago: R. White, *Prehistoric Art* (New York: Harry Abrams, 2003) 16; B. Bower, "First Europeans May Have Painted," *Science News* 182, no. 2 (2012) 15.

It is suspected that Cro-Magnon man in Europe may have been practicing some type of animistic religion by about 30,000 years ago: M. Aldhouse-Green and S. Aldhouse-Green, *The Quest for the Shaman* (London: Thames and Hudson, 2005) 62.

Page 135

The growing and processing of olives and olive oil also began in the Chalcolithic: R. Gonen, "The Chalcolithic Period," A. Ben-Tor, ed., *The Archeology of Ancient Israel*, Ch. 3 (Tel Aviv: Open University of Israel, 1992) 41.

"Ötzi the Iceman" carried a copper axe, which puts him in the Chalcolithic: D. Roberts, "The Ice Man," *National Geographic* 183, no. 6 (1993) 36-67; J. H. Dickson, K. Oeggl, and L. L. Handley, "The Iceman Reconsidered," *Scientific American* 288, no. 5 (2003) 60-69; B. Bower, "Iceman has the world's oldest tattoos," *Science News* 189, no. 2 (2016) 5.

The Natufians were already cultivating wheat and barley ~12,000 to 10,000 years ago: J. Mellaart, *The Neolithic of the Near East* (London: Thames and Hudson, 1975) 29; J. I. Rose, "New Light on Human Prehistory of the Arabo-Persian Gulf Oasis," *Current Anthropology* 31, no. 6 (2010) 862.

Page 136

The Ubaid Period records the first agricultural (non-nomad) occupation of southern Mesopotamia: J. L. Huot, "The First Famers at Oueili," *Biblical Archeologist* 55, no. 4 (1992) 188-195; J. I. Rose, "New Light on Human Prehistory in the Arabo-Persian Gulf Oasis," *Current Anthropology* 51, no. 6 (2010) 850-854, 864.

"Eridu ware" pottery was first in order of appearance; later "Uruk ware" signified the arrival of the Sumerians at Eridu: R. J. Fischer, *Historical Genesis: From Adam to Abraham* (New York, NY: University Free Press of America, 2008) 30-31, 35; D. Fischer, "In Search of the Historical Adam: Pt. 2," *Perspectives on Science and Christian Faith* 46, no. 1 (1994) 48.

The "descent of kingship from heaven" was supposedly related to the first city founded in Mesopotamia; that is, Eridu: T. Jacobsen, *The Sumerian King List*, Assyriological Studies no. 11 (Chicago, IL: Oriental Institute, 1966) 58.

Page 139

For a non-technical discussion of DNA and its relationship to human origins, refer to B. Sykes, *The Seven Daughters of Eve* (New York, NY: W. W. Norton, 2001) 306 pp.

Male-transmitted DNA enhances the ability to distinguish one human population from another: G. Stix, "Traces of a Distant Past," *Scientific American* 267, no. 1 (2008) 56-63.

Some DNA studies show that Mitochondrial Eve and Y Chromosome Adam were roughly contemporaneous: L. S. Whitfield, J. E. Sulston, and P. N. Goodfellow, "Sequence Variation of the Human Y Chromosome," *Nature* 378 (1995) 379-380; E. Wayman, "Y Chromosome Adam Gets Older," *Science News* 184, no. 5 (2013) 14.

Tracing human origins to a single mother and to a larger contemporaneous population at the same time may seem contradictory, but actually it is quite plausible: G. Davidson, "Genetics, the Nephilim, and the Historicity of Adam," *Perspectives on Science and Christian Faith* 67, no. 1 (2015) 24-34.

Page 140

From the fossil record, a very approximate date of 150,000 years ago for Mitochondrial Eve and Y Chromosome Adam is not unreasonable since *Homo sapiens* fossils have been dated back to 195,000 YBP, or even before: I. McDougall, D. F. Brown, and J. G. Fleagle, "Stratigraphic Placement and Age of Modern Humans from Kibish, Ethiopia," *Nature* 433 (2005) 733-736. However, the study of early human populations changes all the time with new discoveries. One new discovery, which is still controversial, puts the oldest humans between 350,000 and 280,000 years ago in Morocco; K. Wong, "The Oldest *Homo sapiens*?" *Scientific American* 317, no. 3 (2017) 12-14.

DNA testing of Arabs and Jews has linked both groups to a common male ancestor several thousand years ago: M. F. Hammer et al., "Jewish and Middle Eastern Non-Jewish Populations Share a Common Pool of Y-Chromosome Biallelic Haplotypes," *Proceedings National Academy of Science* 97, no. 12 (2000) 6,769-6,774.

Y Chromosome studies have conferred a distinct paternal genealogy for Jewish priests: K. Skorecki et al., "Y Chromosomes of Jewish Priests," *Nature* 385 (1997) 32.

The earliest, but relatively minor, migration of humans out of Africa is believed to have occurred about 100,000 years ago: B. Bower, "Stone Tools Hint at Earlier Human Exit from Africa," *Science News* 179, no. 5 (2011) 5-6; J. I. Rose, V. I Usik, A. E. Marks et al., The Nubian Complex of Dhofar, Oman: An African Middle Stone Age Industry in Southern Arabia, PLos ONE 6, no. 11 (2011) e28239, doi.org/10.1371/journal.pone.0028239.

A date of approximately 50,000 YBP seems to have been a time of a burst of creativity in human occupied sites: P. A. Underhill et al., "Y Chromosome Sequence Variation and the History of Human Populations," *Nature Genetics* 26 (2000) 358.

Since the DNA genetic evidence supports humans originating in Africa, if one is looking for Eden based on a Paleolithic Adam and Eve, it will have to be in Africa: D. L. Wilcox, "Genetic Insights for Human Origins in Africa and for Later Neanderthal Contact," *Perspectives on Science and Christian Faith* 66, no. 3 (2014) 140-153; T. H. Saey, "One Africa Exodus Populated Globe: DNA Data Point to Migration Less Than 75,000 years ago," *Science News* 190, no. 8 (2016) 6-7.

Page 142

Over fifty years ago linguists proposed three separate waves of migration into the Americas: J. H. Greenberg and M. Ruhlen, "Linguistic Origins of Native Americans," *Scientific American* 267 (1992) 94-99.

The DNA trail for the second, major wave of human migration can be traced back to the !Kung people of Africa: B. Sykes, *The Seven Daughters of Eve* (New York, NY: W. W. Norton, 2001) 306 pp.

Early first-wave human migration may have spread as far as China by approximately 80,000 years ago: B. Bower, "Reading the Stones," *Science News* 187, no. 7 (2015) 16-21; B. Bower, "Before Europe, Humans Went to Asia," *Science News* 188, no. 4 (2015) 15; B. Bower, "When People Reached Arabia: Fossil Puts Humans on the Peninsula by 86,000 years ago," *Science News* 193, no. 8 (2018) 12.

Interbreeding of *Homo sapiens* with Neanderthal probably occurred during the second wave of human migration ~55,000 to 40,000 years ago: B. Bower, "Fossil Recasts History of First Europeans," *Science News* 187, no. 4 (2015) 5; T. H. Saey, DNA points to more recent Neanderthal interbreeding, *Science News* 187, no. 12 (2015) 11.

The Basque people in Spain may have been the first people group to reach Europe: F. Calafell and J. Bertranpetit, "Principal Component Analyses of Gene Frequencies and the Origin of Basques," *American Journal of Physical Anthropology* 93 (1994) 201-215.

Humans reached the British Isles at least by 12,800 to 12,600 years ago: B. Bower, "Bear Bone Rewrites Human History in Ireland," *Science News* 189, no. 11 (2016) 4.

Humans reached Australia about 50,000 years ago and Tasmania about 30,000 years ago: M. Roberts-Thomson et al., "An Ancient Common

Origin of Aboriginal Australians and New Guinea Highlanders is Supported by α-Globin Haplotype Analyses," *American Journal of Human Genetics* 58 (1996) 1,017-1,024; J. R. Roberts et al., "Optical and Radiocarbon Dating at Jinmium Rock Shelter in Northern Australia," *Nature* 393 (1998) 358-362; M. Temming, "Humans' Arrival in Australia Re-dated," *Science News* 192, no. 2 (2017) 10.

Humans reached Japan by 12,000 years ago (the Ainu), and then later a larger population moved from Korea to Japan, replacing the ancient Ainu people except for the northernmost island: S. Horai et al., "mtDNA Polymorphism in East Asian Populations, with Special Reference to the Peopling of Japan," *American Journal of Human Genetics* 59 (1996) 579-590.

Humans migrated from Siberia into the New World at around 25,000 years ago: B. Bower, "DNA Illuminates Siberian Migration," *Science News* 186, no. 7 (2014) 12.

Page 143

Most of the Native American tribal groups of North America – and all of those in South America – fall into the *Amerind* group: A. Torroni et al., "Asian Affinities and Continental Radiation of the Four Founding Native American mtDNAs," *American Journal of Human Genetics* 53 (1993) 563-590.

The earliest native Americans probably first traveled south along the Pacific coast by boat, reaching the tip of South America by about 18,500 years ago: B. Bower, "A Coastal Route Could Have Led Humans into the Americas," *Science News* 193, no. 11 (2018) 13.

New World peoples reached Central America by about 14,000 to 13,000 years ago and Amazonia by about 10,000 years ago: B. Bower, "Migrants Settle Americas in Tandem," *Science News* 175, no. 3 (2009) 5-6; H. Pringle, "The First Americans," *Scientific American* 305, no. 5 (2011) 37-41; J. C. Chatters, D. J. Kennett, Y. Asmerom, B. M. Kemp, V. Polyak, et al., "Late Pleistocene Human Skeleton and mtDNA Link Paleoamericans and Modern Native Americans," *Science* 344 (2014) 750-754; B. Bower, "Florida Inhabited Surprisingly Early," *Science News* 189, no. 12 (2016) 8; B. Bower, "People may have lived in Brazil more than 20,000 years ago," *Science News* 192, no. 4 (2017) 17.

The last major wave of human migration was into the islands of the Pacific: J. K. Lum et al., "Mitochondrial and Nuclear Genetic Relationships among Pacific Island and Asian Populations," *American Journal of Human Genetics* 63 (1998) 613-624.

Amazingly, the Easter Island Polynesians may have even sailed back and forth to South America, but this is controversial: B. Bower, "Easter Islanders Sailed to America," *Science News* 186, no. 11 (2014) 12; D. Garisto, "Polynesian sailings to America Doubted," *Science News* 193, no. 8 (2018) 18.

Ethnologists have traced the three sons of Noah to the region that roughly surrounds the Mediterranean: E. A. Speiser, *The Interpreters Dictionary of the Bible* 3, Man: *Ethnic Divisions* (New York, NY: Abingdon Press, 1962) 235-242; G. J. Wenham, *Word Biblical Commentary* 1 Genesis 1-15 (Waco: Word Books Publishers, 1987) 216-232; V. P. Hamilton, *The Book of Genesis, Chapter 1-17* (Grand Rapids, MI: Eerdmans, 1990) 330-365; K. A. Kitchen, *On the Reliability of the Old Testament* (Grand Rapids, MI: Eerdmans, 2003) 426-427, 430-438.

Page 144

By about 2400 B.C., the Sumerian language was entirely replaced as a "living language" but still remained as a written language: J. N. Postgate, *Early Mesopotamia – Society and Economy at the Dawn of History* (Routledge: London, 1992) 37; D. O. Edzard, "The Sumerian Language," *in* J. M. Sasson, ed., *Civilization of the Ancient Near East* 4 (New York, NY: Scribners, 1995) 2109; J. Huehnergard, "Semitic Languages," *in* J. M. Sasson, ed., *Civilization of the Ancient Near East* 4 (New York, NY: Scribners, 1995) 2118; A. Kitchen, C. Ehret, S. Assefa, and C. J. Milligan, "Bayesian Phylogenetic Analysis of Semitic Languages Identifies an Early Bronze Age Origin of Semitic in the Near East," *Proceedings of the Royal Society B*, Biological Sciences 276, no. 1668 (2009) 2,703-2,710.

A date of around 2700-2600 B.C. for the dispersion of languages story in Genesis correlates with other ancient Sumerian stories on the same subject: K. A. Kitchen, *On the Reliability of the Old Testament* (Grand Rapids, MI: Eerdmans, 2003) 426.

Page 145

The intent of the Table of Nations in Genesis 10 was to show the relative kinship of all the *known* nations of the world to the people groups who were descended from Noah's sons: M. D. Johnson, *The Purpose of the Biblical Geneologies* (Cambridge: Cambridge University Press, 1969) 77.

The Genesis 10 text was probably written down at the time when narrative writing began (ca 2000 B.C.), the text coming as a tradition with Abraham journeying northwestward from Mesopotamia: K. A. Kitchen, *On the Reliability of the Old Testament* (Grand Rapids, MI: Eerdmans, 2003) 426.

My lineage has been traced based on DNA analysis and genealogy records: Analysis of my Mitochondrial DNA was performed by the National Geographic Society's genographic project and Roy Hill provided the genealogy for the Read, Sommers, and Sinn families.

From the study of haplotypes, it is possible to trace the origins of founder mutations and to track human populations: D. Drayna, "Founder Mutations," *Scientific American* 293, no. 4 (2005) 80.

Page 146

Cheddar Man is the most complete skeleton every found in Britain and has been dated to about 9,000 years ago: B. Sykes, "Cheddar Man Speaks," *The Seven Daughters of Eve*, Ch. 12, (New York, NY: W. W. Norton, 2001) 169-184; S. Lyall, "Tracing Your Family Tree to Cheddar Man's Mum," *The New York Times*, March 24, 1997; http://www.nytimes.com/1997.

Page 149

Some people make the biblical Eve the equivalent to Mitochrondrial Eve and the biblical Adam equivalent to Y Chromosome Adam: H. Ross, *The Genesis Question* (Colorado Springs: NavPress, 1998) 231 pp.

There is substantive evidence that runs counter to the Progressive Creationist's "big bang" model of human advancement and also the mental capacities of the Neanderthal: D. C. Johanson and K. Wong, *Lucy's Legacy: The Quest for Human Origins* (New York, NY: Three Rivers Press, 2010) 259-267; B. Bower, "Reading the Stones," *Science News* 187, no. 7 (2015) 16-21; B. Bower, "Cache of Eagle Claws Points to Neandertal Jewelry-Making," *Science News* 187, no. 8 (2015) 7; B. Bower, "Stone Circles Show Neandertals' Skills: Ancient Stalagmite Structures Found Inside French Cave," *Science News* 189, no. 13 (2016) 7.

Can the "gaps" in the genealogies of Genesis be stretched back to millions of years as has been done by some: G. R. Morton, "Dating Adam," *Perspectives on Science and Christian Faith* 51, no. 2 (1999) 87-97.

Another problem with the Progressive Creationist view is its anthropological universal position that the flood was local but killed all humans on planet Earth except for Noah and his family: H. Ross, "Explaining the Extent of Noah's Flood," *New Reasons To Believe* 1, no. 2 (2009) 14-15; H. Ross, "Exploring the Extent of the Flood," *New Reasons to Believe* 1, no. 3 (2009) 14-15.

Page 150

Anthropological universality is assumed without regard to hundreds, perhaps thousands, of ancient human archeological sites: D. A. Young, "Theology and Natural Science," *The Reformed Journal* 38, no. 5 (1988) 13.

Most damaging to the Mitochondrial Eve view is the DNA-genomics evidence: D. R. Venema, "Genesis and the Genome: Genomics Evidence for Human-Ape Common Ancestry and Ancestral Hominid Population Sizes," *Perspectives on Science and Christian Faith* 62, no. 3 (2010) 166-178.

The (strict) Evolutionary Creationist view considers the people and events in the early chapters of Genesis (before Abraham) to be unhistorical, legendary, or fictional: D. O. Lamoureux, *Evolutionary Creation: A Christian Approach to Evolution* (Eugene, OR: Wipf and Stock, 2008) 319, 367.

Do real people start with Abraham, as maintained by some Evolutionary Creationists?: D. O. Lamoureux, "Lamoureux Response to

Montgomery," *Perspectives on Science and Christian Faith* 63, no. 1 (2011) 72.

Page 151

There were other *Homo sapiens* before Adam, but Adam was made the first "Homo divinus" in the spiritual sense: The term "Homo divinus" was suggested by J. R. W. Stott, *Understanding the Bible* (London: Scripture Union, 1972) 63.

Page 152

The expression "dust of the ground" is a poetic figure of speech, one that always signifies mortality in the Old Testament: R. Clouser, "Reading Genesis," *Perspectives on Science and Christian Faith*, 68, no. 4 (2016) 246.

In a Near Eastern cultural context, among the materials used by the gods for the creation of man was the "clay" of the earth: U. Cassuto, *A Commentary on the Book of Genesis*, Pt. 1 (Jerusalem: Magnes Press, 1972) translated from Hebrew by Israel Abrahams, 104; D. C. Harlow, "The Genesis Creation Accounts," *Origins Symposium*, Calvin College, March 3, 2006, 39 pp.

When Near Eastern texts speak of people being created from "dust" or "clay," they are addressing the nature (mortality) of all humanity: J. H. Walton, *The Lost World of Genesis One: Ancient Cosmology and the Origins Debate* (Downers Grove, IL: InterVarsity Press Academic, 2009) 70; J. H. Walton, *The Lost World of Adam and Eve* (Downers Grove, IL: InterVarsity Press, 2015) 76.

The Sumerian word for "rib" could for the ancients alternately mean "life," and in Sumerian literature the "lady of the rib" came to be identified with the "lady who makes live": S. N. Kramer, *History Begins at Sumer* (London: Thames and Hudson, 1961) 209-210; J. M. Sasson, *Civilizations of the Ancient Near East* (New York: Scribners, 1995) 1,818; D. O. Lamoureux, *Evolutionary Creation: A Christian Approach to Evolution* (Eugene, OR: Wipf and Stock, 2008) 257.

Genesis gives many hints of other people besides Adam and Eve that were alive at the same time they were: Refer to the book by D. Fischer, *The Origins Solution* (Lima, OH: Fairway Press, 1996) 382 pp.

Page 153

Another translation of Genesis 4:14 could be: "…and Cain dwelt in the land of nomads": D. J. Fischer, *Historical Genesis: From Adam to Abraham* (Lanham, MD: University Free Press of America, 2008) 51-52.

To most biblical scholars it seems certain from Hebrew syntax that this "generations of" formula in Genesis 2:4 is intended to be a superscription to what follows in Genesis 2: K. A. Kitchen, *On the Reliability of the Old Testament* (Grand Rapids, NY: Eerdmans, 2003) 427-428; R. Clouser, "Reading Genesis," *Perspectives on Science and Christian Faith*, 68, no. 4 (2016) 246-248.

The majority of scholars over the years have favored a parallel structure between Genesis 1 and 2, with Adam and Eve of Genesis 2 being the specific "male" and "female" of Genesis 1: P. J. Wiseman, *Ancient Records and the Structure of Genesis – a Case for Literary Unity* (Nashville, TN: Thomas Nelson, 1949; 1985) 148 pp; H. F. Blank, "On the Structure of Genesis," *Perspectives on Science and Christian Faith* 56, no. 1 (2004) 74-75.

However, the second creation account can also be viewed as a sequel to the Genesis 1 account: J. H. Walton, *The Lost World of Adam and Eve* (Downers Grove, IL: InterVarsity Press, 2015) 63, 64.

Alternately, a compromise position is that while Genesis 2:4 is certainly a superscription to which follows in Genesis 2, it is not two creation accounts, but one creation account (Genesis 1) followed by another account of the beginning of redemption (Genesis 2): R. Clouser, "Reading Genesis," *Perspectives on Science and Christian Faith*, 68, no. 4 (2016) 248.

In 1655, Isaac de La Peyrè suggested that pre-Adamite people had been created in the first phase of creation recounted in Genesis 1, while Adam and Eve did not appear until the next phase, recounted in Genesis 2: D. A. Young, *The Biblical Flood* (Grand Rapids, MI: Eerdmans, 1995) 51-53.

Page 155

The true sense of Genesis 1:27-28 is that God created *them* (in the plural): U. Cassuto, *A Commentary on the Book of Genesis*, Pt. 1 (Jerusalem: Magnes

Press, 1972) translated from Hebrew by Israel Abrahams, 58.

From the worldview of the ancient world, people were considered to be in the "image of God" when they embodied his qualities and did his work: D. O. Lamoureux, "Beyond Original Sin: Is a Theological Paradigm Shift Inevitable?" *Perspectives on Science and Christian Faith* 67, no. 1 (2015) 35-49.

It is the spirit nature that survives the death of the body; it is a special creative act of God, and it differentiates us from every other creature: J. H. Walton, *Ancient Near Eastern Thought and the Old Testament* (Grand Rapids, MI: Baker Academic, 2006) 210-214.

Page 156

The Old Testament never uses the word "fall" as a term: J. H. Walton, *The Lost World of Adam and Eve* (Downers Grove, IL: InterVarsity Press, 2015) 192-193.

The doctrine of sin being passed down from generation to generation was first formulated by Augustine: J. H. Walton, *The Lost World of Adam and Eve* (Downers Grove, IL: InterVarsity Press, 2015) 142, 145.

Adam, as a human being, was "doomed to death" (the meaning of the Hebrew phrase in Gen. 2:17): J. H. Walton, *The Lost World of Genesis 1: Ancient Cosmology and the Origins Debate* (Downers Grove, IL: InterVarsity Press Academic, 2009) 101.

Chapter 10: Putting It All Together

Page 159

From his infinite time frame, God condescends to enter the finite realm of human history: P. Enns, *Inspiration and Incarnation* (Grand Rapids, MI: Baker Academic, 2005) 160-161.

Page 160

The Old Testament is Jewish history, not human history: D. Fischer, "Historical Adam?," *Perspectives on Science and Christian Faith* 67, no. 2, 159.

Page 161

Creation of the physical world of matter and energy could have begun by direct fiat of God, and then could have evolved under God's direction to its present form. A good simple book on cosmology written for the lay public is: N. deGrasse Tyson, *Astrophysics for People in a Hurry* (New York, NY: W. W. Norton, 2017) 222 pp.

A Worldview Approach favors a theistic evolution position based on the strong genetic evidence for evolution that has emerged over the past few decades: F. S. Collins, *The Language of God* (New York: Free Press, 2006) 294 pp; D. R. Venema, "Genesis and the genome: Genomics evidence for human-ape common ancestry and ancestral hominid population sizes," *Perspectives on Science and Christian Faith* 62, no. 3 (2010) 166-178.

Page 162

By the time of Adam, all humans had attained a religious consciousness: R. A. Clouser, "Genesis on the Origin of the Human Race," *Perspectives on Science and Christian Faith* 43, no. 1 (1991) 2-13.

Page 163

A new world order commenced with Adam when God decided that this was the time in human history to impart his Holy Spirit into a specific human, Adam: B. Enick, *Evolving in Eden* (Pittsburg, PA: AdamEve Publishing, 2007) 326 pp.

Although humans and sin were in the world before Adam, the manner of sin was in the form of offenses against nature and all died a natural death: R. J. Fischer, *Historical Genesis: From Adam to Abraham* (Lanham, MD: University Free Press of America, 2008) 27.

Page 164

The garden of Eden is considered a historical place, with the four rivers of Eden pinpointing its location at the head of the Persian Gulf in today's Iraq: C. A. Hill, "The Garden of Eden: A Modern Landscape," *Perspectives on Science and Christian Faith* 52, no. 1 (2000) 31-46.

In ancient Near East writings, serpents and serpentine creatures played a prominent role as adversaries of both humans and gods: W. C. Kaiser, senior editor, *The NIV Archaeological Study Bible* (Grand Rapids, MI: Zondervan, 2005) 8.

Serpents similarly oppose humans and gods in the Mesopotamian *Enuma Elish* myth and the Egyptian pyramid texts: J. H. Walton, *The Lost World of*

Adam and Eve (Downers Grove, IL: InterVarsity Press, 2015) 129.

Page 165

For Egyptian creation myths refer to: J. V. Miller and J. M. Soden, *In the Beginning…We Misunderstood: Interpreting Genesis 1 in Its Original Context* (Grand Rapids, MI: Kregel Publications, 2012) 220 pp.

The invention of cuneiform writing happened around 2000 B.C., and narrative writing didn't commence until 2500-2000 B.C.: J. Friberg, "Numbers and Measures in the Earliest Written Records," *Scientific American* 250, no. 2 (1984) 110-118.

The Genesis authors/scribes were using common literary motifs to convey truths about humanity that were familiar topics in their ancient world: J. H. Walton, *The Lost World of Adam and Eve* (Downers Grove, IL: InterVarsity Press, 2015) 110.

Near Eastern creation myths are a clear product of the myth-based polytheistic Mesopotamian culture: W. C. Kaiser, senior editor, *The NIV Archaeological Study Bible* (Grand Rapids, MI: Zondervan, 2005) 888; K. A. Kitchen, *On the Reliability of the Old Testament* (Grand Rapids, MI: Eerdmans, 2001) 424-425; B. S. Childs, *Myth and Reality in the Old Testament* (London, SCM Press Ltd, 1960) 109 pp.

The Hebrew language that we know from the Old Testament did not exist in the second millennium B.C.: P. Enns, *Inspiration and Incarnation* (Grand Rapids, MI: Baker Academic, 2005) 50.

Page 166

A Worldview Approach does not concord with the (strict) Evolutionary Creationist position that virtually all of the narrative details in Genesis 2-8 are borrowed from mythology but transformed to craft new stories with a decidedly different theology: D. C. Harlow, "After Adam: Reading Genesis in an Age of Evolutionary Science," *Perspectives on Science and Christian Faith* 62, no. 3 (2010), 194.

It should be expected that the Israelites held many concepts in common with the rest of the ancient world, but this is far different from suggesting that their literature was borrowed or copied from other Near Eastern literature: J. H. Walton, *"The Lost World of Genesis One: Ancient Cosmology and the Origins Debate* (Downers Grove, IL: InterVarsity Press, 2009) 13.

Page 167

Note on Figure 10-3: The Library of Congress supplied the following information on Fig. 10-3: "Currier & Ives entitled this lithograph 'The tree of life: on either side of the river there was the tree of life which bare twelve manner of fruits – Rev. ch. XXII, 2.' Actually Currier & Ives may have either hired an artist to create the image or used an existing image in creating theirs. However, the Catalog of Copyright Entries does not assign an artist's name to the piece, nor do Currier & Ives acknowledge an artist in their text on the image."

The overall purpose of the Old Testament was to chronicle the history of God's covenants with this one people group: R. J. Fischer, *Historical Genesis: From Adam to Abraham* (Lanham, MD: University Free Press of America, 2008) 117, 141.

Page 168

Noah's flood was considered to be the "primitive flood catastrophe" (*abubu*) that wiped out the entire antediluvian world: Y. S. Chen, *The Primeval Flood Catastrophe: Origins and Early Development in Mesopotamian Traditions* (Oxford: Oxford University Press, 2013) 100.

A Worldview Approach places Noah's flood in Mesopotamia in the time frame of about 2900 B.C. when large boats were being built using the tools then available for ship building: C. A. Hill, "A Time and Place for Noah," *Perspectives on Science and Christian Faith* 53, no. 1 (2001) 37-38.

The geological and paleontological geosciences severely disagree with the Genesis flood being global, which position changes the text from a historically based account into one fully unsupported by the findings of modern science: C. Hill, G. Davidson, T. Helble, and W. Ranney, eds., *Grand Canyon: Monument to an Ancient Earth – Can Noah's Flood Explain the Grand Canyon?* (Grand Rapids, MI: Kregel Publications, 2016) 240 pp.

A fascinating detective story about the search for the Adites' "lost city of Ubar" is Nicholas Clapp's book: *The Road to Ubar: Finding the Atlantis of the*

Sands (Boston: Houghton Mifflin, 1998) 342 pp. Also see Wikipedia, "Atlantis of the Sands," for an ongoing debate on this topic.

Page 169

The Mesopotamian stories of the flood came down to us sometime between the Early Dynastic Period III and Old Babylonian Period: For a detailed discussion of Mesopotamian flood traditions, refer to Y. S. Chen, *The Primeval Flood Catastrophe: Origins and Early Development in Mesopotamian Traditions* (Oxford, Oxford University Press, 2013) 313 pp. Chen makes no attempt to correlate these primeval flood traditions with the Genesis flood.

Page 170

The assimilation process of the Noah flood stories developed gradually, often by the accretion of different versions of the story based on the political ideology of each different Mesopotamian region: Y. S. Chen, *The Primeval Flood Catastrophe: Origins and Early Development in Mesopotamian Traditions* (Oxford, Oxford University Press, 2013) 3, 5, 196.

The differences between Genesis and mythological versions are so numerous as to preclude them from being copied directly from each other: K. A. Kitchen, *On the Reliability of the Old Testament* (Grand Rapids, MI: Eerdmans, 2006) 425.

It has not been shown that the Hebrew version was borrowed, even indirectly, from the Babylonians: A. R. Millard, "New Babylonian 'Genesis' Story," *Tyndale Bulletin* 18 (1967) 3-18; A. Millard, "The Babylonian Flood Story – A New Tablet and an Old One Misrepresented," *Faith & Thought* 60 (2016) 25-30.

The Genesis account would have had the advantage of being written by the covenant line descended from Noah, where the flood story would have been faithfully transmitted orally, and then in written in cuneiform to succeeding generations: W. C. Kaiser, senior editor, *The NIV Archaeological Study Bible* (Grand Rapids, MI: Zondervan, 2005) 448. *The NIV Archeological Study Bible* attests to the faithful transmittal of the biblical text over time, and that only in a very few places, and only in minor word cases, does it appear that a scribe intentionally altered a sacred manuscript.

Page 172

The names Terah (Abraham's father), Nahor (Abraham's grandfather), and Serug (Abraham's great-grandfather) have survived from antiquity as names of towns in the vicinity of Haran: W. C. Kaiser, senior editor, *The NIV Archaeological Study Bible* (Grand Rapids, MI: Zondervan, 2005) 22.

DNA testing of Arabs and Jews has linked both groups to a common ancestor several thousand years ago: M. F. Mammer et al., "Jewish and Middle Eastern Non-Jewish Populations Share a Common Pool of Y-Chromosome Biallelic Haplotypes," *Proceedings of the National Academy of Science* 97, no. 12 (2000) 6,769-6,774.

The tradition of exaggerated "long reigns" for gods and kings seems to have been a common religious tradition for the peoples of the ancient Near East: Diodorus Siculus, *Diodorus on Egypt*, trans. Edwin Murphy (London: McFarland, 1985) 32-33; J. A. Hoffmeier, *Israel in Egypt: The Evidence for the Authenticity of the Exodus Tradition* (Oxford: Oxford Press, 1996) 41.

Some (strict) Evolutionary Creationists consider all of the patriarchs from Adam down to, but not including, Abraham to be non-historical persons: D. O. Lamoureux, *Evolution: Scripture and Nature Say YES!* (Grand Rapids, MI: Zondervan, 2016) 120.

The minimalist movement started in the 1870s with Julius Wellhausen and others and continues up until today in a number of different forms and schools: For a thorough critique of minimalism, from the 1800s until today, refer to K. A. Kitchen, *On the Reliability of the Old Testament* (Grand Rapids, MI: Eerdmans, 2003) 449-500.

Page 173

Wellhausen proposed – based mainly on word studies – that a number of different biblical authors (J, E, D, P) had actually written Genesis much later than the patriarchal Period, during the Monarchy/Exile: W. C. Kaiser, senior editor, *The NIV Archaeological Study Bible* (Grand Rapids, MI: Zondervan, 2005) 15.

The world of Genesis 12-50 is certainly *not* that of the Monarchy period. It fits only the period before that, the twentieth to seventeenth centuries (ca 2000-1600 B.C.): K. A. Kitchen, *On the Reliability of the Old Testament* (Grand Rapids, MI: Eerdmans, 2003) 462, 487, 495.

Different sources do not imply that these stories were made up during the much later Monarchy/Exile Period: R. Clouser, "Reading Genesis," *Perspectives on Science and Christian Faith*, 68, no. 4 (2016) 248.

Some scholars have attributed the parallel lines of Cain (Gen. 4:17-18) and Seth (Gen. 5:6-25) to their close geographical proximity to each other: R. J. Fischer, *Historical Genesis: From Adam to Abraham* (New York, NY: University Press of America, 2008) 53-54.

Page 174

DNA evidence has confirmed a distinct paternal genealogy for Jewish priests: K. Skorecki et al., "Y-Chromosomes of Jewish Preists," *Nature* 385 (1997) 32.

We might view Moses as an editor/historian who drew together the details of the family histories of Abraham and his descendants into a single text: W. C. Kaiser, senior editor, *The NIV Archaeological Study Bible* (Grand Rapids, MI: Zondervan, 2005) 2.

Page 175

Abraham could have taken this early Genesis account with him when he left Ur for Haran, either in oral or written form: U. Cassuto, *A Commentary on the Book of Genesis*, Pt. 1, translated from Hebrew by Israel Abrahams (Jerusalem: Magnes Press, 1972) 273.

Moses could have compiled these stories into a single chronological narrative sometime between the fifteenth and thirteenth centuries B.C.: K. A. Kitchen, *On the Reliability of the Old Testament* (Grand Rapids, MI: Eerdmans, 2003) 366-367.

Historical "memories" and geographical information, as well as old words that had disappeared from the living language before the time of Moses, attest to these stories being handed down from more ancient times: C. K. Keil and F. Delitzsch, *Commentary on the Old Testament* (Grand Rapids, MI: Eerdmans, 1975) 32; U. Cassuto, *A Commentary on the Book of Genesis*, Pt. 2, translated from Hebrew by Israel Abrahams (Jerusalem: Magnes Press, 1972) 252.

The Genesis text could have been put together in a logical chronological sequence and converted by authors/scribes into classical Masorite Hebrew – a form of the Hebrew language that did not even exist at the time of Moses: D. C. Harlow, "The Genesis Creation Accounts," *Origins Symposium*, Calvin College (March 3, 2006) 6.

Page 176

Part of the problem of finding archeological evidence for the Exodus is that biblical scholars cannot agree on the date or the route of the Exodus: W. C. Kaiser, senior editor, *The NIV Archaeological Study Bible* (Grand Rapids, MI: Zondervan, 2005) 98, 108-109.

Egypt's East Nile delta is where the Hebrews resided and is essentially an alluvial fan of mud which was made into mud-brick structures that were repeatedly leveled, replaced, and which quickly turned back into soil. This is why no trace of the Hebrews have been found in Egypt: it would not be logical to expect these residential structures to exist after three thousand years: K. A. Kitchen, *On the Reliability of the Old Testament* (Grand Rapids, MI: Eerdmans, 2003) 246.

Page 177

Satellite images of the western Nile delta part of Egypt are now being used to trace the most likely journey of the Exodus: J. A. Hoffmeier, *Israel in Egypt: The Evidence for the Authenticity of the Exodus Tradition* (Oxford: Oxford University Press, 1996) 240 pp; J. A. Hoffmeier and S. O. Moshier, "New Paleo-Environmental Evidence from North Sinai to Complement Manfred Bietak's Map of the Eastern Delta and Some Historical Implications," *in* T. E. Levy, T. Schneider, and W. H. Propp, eds., *Israel's Exodus in Transdisciplinary Perspective*, Quantitative Methods in the Humanities and Social Sciences (Switzerland, Springer International, 2006) 244 pp; S. O. Moshier and J. K. Hoffmeier, "Which Way Out of Egypt? Physical Geography Related to the Exodus Itinerary," *in* T. E. Levy, T. Schneider, and W. H. Propp, eds., *Israel's Exodus in Transdisciplinary*

Perspective, Quantitative Methods in the Humanities and Social Sciences (Switzerland: Springer International 2015) 101-108.

A known solar eclipse that occurred in 1207 B.C., possibly in conjunction with Joshua's victory at Gibeon, gives credence to the late-date view of Exodus: C. J. Humphreys and W. G. Waddington, "Solar Eclipse of 1207 helps to Date the Pharaohs," *A&G (Astronomy and Geophysics*, published by the Royal Astronomical Society) 58 (2017) 5.39-5.42.

The "death through sin" that Paul is talking about in Romans 5:12 is *spiritual* death, not physical death: R. J. Fischer, *Historical Genesis: From Adam to Abraham* (Lanham, MD: University Free Press of America, 2008) 27.

Page 178

According to (strict) Evolutionary Creationist Daniel Harlow, the doctrine of original sin does not require that Adam and Eve be historical figures: D. C. Harlow, "After Adam: Reading Genesis in an Age of Evolutionary Science," *Perspectives on Science and Christian Faith* 62, no. 3 (2010) 179; D. C. Harlow, *Adam and Eve as Symbolic Figures in Biblical Literature*, Abstract, 2009 Annual Meeting, American Scientific Affiliation, Baylor University, Waco, TX: 18.

The Last Supper by Pascal Dagnan-Bouveret, 1896. *Wikimedia Commons*

INDEX

Bold = names on, or in figure captions of, photos, artwork, drawings, tables, charts, or boxes

A

Aaron, 174
Abel, 135, 153, 170, 174
Abraham, 46, **47**, 47, 48, **49**, **50**, 52, 53, 58, **133**, 137, **137**, 140, 145, **148**, 150, 156, **162**, 163, 164, 172, 174, 175, **176**, 177
Abridgement ages, 51, 135, **136**
Abrahamic (Sinaic) covenant, **162**, 174
Accommodationist position, see Evolutionary Creationist position
'Ad, **168**, 168, 169
Adah, 59
Adam and Eve, 6, **28**, **34**, 38, 41, 47, 48, **48**, 50, **50**, 52, 53, 58, 59, 90, 97, 128, 131, **133**, 134, 135, **136**, 136, 137, **137**, 1338, 140, 145,146, 147, **148**, 150, 152, 153, 155, **157**, **162**, **163**, 163, 164, 172, 177, 179
  Adam and Eve and origins, 15, **95**
  Historical Adam, 7, 47, **148**, 150,151
  How the story of Adam and Eve was transmitted, 165, **165**
  Were Adam and Eve real persons? 164
  When did Adam and Eve live? **136**, 139
Adamic covenant, **162**, 166
Afghanistan, 63
Africa, 132, 134, **135**, 140, **141**, 142, 143, 145, 147, **148**, 148, 149, 153
After its kind, 90, **108**, 111, 112, 113, 125, **125**
Age of how long ago *Homo sapiens* lived, 140
Age of planet Earth, 18, 19, 39, 41, 42, 53, **89**, 91, 93, **94**, 94, **95**, 117
Age of universe, 18, **19**, 42, 53, **89**, 93, **95**
Agri Dag (Mount Ararat), 78, 79, **98**
Akkad, **33**, 46, 61, 166
Akkadian language, 138, **169**
Aleut-Eskimo migration, 143
Al Hatifah, 30
Alps, 83
Al'Ubaid, **33**, 48, 138
Amanus Mountains, **33**, 62, 64, **65**, 66, 75
Amazonia, 143
Amenhotep II, 176
Amerind migration, 142, 143
Amram, 50, **50**
Animism, 8, 134, 135, **162**, 162
An Nasiriyah, **33**, 34, 35
Anthropology, 131, 132, 136, 143
  Anthropological evidence of humans, **141**, 142, 145, 148, 149, 160, 164, 173, 179
  Anthropological time periods, 132, **133**
  Anthropological universality, 148, 149, 150, **171**
Appearance of age, 125, 127
Arabia, 29, 31, 36, 60, 62, **63**, 142, 143, 144, 169
Arabian Highlands, **33**, 66
Arabian Peninsula, **30**
Arabian Sea, 66, 67
Arabian shield, 39
Aram, 169
Ararat (Urartu) region, 62, 69, 73, 79, 80
Archaeopteryx, **119**, 119
Archean Era, **95**, 117
Archeology, 131, 136
  Archeological evidence of humans, **138**, 140, 142, **144**, 145, 148, 149, 160, 173, 179
  Archeological time periods in Mesopotamia, **34**, **84**, 132, **133**, **137**, 138
Ark, Noah's, 56, 57, 64, **65**, **70**, 74, **75**, 77, 78, 90, 126, 149
  Size of ark, 56, 76, 85
  Color of ark, 76
  Decks on ark, 77
  Door of ark, 77
  Landing place of ark, 78
  Rooms in ark, 75
  Shape of ark, 76
  Waterproof covering of ark, 77
  Window of ark, 77
  "Window" (skylight) of ark, 77
  Wood for ark, 62, 74, 75

Ark of the covenant, 78
Armenia, **33**, 52, 66, **79**, 79, **98**
Arphaxed, **49**, **50**, 52
Ash Shamiyah, **33**, 34
Asher, 51
Asia, 82, 132, 134, 135, 147
Ashur, **33**, 34, **38**
Assyria, **33**, 34, 38, 61, 75, 78, **78**, 81, 144
Astrometrical hypothesis for long patriarchal ages, 47
Atlantis, 78
Atra-hasis (Akkadian name equivalent for Noah) epic, **59**, 60, 69, 169, **169**
Augustine, 47, 148, 156
Austria, 135
Australia, 72, 117, 135, **135**, 142, 147, 163, 110

B

Babylon, **33**, 61, 68, **130**
Babylonia, **33**, **59**, 61, 62
Baghdad, **33**, 34, 39, 61, 69, 76, 86
Bahrain, **30**, 37
Baliki River, **33**
Basque people, **141**, 142
Basra, **33**, 34, 35
Bdellium, **30**, 31, 38
Ben Halafta's age for creation of Adam, 41
Bering Strait, 142
Berossus, 78
Beytişebab, 69
Biblical evidence for old Earth, 92, 93
Bitumen (pitch), **33**, **38**, 39, 62, 63, 64, **65**, 76, 138
Black Sea, 61, 66, 81, **135**
British Isles, 142
Bronze, **63**
Bronze Age, 48, 60, **133**, 135, 137
Buddhism, 8
Burgess shale, 118, **118**

C

Cain, 26, 58, 62, 90, **133**, 135, 151, 152, 153, 170, 173
Calneh, 61
Cambrian explosion of life, **95**, 118, 126
Canals, 62, **63**, 63, 64, **65**, 65, 138
Canaan, 153, 177
Canopy theory, 48, 89, 90
Carnelian, 31, 62, **63**
Caspian Sea, **33**, 66, 81
Catastrophic plate tectonics, 90, 91, 102
Caucasus Mountains, 79
Cave adaption of species, **121**
    Amblyopsidae fishes, 122, **123**
Cedar wood, 62, 64, 75
Cenozoic Era, **95**, 102
Central America, 143
Chalcedony, 31, **63**
Chalcolithic Period, **133**, 135, 173
Cheddar Man, **133**, 146, **147**
Cherubim, 153
China, 118, 132, 135, 142, 172
Christ, 47, 87, 128, 153, 159, 163, 167, 174, 177
Chronologies of Genesis, 18, 39, 41, 42, 48, **48**, 51, **53**, 90, 91, 136
Cizre, **33**, 69, 80, 82, 86
Clan ages, 47, **136**, 137
Common descent, 110, 114, 116, 127, 128, 155
Common design, 126, 127
Comparative anatomy, 112, 114, 115
Comparative genetics (DNA), 115
Concordist position, see Progressive Creationist position
Continental Divide, 84
Convergent evolution, 117, 122
Copernicus, 95
Copper, 62, **133**, 135
Cosmic-background microwave radiation, 18, **19**
Cosmology of the Near East, 3, 9, **10**, 22, 65
    Three-tiered cosmology, 10, 46
Covenants with Adam's line, 87, 92, 144, **144**, 151, 157, 160, 162, 164, 167, 170
Creation by fiat, **108**, 120, 125
Creation science, see Young-Earth Creationist position
Crete, 143
Critical scholars, 26, 32, 172, 173
Cro-Magnon man, **133**, 134
Cubit (Mesopotamian), 56, 73, 76, 77, 85
Cudi Dag (Jabel Judi) region, 69, 80, 82, **98**
Culture, 4, 8, 9, 58, 64, 72, **112**, 144, 151, 160, 167, 170
Cultural interaction between Adamite and non-Adamite lines, 164

Cuneiform writing on tablets, 43, **43**, **47**, 47, 58, 59, 64, 138, 151, 165, **165**, 175
Cuneiform tablet of the Sumerian flood story, **59**
Cuneiform tablet of the Sumerian king list, **59**
Cuneiform tablets depicting tokens and pictographic script, **64**
Cush, 31, 32, **38**
Cyclonic storms, 66, 67, 68, 86
Cyprus, **135**, 143

D

Damman Formation, 37
Daniel, 73, 153
Darwin, 109, 110, 114, 115, 117, 119, 120, 124, 126
Dating of archeological material, 85
By pottery type, 85
By radiocarbon dating, **34**, 61, 80, 85
David, **47**, **133**, 137, 163
Day-Age view, **17**, 20, 21, 22, 126
Days of Genesis 1, 4, 19
Day 1, **2**, 20, 23, **23**
Day 2, **23**, 24
Day 3, **23**, 24
Day 4, 23, **23**
Day 5, **23**, 24
Day 6, **23**, 24
Day 7, **23**
Dead Sea Scrolls, 53
Death of animals due to "original sin," 90
Decimal (base 10) numbers, 43, 46, **47**, 51, 111
Dhofar region, **30**, 169
Diluvialist school of thought, 83, **83**
Dinosaurs, **95**, 96,
Tyrannosaurus rex, **97**, 97
Divergent evolution, 111, 122
Divine accommodation, see Evolutionary Creationist position
Diyala River, **33**
Diyarbakir, **33**
DNA genomics, 112, 115, **125**, 126, 127, 134, 138, 139, 140, 142, 143, **144**, 145, **147**, 149, 172, 173, 174
DNA testing of Arabs and Jews, 140, 150
DNA testing of Jewish priests, 140, 174
Doğubayazit "ark," 80, **80**
Domestication of animals, 59
Doves, **57**, 78, 81, **82**
Domesticated *Columba livia*
Dual revelation, 6
Dynasty of Akkad Period, **34**, **137**

E

Early Dynastic I Period, **34**, 84, **84**, **137**, 171
Early Dynastic II Period, **34**, 137
Early Dynastic III Period, **34**, 137, 169
Easter Island (Rapa Nui), 143
Eber, **49**, **50**, 52
Eden, 91, 93, 140
Ediacaran life, **95**, 118, 119
Egypt, 30, 45, 46, **47**, 62, 63, 73, 76, 77, 93, 135, 143, 153, 174, 176
Elam, 31, 32, **33**
El-Oueili, 136, 138, **138**
Emar, 45, **47**
Embryonic organs, 115
Empty Quarter, **30**
England, **83**, 146, **147**
Enoch, 47, **49**, **50**, 135, 173, **174**
Enosh, **49**, 52
Enuma Elish, 164, **165**
Epigenetics, 116, 117, 124, 139
Eretz (earth, not Earth), 72, **72**
Eridu, **30**, **33**, **34**, 35, 36, 37, 38, **59**, 136, **137**, 138
Eridu-ware pottery, **138**
Erratic glacial boulders, 83, **83**
Ethiopia, 31, 143, 144
Ethnology, 143, 168
Euphrates River, **30**, 31, **33**, 34, 36, 37, **38**, 39, 61, 62, 64, 66, 67, 68, 72, 75, 76, 79, 84, 85, 86, 138
Europe, 46, 82, 83, 132, **133**, 134, 136, 142, 145, 146, 147, 163, 170
Eve, see also Adam and Eve, 130, 148, 149, 150, 151, 152, 157, 164
Evolution, 6, 109, 114, 126, 131
Evolution and the new genetics, 15, 109
Evolution in fossil record, 117, 118, 119
Evolutionary convergence, 117
Natural genetic engineering, 117, 124
Naturalistic evolution, 125, 129
Theistic evolution, 111, 125, 128, 129

Evolutionary Creationist position, 4, 6, 15, 42, 55, 125, **125**, 128, 147, **148**, 150, 151, 179
Divine accommodation, 6, 15
(Strict) Evolutionary Creationist position, 7, 14, 29, **148**, 150, 151, 166, 172, 178, 179
Exile, 172, 173, 175
Exodus, 137, 172, 176
Archeological evidence, 176, 177
Early and late dates for Exodus, 176, 177
Ezekiel, 153

F

Fara (mound of Shuruppak), 60, 84, 171
Figurative language, 24, 152, 160, 164, 165
Fiji, 143
Firmament, 3, 9, 56, 65
Fixity of species, **108**, 111, **113**
Flood geology, 15, 38, 39, 42, 55, 69, 71, 83, 87, 89, **89**, 90, 91, **92**, 93, 95, 96, 97, 98, 99, 100, 102, 106, **107**, 10, **144**, 167
Flood legends around world, 93
Flood sediments, deposition and erosion, 82, 85
Folk taxonomy, 112
Foundations of the deep, 9
Fossil record, 95, **95**, 96, 97, 114, 117, 126, 132, 156
Fossils, 71, 83, 89, **89**, 90, **92**, 106
"Living fossil," 120, **120**
Framework view, see Literary view
France, 134, 135, 142
Frankincense, **30**, 31

G

Gad, 51
Galápagos Islands, 111
Galileo, 95
Gap Theory view, **17**, 20
Garden of Eden, 6, 15, **28**, 29, **34**, 36, **36**, 37, **37**, 38, **38**, 39, 6, 91, 93, 131, 136, **137**, 138, **148**, 149, 155, 156, 164, 167
A spring rises in Eden, 36
Garden of Eden on a modern landscape, **38**
Location of Garden of Eden, Genesis 2 vs. flood geology, **38**
Where the four rivers of Eden meet, 35, 61
Genealogies of Genesis, **49**, **50**, 51, 53, **53**, 58, **59**, 59, 150, 151, 155, 160
Abridgment of genealogies, 51, 135, **136**
Gaps in genealogies, 51, 137, 148, 149
Ten-generational pattern of genealogies, 51
Three genealogical numbering systems, 51
General revelation, 9
General Theory of Evolution, see Macroevolution
Genesis 1, 3, 18, 20, 22, 23, 24, 25, **27**
Genesis 1, creation controversy, 7
Genesis 1, Gutenberg Bible, **16**
Genetics
Genetic sequence identity, 115, **116**
Modern genetics, 114, 148, 179
The "new genetics," 15, 116, 120
Gentiles, 160, 162, 167, 174, 177, 178
Gideon, 163
Gihon River, 31, 32, 34, 36, **38**, 39
Geologic column, **94**
Geologic evidence for old Earth, 93, **95**
Genome, 116, 129, 134
Genome Project, 116
Germany, 146
Gilgamesh (Assyrian equivalent name for Noah), **59**
Gilgamesh epic, 60, 61, 164, 169, **169**
Glacial theory, 83
God-of-the-gaps theology, 126
Gough's Cave, 146
Gold, **30**, 31, 62, **63**
Gopher wood, 39, 74, 80
Grand Canyon, 83, 85, 88, 92, 94, 99, **99**, **100**, 101, 102, 103, 106
Grand Staircase, 102
Great Britain, **133**, 142, 145, 146
Greece, 143
Greenland, 117, 143
Gulf of Aden, **30**
Gutenberg Bible, **17**

H

Habuda, **33**, 62
Ham, 143, 168
Haplotype, 145
Haran, **33**, 46, 48, 172, 175
Hassuna pottery, 138
Havilah, 29, 61
Hawaii, 124, 143
Hawaiian planthopper evolution, 123

Hebrews, 51
Hermeneutics, 3, 5, 7, 18, 106
Hiddekel (Tigris) River, **33**, 34
Historical Adam view, 7, 147, **148**, 150, 151
Historical memory, 26, 30, 31, 32, 56, 69, 152, 170, 173
Historicity and antiquity of Genesis, 13, 14, 31, 32, 56, 78, 136, 168, 173
Hit, **33**, 34, **38**, 39, 62, 64, **65**, 69, 76
Holy Spirit, 18, **162**, 163, 164, 179
Homo, 132
  Homo denisovan, **133**
  Homo erectus, 132, **133**
  Homo floresiesis, 132, **133**
  Homo naledi, **133**
  Homo neanderthalensis, **133**
  Homo sapiens, 19, **133**, 134, 138, 139, 140, 142, **148**, 148, 149, 150, 153, 161, **162**
Homo divinus, 151
Homoinids, 114, **133**, 134, 149
Horizontal DNA transfer, 116, 124
Hydrology of the flood, 65, **67**, 68, 85

I

Ice Ages, 83, **83**, 142
Igneous rock, **101**, 101
Image of God, 125, 155
Incense routes of old, **30**, 31
India, 62, 142
Indian Ocean, 31, 66, 67
Indonesia, 142, 143
Inerrancy, 5, 14
Intelligent Design, 125, **125**, 126, 127
Iran, **33**, 39, 61, 62, 66, **98**, 143
Iraq, 33, 34, 35, **38**, 39, 43, 48, 58, 61, 62, 66, 67, 68, 79, 81, 86, 144, 164
Iron Age, 60, **133**, 137
Isaac, 47, 50, **50**, 137, 174, 175, **176**
Islam, 8
Israel, 93, 132, **135**, **144**, 145, **162**, 177
Israelites, 48, 87, 153, 154, 166, 174, 175, 177
Italy, 135, 143

J

Jabel, **34**, 51, 58, 59, 60, **137**, 153
Jabel Judi Mountains, **33**, 69, 78, 79, 80, 81, 82, 86
Jacob, 47, 50, **50**, 137, 172, 175, **176**
Japeth, 143, 168
Jared, **49**, **50**, 173
Java man, **133**
Jemdet Nasr, **33**, 43, **65**, **84**
Jemdet Nasr Period, **34**, 60, 61, 64, 75, 84, **137**, 170
Jericho, 87, **135**
Jerusalem, 38, 72, 161
Jesus, 51, 53, 58, 87, 153, 172, 174, 177
Jezira desert region, 69
Joktan, 29
Joram, 51
Jordan, **30, 33**, **135**
Jordan River, 87, 129
Joseph, 45, **47**, 47, 73, 137, **137**, 163, 172
Joshua, 45, **47**
Joshua's long day, 10, 87, 129
  Celestial omens interpretation, 10, **12**
  Solar eclipse interpretation, 11, 12, **13, 87**, 177
Jubal, **34**, 51, 58, 59, 60, **137**

K

Karan River, **30**, 31, **33**, 35, 39
Karkheh River, 32
Kassites, 31, 32
Kazumura lava tube, Hawaii, 124
Kebanan, **133**
Kenan, **49**, **50**, 52
Khabūr River, **33**, 68
Kish, **33**, 62, 68, 84, **84**, 85
Kohath, 50, **50**
!Kung people, **140**, 142
Kurdistan, **33, 66**
Kuwait, 30, **30**, 35, 37, 39

L

Lagash, **33**, 84, **84**, 85
Lamarkian evolution, 117, 124
Lamech, **49**, 50, **50**, 52, 59, 170
Lake Urmia, **33**, 79
Lake Van, **33**, 79
Land of Havilah, 29, **30**
Lapis lazuli, 62, **63**
Larsa, **33**, **59**
Last Glacial Maximum, 142
Law (of Israel), **162**, 163

Lebanon, 143
Levi, 50, **50**
Libya, 143
Linguistic evidence for human migration, 142
Linnaean classification system, 111, **111**, 112, 114, 115
Linnaeus, Carolus, 114
Literary view, **17**, 22, 23, 24, 154
    Objections to the Literary view, 26
Little Zab River, **33**

M

Macroevolution, 110, 119, 122, 124, 128
Mahalalel, **49**, **50**
Mahd and Dhahab gold mine, **30**, 31, **38**, 39, **63**
Mammoth Cave, Kentucky, 121
Marib, **30**, 31
Marshland of Persian Gulf, **33**, 34, 35, 62, 76, 86
Masoretic Hebrew Bible, 25, **25**, 41, 52, 53, 111
Mathematical model of ark travel to mountains of Ararat, 85
Mediterranean, 3, 66, 73, **135**, 143, 145, 146
Melanesians, 143
Mesolithic Period, **133**, 134, 136, 149
Mesopotamia, 30, 31, **31**, **33**, 34, 35, 37, 38, 39, 46, **47**, 48, 56, 59, 61, 62, 65, 66, 67, **67**, 68, 69, 73, 74, 75, 76, 77, 81, 84, 91, 93, 111, 113, 135, 138, 140, 145, **148**, 148, 149, 152, **162**, 164, 169
    Climate, 62
    Natural resources, 62
Mesopotamian
    Agriculture and animal husbandry, 62, 112, 113
    Architecture, 42, 43, 63, **65**
    Astronomy, 43, 63
    Canals, 62, **63**, 64, 65, **65**, 138
    Cities, 62, 63, 153
    Concept of numbers, 42, 43, 44, 45, 46, **48**
    Export-import trade routes, 62
    Mathematics, 42, 43, 50, 63, 64, **65**
    Metal and jewelry working, **63**
    Music, **60**, 60
    Occupations, 59, 60
    Religion, 50, 64, **65**
    Worldview, 56, 58
    Writing, literature, legal, 63, 152
Mesopotamian alluvial plain and hydrologic basin, **33**, 56, 61, 62, 63, 65, 68, 69, 72, 82, 84, 86, 144
Mesozoic Era, **94**, **95**, 96, 100, **100**, 102
Metamorphic rock, **101**, 101
Meteorology of Noah's flood, 65, 86
    Accordance with biblical account, 67
    Cyclonic storms, 66, 67, 68
    Precipitation, 66
    Prolonged flooding, 68
    Wind, 66, 67
Methuselah, **40**, 41, 47, **49**, **50**, 52
Microevolution, 110, 122
Micronesians, 143
Middle Ages, 4
Middle East, 135, 135, 137, 140, 142, 146
Migration of humans around world, 140, **141**
Minimalists, see Critical scholars
Miracles
    Nature miracles, 86, 87, 129
Mitochondrial DNA, 139, 145, 148
Mitochondrial Eve, **133**, 139, 140, 145, 149, 150
Modern geology, 89, **89**, 92, 93, 96, 100, **100**, 107
Monarchy, 172, 173, 175
Mongolia, 142, 143
Monotheism, **162**
Moses, 45, **47**, 50, **50**, 137, **158**, **165**, **169**, 172, 174, 175, 177
    Historian/compiler author of Genesis, 174, 176
    How the story of Moses could have been transmitted, **176**
    Mosaic (Sinaic) covenant (the Law), **162**
    Was Moses a real person and Exodus a real event?, 176
Mosul, **33**, 66, 69, 86
Mountains ("highlands") of Ararat, 9, **57**, 61, 71, 77, 78, **78**
Mount Ararat, **33**, 73, 74, 78, 79, **79**, 80, 81, 82, 85, 87, 90, 98, **98**, 99, 102, **149**
Mount Nisibis, **33**, 78
Mount Nisir, **33**, 78
Mudcracks, **104**, 104, 106, 139
Mutations, genetic, 110, 145
Myrrh, 31

N

Naamah, 59
Nahor, **49**, **50**, 172
Native Americans, 143

Natufian, **133**, **135**
Naturalistic, 125, **129**
Na-Dene migration, 143
Navarra expedition to Mount Ararat, 80
Neanderthal (Neandertal) humans, **133**, 140, 142, 149
Near East, 135, **135**, 145, 150, 152, 159, 164, 165, **166**, 168, 172, 176
Neolithic Period, 78, 133, 149, 173
  Early Neolithic (pre-pottery), **133**, 134, **135**, 138
  Late Neolithic (pottery) **133**, 136, 138
Nephilim, 154, 170
New covenant, **162**, 177, 178
New Guinea, 142, 143
New World, 143
New Zealand (Maoris), 143
Nile Delta, 176
Nimrod, **34**, 34, 61, **137**
Nineveh, **33**, 34, **63**, 81, 82, 84, **169**
Nippur, **33**, 35, 62, **169**
Noachian covenant, 92, **164**
Noah, 6, 26, **34**, 47, 48, **48**, **49**, **50**, 52, 56, 67, 90, **133**, 137, **137**, 143, 144, **144**, **162**, 163, 169, 170
  A time for Noah, 58, 60
  A place for Noah, 58, 61
  Noah's worldview, 64
  Was Noah a real person?, 58, 168
Noah's flood, 6, **34**, **38**, 39, 48, **54**, 55, 57, 65, 69, 83, 84, 85, 87, 90, 96, 99, 100, 102, 103, 106, **107**, 125, 137, **137**, 144, 162
  Global or local?, 15, 38, 55, 69, 71, 73, 74, 78, 81, 82, 83, 85, 87, 89, 90, 91, **92**, 93, 95, 96, 101, **144**, 148, 167, 168, 170, 171, **171**
  Historical or mythological?, 14, 15, 78, **79**, 169
Nomads, 134, 138, 153
North America, 72, 135, 142, 143, 147, 170
Numbers of Genesis, 23, 41
  Concept of numbers change over time, 46, **47**
  Cryptographic (gematria) numbers, 50
  Exaggerated numbers, 41, 44, 47, 50, 56, **59**, **136**, 137, 172
  Numerical numbers, 44, 45, 48, **48**, 56, 61, 64, 78, 86, 91
  Numerological numbers, 44, 45, 46, 48, **48**, 51, 53, **53**, 56, 61, 64, 76, 78, 86, 91, 111
  Preferred or figurative numbers, 44, 45, 46, **47**, **48**, 50, 56
  Sacred symbolic numbers, 44, 46, **47**, 48, **48**, 50, 76, 91
  Symmetry and harmony of numbers, 25, 26, 50, 51, 91, 111, 136, 165, 172

O
Obsidian, 62
Old Babylonian Period, 31, 32, **34**, **137**, 169, 170
Old-Earth Creationists, 42, **42**, **53**, **89**
Olive trees, **57**, 81, 86
  Olea europa, 81
  Oman, 86
Onyx, **30**, 31, 62
Ophir, 61
Ophir gold mine, 31
Original sin, 90, 150, 155, 178
Origins, human, 15, **95**, **130**, 130, 140, 147, **148**, 149, 156, 160
Ötzi the Ice Man, **133**, 135, 136, 146

P
Pacific Islands, 143
Paleolithic Period (Old Stone Age), **133**, 134, 136, 140
  Lower Paleolithic, **133**, 134
  Middle Paleolithic, **133**, 134
  Upper Paleolithic, **133**
Paleozoic Era, **95**, 96
Paluxy "man tracks," 96
Patriarchs, 47
  Long patriarchal lifespans, 47, 48, **48**, 50
  Overlap of patriarchal lifespans, 52
Palestine, **30**, 46, **47**, 48, 66, **72**, 72, 73, 93, 112, 143, 164
Panspermia, 117
Paul, Apostle, 97, 177, 178
Peking man, **133**
Peleg, **49**, **50**
Pentecost, **164**
Persian Gulf, 29, 30, **30**, **33**, 34, 35, 38, **38**, 39, 57, 61, 62, 66, 67, 68, 74, 82, 84, 85, 86, 134, 153, 164
Phenotypic plasticity, 116, 124
Pictographic writing, 63, **64**
Pishon River (Wadi al Batin), 29, **30**, 34, 35, 36, **38**, 39, 84, 173

Pitch, see Bitumen
Plate tectonics, 92, **95**
Poland, 146
Polynesia, 143
Polytheism, 63, 64, 163, 164, 166, 170
Portugal, 134
Pottery, 63, 138, **138**
Pre-Adamic Ruin and Restoration Theory, see the Gap Theory view
Pre-Adamite humans, 131, 151, 152, 154, 161, **162**, 162, 163
Pre-flood and post-flood kings, 36, **59**
Precambrian rock, 96, **100**, 106
Pre-Paleolithic Period, 132, **133**
Pre-pottery, 138
Principle of faunal succession, 114
Principle of superposition, 94
Principle of uniformitarianism, 103
Progressive Creationist position, 4, 5, **5**, 6, 20, 21, 125, **125**, 126, 145, 146, 147, **148**, 148, 149, 150, 151, 172, 179
Progressive Revelation, 109, 161, **162**, 162, **163**
Proterozoic Era, **95**, 96, 100
Pseudogenes, 116
Punctuated equilibrium, 120

Q
Qingjiang biota, 118
Queen of Sheba, 31, **32**

R
Radioactive decay, 139
Radiocarbon (C-14) dating method, **34**, 61, 80, 85, **137**
Radiometric (absolute) dating method, **94**, 94, 139
Rafts, 75, **75**
Rainbow, 87, **89**, 90, 91, 171, **171**
Raindrop prints, **105**, 105, 106
Ramesses the Great, 137, 176, 177
Ras-el-ain spring, **33**, 68
Raven, **57**, 78
Red Sea, **30**, 31, 66, 142
Reeds, use of, 39, 62, 63, **64**, 75, 76, 78
Regressive evolution, 121, 122, 124
Relative dating method, **94**, 94
Reu, **49**, **50**
Rhodes, 143
Rib of Adam, 152, 164
Ripple marks, **104**, 104, 106
Royal Cemetery of Ur, 60, **63**, 84

S
Samaritan Bible, 41, 52, 53
Samaritan pottery, 138
Samoa, 143
Sardinia, 143
Sargon II, 69, 75
Satan, 20, 164
Saudi Arabia, 29, 30, **30**, **33**, 37, 38, 39, 61, 78
Science-Scripture debate, 3, 4, 5, 6, 7, 9, 15, 83, 131, 157, 159, 160, 179
Scientific concordist position, see Progressive Creationist position
Sedimentary rock, 71, 83, 89, **89**, 90, **92**, 93, **95**, 96, **98**, 99, **101**, 101, 102, 106, 114, 119
Sedimentary structures, 103, 106
Semitic-Hamitic languages, 144, 169
Septuagent Greek Bible, 41, 52, 53
Serpent in Eden, 164
Serug, **49**, **50**, 173
Seth, 48, **48**, **50**, 58,173
Seven stanza templates, 45
Sexagesimal (base 60) numbers, 43, 45, 46, **47**, 48, **48**, 50, 64, 76, 11
Shalmaneser's Cave, 69
Shamal wind, **33**, 62, 66, 67, **67**
Sharqi wind, **33**, 62, 67, **67**, 69, 86
Shelah, **49**, **50**, 52
Shem, **49**, **50**, 52, 143, 153, 168, 169
Ship building, 64, 168
Shuruppak, **33**, **34**, 60, 61, 62, 64, **65**, 76, 84, **84**, 85, 86, 137, 169, 171
Shamanism, 7, 134, **135**, **162**, 162
Sheol, 9, **10**
Sibera, 142
Silver, 31, 62
Sinaic (Mosaic) covenant, 117
Sippar, **33**, 62, 68
Six days of creation, 15, 17, 18, 19, 20, 22, 23, 39, 41, 42, 90, 91, 125
Somalia, 31
Solomon, 31, **32**, **47**, **133**, 137

Sons of God, 153
South America, 72, 135, 143, 147, 170
South Africa, 117
Spain, 134, 142, 143
Special revelation, 9
Special Theory of Evolution, see Microevolution
Species, 110, 132
Storm surge (*abubu*), 69
Sudan, **30**
Suhaili wind, **33**, 62, 67, **67**, 69, 86
Sumer, **33**, 61, **64**, 166, 169
Sumerians, 46, 47, 73, 138
  Sumerian Enki myth, 164, **164**
  Sumerian flood story, **59**, 169, **169**
  Sumerian Gilgamesh epic, 69, 164
  Sumerian King List, 46, **59**, 60, 61, 64, **65**, 169, **169**
  Sumerian Nippur tablets, 69
Sumerian language and literature, 64, **65**, 138, 144, 148, 151, 152, 166
Susa, 31, 32, **33**, 63, **64**
Susiana, 31
Syria, **33**, 61, 66, 68, 93, 135, 144

T

Table of Nations, 29, **141**, 143, 144, **144**, 145, 146, 147, 149, 168
Tahiti, 143
Tasmania, 142
Taurus Mountains, **33**, 34, 66, 68, 82, **135**
Temples, **63**, 63, **64**, 73, 77, 85, 138
Terah, **49**, **50**, 52, 172
Theistic evolution, 111, 125, 128
Third Dynasty of Ur, **34**, 137
Tigris (Hiddekel)River, **30**, 31, 32, **33**, 34, 35, 36, **38**, 39, 61, 62, 66, 67, 68, 73, 78, **78**, 80, **80**, 81, 84, 85, 86, 138
Timber, 62
Timeline of biblical history, 15, **15**, 160, **160**, 179
Tin, 62, 137
Tower of Babel, **34**, 38, 61, 130, 137, 143, 144, 171
Tracks, **105**, 105, 106
Transitional fossils, 119, 120
Transposons ("jumping genes"), 116, 124
Tree of life (knowledge of good and evil), 156, 167, **167**
Troglobites, **121**, **123**, 123, 124
Trogloxenes, 121
Troglophiles, 121, 123
Tubal-Cain, **34**, 58, 59, 60, **133**, 135, **137**
Turkey, **33**, 61, 62, 66, 68, 69, 78, **79**, 80, 82, 98, **98**, 99, **135**, 143
Twenty Four-Hour Day view, **17**, 17, 18, 22, 90

U

Ubaid Period, **34**, 42, 62, 76, 78, 84, **84**, **135**, 143
Ulhu, 69
Umm Qasr, 30, **30**, 35
Ur, **30**, **33**, 35, 46, 61, **63**, 64, 84, **84**, 85, 164, 175
Urartu, **33**, **57**, 61, 62, 68, 69, 78, **78**, 80, 82, 86, 99
Uruk, **30**, **33**, 43, 59, 60, 61, 62, **64**, **84**, 85, 113, **137**, 153
Uruk Period, **34**, 39, 60, 61
Uruk-ware pottery, 138
Ussher's calculation for age of world, 41
Ut-napišhtim (Assyrian name for Noah), 60
Uz, 169
Uzziah, 51

V

Vapor canopy, **89**, 90, 91
Vestigial structures, 114, **115**, 116, 127
Vineyards, 81, 86
  Vitisvini Cera, 81, 82

W

Wadi al Batin (Pishon River), 30, **30**, **33**, 35, 84
Wadi al Jarir, **30**, 31
Wadi Aqia, **30**
Wadi Rimah, 30, **30**, 31
Wadi Qahd, **30**, 31
Wellhausen, Julius, 172, 173
Western world, **47**
Worldview
  Worldview and revelation, 14, 15
  Worldview is an approach, not a position, 3, 7, 8
  Worldview is more than culture, 4
  Worldview is subjective, **7**, 8
  Worldview of the Genesis authors/scribes, 9

Worldview Approach
Basic premise of the Worldview Approach, 12, 13, 55
Worldview Approach to Adam and Eve, 164, 166
Worldview Approach to Christ and the church, 177
Worldview Approach to creation and creation myths, 161, 166
Worldview Approach to figurative language, 152
Worldview Approach to evolution, 113
Worldview Approach to extent of "earth" over time, 72, 73
Worldview Approach to garden of Eden, 37
Worldview Approach to Genesis 1, 22, 24, 26
Worldview Approach to Genesis chronologies, 53, **53**
Worldview Approach to Moses and the Monarchy/Exile, 174
Worldview Approach to Noah's flood, 55, 56, 61
Worldview Approach to Origins, 151
Worldview Approach to patriarchs, 172

Y
Y Chromosome Adam, 139, 149, 150
Y Chromosome DNA, 139
Yemen, 29, **30**, 31, **38**
Young-Earth Creationist position, 4, 5, **5**, 6, 7, 29, 38, 42, **42**, 48, **53**, 55, 69, 71, **71**, 73, 74, 78, 82, 85, 89, **89**, 92, **92**, 100, 106, 111, 125, **125**, 126, 127, **135**, **144**, 145, 147, **148**, 151, 152, 156, 167, 168, **171**, 172, 179

Z
Zagros Mountains, 9, **33**, 66, 69, 81, 98
Ziggurats, 37, 38, 61, **63**, 73, 85, **130**, 138, 172
Zillah, 51, 59, 63, **63**
Zilpah, 51
Ziusudra (Sumerian name equivalent to Noah), 36, **59**, 60, 61, 69, **169**

Jesus on the Cross

The Lenten Positive Act Challenge, Dollar Photo Club.

FULL ENDORSEMENTS

"Carol Hill's book is terrific and I am happy to endorse it. A Worldview Approach to Science and Scripture *provides an excellent integration of God's revelation in His written word in Genesis to His revelation in His created world as seen through the eyes of science. Her presentation is both scholarly and accessible to laymen. This book is extremely timely, given the recent Barna surveys (2018) that indicate that the most common reason that students today are losing their faith is over perceived conflicts between faith and science."*

— **Walter Bradley,** Emeritus Professor, Texas A&M University and Baylor University, Past President of The American Scientific Affiliation, and author of *The Mystery of Life's Origins*

"This book does a much-needed service for all Christians troubled by the relation of their faith to science. While other writers have also taken the position that Genesis is to be understood from the standpoint of the worldview of its own time, Hill has spelled out just how to do that in nontechnical language and convincing detail."

— **Roy Clouser,** Emeritus Professor, The College of New Jersey, author of *The Myth Of Religious Neutrality*

"Carol Hill takes both Scripture and science seriously, affirming the inspiration of the Bible and the evidence for biological evolution. She shows that the evolution of Homo sapiens *as a group 200,000 years ago can fit with a historical Adam and Eve living around 5000 B.C. This view is an important option for the church to consider alongside other recent work on human origins. Hill provides a beautiful presentation in this full-color book."*

— **Deborah B. Haarsma,** President of BioLogos, Former Professor of Physics & Astronomy at Calvin University, and co-author of *Origins: Christian Perspectives on Creation, Evolution, and Intelligent Design.*

"In most discussions or debates about cosmic and human origins and how the Book of Genesis figures in such conversations, we are accustomed to looking back from our vantage point to figure things out. But what if we tried to put ourselves in the shoes of the people who produced a text like Genesis? What if we tried to see the universe from inside their worldview and look out onto the world then through this lens instead of only looking back from now? Carol Hill helps us to see that taking such a vantage point may be a key way to unlock the meaning of an ancient witness like Genesis, and readers will benefit from this unique perspective on the larger issues at hand in all such considerations."

— **Scott Hoezee,** Director of The Center for Excellence in Preaching, Calvin Theological Seminary, and author of *Proclaim the Wonder: Preaching Science on Sunday*

"YOUR FAITH SHOULD BE INFORMED BY SCIENCE.

FULL ENDORSEMENTS

"In this book, Carol Hill, a seasoned geologist, combines her expertise and her knowledge and love for the Bible to address some of the most vexing issues of our time regarding the origins of the earth, animal and human life, offering some guidance on how to read Genesis in a responsible way in the light of the latest scientific developments. Her worldview or contextual approach brings the reader face to face with archeological, biblical and scientific data that enable one to gain a new appreciation for what the Bible is trying to teach. This is a very helpful tool!"

— **James K. Hoffmeier,** Emeritus Professor of Near Eastern Archaeology & Old Testament, Trinity Evangelical Divinity School, and author of *Israel in Egypt* and *Ancient Israel in Sinai*

"As the title indicates, this book stakes out the claim for a 'worldview' approach to synthesizing the biblical witness with the findings of the natural sciences. In so doing it occupies the terrain somewhere between old-earth creationism and evolutionary creationism. Hill's approach accepts the evolutionary paradigm while also strongly affirming the historicity of Genesis 1-11. She writes with remarkable clarity and charity. Without agreeing with all the positions set forth, I have to say this is one of the best one-volume works on the creation/evolution dialogue in print."

— **Kenneth Keathley,** Senior Professor of Theology, Southeastern Baptist Theological Seminary, and author of *40 Questions About Creation and Evolution*

"Building on her expertise in geology, Carol Hill argues attractively for a 'worldview' approach to the book of Genesis. Her common-sense approach shows the errors of many common views, setting out the context for the biblical records."

— **Alan Millard,** Emeritus Rankin Professor of Hebrew and Ancient Semitic Languages, University of Liverpool, and author of *Discoveries From Bible Times*

"A Worldview Approach to Science and Scripture *makes an important contribution to the discussion of the compatibility of scientific and biblical understandings which will be helpful in both the Christian community and general culture. Professional geologist Carol Hill draws the reader into the world of ancient Mesopotamia and the land of Eden, both culturally and scientifically, in order to provide context for interpreting the early chapters of Genesis. Through the plentiful use of photographs, maps, and tables, the book offers rich new information and insights for addressing some very old questions – regarding origins of the living world, a global flood, and more."*

— **Michael L. Peterson,** Professor of Philosophy and Religion, Asbury Theological Seminary, and author of *Reason and Religious Belief*

IT SHOULD NOT BE REPLACED BY SCIENCE." — William Newsome

ABOUT THE AUTHOR

Carol Hill is a geologist who has worked in the Grand Canyon and Carlsbad Cavern (and other caves) for over 40 years. She has published her research in books and in *Science* and other professional publications, and this work has been featured on NOVA and National Geographic's *Naked Science* program. Carol is a fellow of the American Scientific Affiliation (ASA) and author of a number of apologetics articles in *Perspectives on Science and Christian Faith*; she is also the senior editor and one of eleven authors of Kregel Publications' book *Grand Canyon: Monument to an Ancient Earth, Can Noah's Flood Explain the Grand Canyon?* Carol and her husband, Alan, of 59 years are long-time members of Heights Cumberland Presbyterian Church in Albuquerque and have taught science-Scripture classes there. A graduate of the University of New Mexico, Carol is currently an Adjunct Professor in the UNM Continuing Ed Department, where she teaches geology and climate science.

The author inspecting an exquisite helictite spray in Kartchner Caverns, Arizona, when she was a consultant for Arizona State Parks before it was open to the public. *Photo by David Jagnow.*

Author climbing up rope to a cave entrance in the Grand Canyon (done with the Grand Canyon National Park's permission). *Photo by Bob Buecher.*

Author hiking in the remote Hualapai Plateau section of Grand Canyon (hiking permit issued by the Hualapai tribe). *Photo by Bob Buecher.*

Acts 1:9. *After he [Christ] said this, he was taken up before their very eyes, and a cloud hid him from their sight.* (NIV)

Christi Himmelfahrt by Gebhard Fugel, c. 1893.